Study Guide

College Algebra
Concepts and Contexts

James Stewart
McMaster University and University of Toronto

Lothar Redlin
The Pennsylvania State University

Saleem Watson
California State University, Long Beach

Phyllis Panman

Prepared by

Florence Newberger
California State University Long Beach

BROOKS/COLE
CENGAGE Learning

Australia • Brazil • Japan • Korea • Mexico • Singapore • Spain • United Kingdom • United States

For product information and technology assistance, contact us at
**Cengage Learning Customer & Sales Support,
1-800-354-9706**

For permission to use material from this text or product, submit all requests online at **www.cengage.com/permissions**
Further permissions questions can be emailed to
permissionrequest@cengage.com

ISBN-13: 978-0-495-38791-6
ISBN-10: 0-495-38791-6

Brooks/Cole
20 Davis Drive
Belmont, CA 94002-3098
USA

Cengage Learning is a leading provider of customized learning solutions with office locations around the globe, including Singapore, the United Kingdom, Australia, Mexico, Brazil, and Japan. Locate your local office at: **www.cengage.com/global**

Cengage Learning products are represented in Canada by Nelson Education, Ltd.

To learn more about Brooks/Cole, visit
www.cengage.com/brookscole

Purchase any of our products at your local college store or at our preferred online store
www.cengagebrain.com

Printed in the United States of America
2 3 4 5 6 7 14 13 12 11 10

TABLE OF CONTENTS

To the Instructor

A STUDY GUIDE AND WORKBOOK

As described in the list, 'How to Use Your Textbook and Study Guide,' below, I organize this study guide with the intent to help the students get the most out of their textbooks, using short directed reading assignments, followed by immediate applications of the ideas and techniques. When I teach college algebra, I also like to have the students work on their own during class, applying the ideas we discuss. In this study guide, I include the full text of many exercises from the textbook, with space to write a solution, so that **the study guide may also act as a workbook, to be used in-class**, to follow up on discussion. (Tip: When the exercise involves a scatter plot, I often ask the students to complete the plot at home before class, to save time.)

A list of the exercises included here can be found in this preface; with few exceptions, this list coincides with the exercises in the textbook that are marked with a pencil icon. Note that the study guide also includes complete solutions to these exercises.

To the Student

HOW TO USE YOUR TEXTBOOK AND STUDY GUIDE:
Eleven steps to enjoying success in College Algebra: Contexts and Concepts

1. Start with the textbook. If the section you are studying has one, use the ✓ Algebra Checkpoint to see if you need to review any prerequisite skills; if so refer to the appropriate Algebra Toolkit in the appendices of the textbook.

2. Read the section in the textbook.

3. Read and think about the answers to the questions entitled, 'Think About It,' in the Concepts section of the exercises at the end of the section in the textbook.

4. Go to class, participate, and take notes. Write in-class work neatly, and make notes (in sentences) that explain how you solve the problems.

5. Turn to the section in the study guide. The study guide directs you to do the following:

    *   📖 **Read** short segments from the text.

    *   Reflect on what you have read, by answering the brief *follow-up questions* after each reading assignment.[1]

    *   Solve the given ✐ **Exercises**. Write your solutions neatly, filling the space from left to right, and then from top to bottom. If you work meanders around the allotted space, draw arrows and number your steps so that you can read it again when you study. Always answer word-problems with word-answers, using sentences. The more effort you make to communicate the ideas, the more likely you will be able to use them to solve new problems, and apply them even outside the mathematics classroom.

    *   Read the Hints and Tips , which help you connect the symbols and numbers with the contexts and graphs. Look for applications of the Hints and Tips in the explained

---

[1] Check your answers in the footnotes.

examples from the textbook (marked with, '$\Rightarrow$'), the *follow-up questions,*[2] and the **Solutions to** ✎ **Exercises**, found at the end of each section in the study guide. (See the Index of Hints and Tips in this preface.)

- Read the **Solutions to the** ✎ **Exercises**, found at the end of each section in the study guide, to reinforce your understanding.

- Read the checklist of OBJECTIVES at the beginning of the section in the study guide. Make sure each objective is familiar to you. Mark the boxes of the objectives that you meet and revisit the exercises for those that you want to study further.

6. Once you have completed the section in the study guide, solve other problems from the many included in the SKILLS and CONTEXTS sections of the exercises in the textbook. Notice that the solutions to the examples in the book and the ✎ **Exercises** in the study guide include words of explanation as well as computation. Most computations are even annotated with more words of explanation. When you write a solution, such as on your notes, homework or exams, write your solution like the ones you have been reading, with complete sentences and plenty of explanation.

7. Give yourself time to reflect on the big picture; consider the exercises you solve, and their similarities and differences. Follow up on these ideas by answering the questions entitled 'Fundamentals' in the CONCEPTS part of the exercises at the end of the section in the textbook.

8. Read the summary of the section in the Concept Check at the end of the chapter (in the Chapter Review in the textbook).

9. Return to the questions entitled, "Think About It," in the CONCEPTS section of the exercises at the end of the section in the textbook. Discuss the questions and your answers with your peers and your teachers, using correct vocabulary and referring to and applying the techniques and ideas at hand.

10. Complete activities from the Explorations, found at the end of each chapter in the textbook. These guide you to experiment with the ideas from that section in a hands-on application, and provide you with another means to reflect on and reinforce the ideas that you are studying.

11. Find uses for the CONCEPTS you study in algebra in CONTEXTS outside your mathematics classroom. Use the precise writing style that you practice in algebra to hone your communication skills in all subjects. Use algebra, symbolically, numerically, verbally and graphically to describe the real world, and become a life long learner of mathematics.

---

[2] Look for uses of the Hints and Tips in the footnotes as well.

Acknowledgements

I would like to thank the authors, James Stewart, Lothar Redlin, Phyllis Panman, and particularly Saleem Watson, for encouraging me to work on this project, and offering me invaluable feedback. I would also like to thank Amy Mulgrew who wrote the solutions to more than half of the exercises in the study guide, carefully following my instructions and matching the style that the authors use in the textbook. My thanks also to accuracy checker Michael Rosen for his careful reading of the Study Guide. I would like to thank Cynthia Ashton, Assistant Editor and my contact at Cengage, for her support as I slowly worked my way through the chapters of the text. Lastly, I would like to thank my family for letting me work on this study guide for far more hours than I expected to spend. I especially appreciate my sons Benjamin and Stuart, for their patience, in waiting for the promised trip to Disneyland, to take place once the task is complete.

✎ Exercises

The following exercises are included in this study guide, with space to write your solutions. Find detailed solutions to each of these exercises at the end of the corresponding sections.

Section	Exercises
1.1	15, 21, 25, 27, 29
1.2	15, 29, 35, 39
1.3	7, 11, 13, 23
1.4	7, 9, 11, 25, 35, 53, 55, 57, 61
1.5	9, 13, 17, 27, 29, 39, 49, 51, 55, 61, 67, 69
1.6	13, 15, 19, 23, 31, 33, 35, 41, 45, 49, 55
1.7	7, 11, 19, 27, 37, 41, 43, 49
1.8	1, 2, 17, 19, 21, 27, 29
1.9	7, 17, 21, 23, 27, 33
2.1	9, 11, 13, 25, 27, 29
2.2	9, 11, 13, 17, 21, 25, 45, 53, 55, 59
2.3	11, 19, 23, 31, 53, 57, 61, 63
2.4	9, 13, 19, 27, 47, 49, 51, 53
2.5	17, 18
2.6	21, 27, 29, 31
2.7	5, 27, 33, 35
3.1	17, 19, 21, 25, 27, 29, 31, 33, 39, 41, 45
3.2	5, 9, 13, 37, 41, 43, 45, 47
3.3	9, 13, 17, 19, 29, 31
3.4	5, 6, 15, 17, 19, 29, 31, 33, 35
3.5	9, 13, 19
4.1	7, 11, 17, 21, 27, 41, 51
4.2	9, 11, 15, 19, 25, 29, 33, 39
4.3	7, 11, 13, 15, 17, 21, 23, 25
4.4	9, 11, 13, 27, 29, 31, 43, 49, 53
4.5	7, 15, 21, 29, 35, 37, 49, 55, 57, 65
4.6	15, 17, 35, 37, 41, 57, 61, 63, 71, 83
5.1	7, 9, 13, 33, 35, 37, 39, 45, 49, 57, 63, 65, 67, 69
5.2	7, 9, 13, 21, 23, 29, 33, 39, 43
5.3	9, 11, 23, 25, 27, 33
5.4	7, 17, 27, 31, 33, 45, 46, 47, 49, 53, 63, 65, 67
5.5	5, 9
6.1	13, 15, 21, 33, 35, 39
6.2	7, 8, 9, 17, 23, 35, 37, 39
6.3	7, 9, 19, 23, 27, 47
6.4	11, 15, 21
6.5	5, 7, 13, 15, 25, 31, 37
6.6	13, 23, 29, 31, 37, 41
7.1	17, 21, 29, 31, 51, 53, 55
7.2	5, 13, 17, 23, 27, 31
7.3	1, 11, 15, 21, 25, 27, 41
7.4	7, 11, 13, 15, 17
7.5	11, 13, 15, 27
7.6	3, 7, 13, 15, 23, 27

Index of Hints and Tips

This index lists the practical hints and tips that we enumerate in this study guide. The numbers refer to the chapter and section in which the hint appears. To better understand how to apply the Hints and Tips, refer to the Solutions to the ✐ Exercises, in which we explicitly demonstrate how the Hints and Tips apply in context. We often also ask you to practice using the Hints and Tips in the italicized follow-up questions, and refer to them in the footnotes.

1.1 Making Sense of Data

OBJECTIVES

☑ *Check the box when you can do the exercises addressing the objective, such as those included in this study guide, which are listed here.*

Exercises

One-variable data

☐ Calculate the average and the median. **15, 21, 25, 27**

☐ Determine if the mean or median is a good measure of central tendency. **21, 27**

Two-variable data

☐ Determine meaningful values for each variable independently. For example, find the greatest value or the median of the data. **25, 27**

☐ Describe relationships between one variable's meaningful values (like greatest or least) and the other variable. **25, 27, 29**

☐ Read information from a table of two-variable data. **25, 27, 29**

☐ Find the amount by which one variable changes when you change the other variable. **29**

SKILLS

One variable data

The exercises in this section ask you to calculate the average and the median, two measures of central tendency, or numbers that reflect the "center" of the data as a whole. Find the definitions of the average and the median in blue boxes in the textbook.

📖 **Read Section 1.1 Example 1**, in which the authors calculate the average age of preschoolers in a class.

📖 **Read Section 1.1 Example 2**, in which the authors calculate the average and median of a list of incomes.

📖 **Read Section 1.1 Example 3**, in which the authors calculate the average and median of the selling price of houses in 2007.

[1]*The average of a list of n numbers is their sum, divided by n.*

What is n in Example 1? _____

What is n in Example 2? _____

What is n in Example 3? _____

What is the first step the authors do when calculating the median?
Answer in your own words:

---

[1] In Example 2, $n = 5$. In Example 3, $n = 8$. To find the median, first put the numbers in order.

✎ **Section 1.1 Exercise 7** A table of one-variable data is given.

(a) Find the average of the data.
(b) Find the median of the data.
(c) How many data points are greater than the average? How many are greater than the median?

Solution

A
113
21
16
16
19
29
21

CONTEXTS

One-variable data

Exercises about one-variable data in context ask you to calculate measures of central tendency, and discuss how well these measures reflect the "center" of the data as a whole.

📖 Read **Section 1.1 Example 2(d)**, in which the authors compare the usefulness of the average and the median as measures of central tendency of the list of incomes.

📖 Read **Section 1.1 Example 3(c) and the paragraph that precedes Example 3**, in which the authors calculate the average and median selling price of a house, and compare the results to those of Example 2.

²*What is an outlier?*
Answer in your own words:

When data have outliers, the _____ (median / average) tends to be a better measure of central tendency.

✎ **Section 1.1 Exercise 21 Investment Seminar** The organizer of an investment seminar surveys the participants on their yearly income. The table below shows the yearly incomes of the participants.

(a) Find the average and median income of the participants.
(b) How many participants have an income above the average?
(c) A new participant joins the seminar, and her yearly income is $500,000. Now what are the average income and median income, and how many participants have an income above the average? Is the average or the median the better measure of central tendency?

Solution

Yearly income
56,000
58,000
48,000
45,000
59,000
72,000
63,000

---

² An outlier is a number that is much larger or much smaller than most of the other data. When data include outliers, the median tends to be a better measure of central tendency.

Two variable data

Exercises involving two-variable data include questions about each variable independently. For example, you may determine the largest and smallest, or calculate measures of central tendency, like the median and the average. Exercises involving two-variable data also include questions about the correspondences between the two variables and how the variables change relative to one another. In this section, you use tables of data to answer questions about two-variable data. (In future sections, you will use functions, equations and graphs in addition to tables, to reveal even more information about two-variable data.)

📖 **Read Section 1.1 Example 4,** in which the authors discuss whether or not the tallest students are the oldest students in a preschool class.

[3] *How tall is the oldest student? _____ inches tall. How old is the tallest student? _____ years old.*

✎ **Section 1.1 Exercise 25 Pets per Household** The homeowners association of a small, gated community conducts a survey to determine trends in the number of pets in their neighborhood. The number of persons in a household and the number of pets they own are recorded. The table below shows the results of the survey.

(a) What is the average number of pets in a household in this community?
(b) Does the largest family have the most pets?
(c) How many people are in the family with the most pets?

Number of persons	2	1	5	3	2	3
Number of pets	5	11	2	0	15	1

Solution

📖 **Read Section 1.1 Example 5,** in which the authors determine the coolest time of day, from a table of two-variable data showing temperatures at various times throughout the day.

[4] *When was the highest temperature recorded?_____ What is the lowest temperature recorded?_____*

---

[3] The oldest student is 45 inches tall. The tallest student is 4 years old.

[4] The highest temperature occurred at 10:00 a.m. The lowest temperature is 58° F.

✎ **Section 1.1 Exercise 27 Snowfall** The Sierra Nevada mountain range is well known for its large snowfalls, especially in the area of Lake Tahoe. This area usually gets over 200 inches of snow a year. The table below gives the snowfall for a each month in 2006.

(a) What was the snowfall for the month of February in 2006.
(b) What month had the highest snowfall? What month had 49 inches?
(c) Find the average monthly snowfall for 2006.
(d) Find the median monthly snowfall for 2006.

Month	1	2	3	4	5	6	7	8	9	10	11	12
Snowfall (in)	49	12	83	42	2	0	0	0	0	0	8	14

Solution

[5]*Does the median give the impression that the Sierra Nevada mountain range receives considerable snowfall? What about the average?*
Answer in your own words:

📖 **Read Section 1.1 Example 6**, in which the authors calculate the amounts by which deep sea pressure changes as depth increases.

[6]*By how much does the pressure change as the depth changes from 10 ft to 30 ft? _____ lbs/in^2
What does the word "steadily" refer to in the answer to part (d)?
Answer in your own words:*

---

[5] The median gives the impression that there is very little snowfall. The average is higher than the median, but it still fails to indicate the great quantity of snow, since the months in which it does not snow affect the average a great deal.

[6] As the depth increases from 10 ft to 30 ft, the pressure changes by 9 lbs/in^2. In part (d), the word "steady" means that the fall-time increases by a constant amount for each 5 foot increase in distance.

✎ **Section 1.1 Exercise 29 Falling Watermelons** The fifth grade class at Wilson Elementary School performs an experiment to determine the time it takes for a watermelon to fall to the ground. They have access to a five-story building for their experiment. Dropping a watermelon out of a window from each story of the building, the class records the time it takes for the watermelon to fall to the ground. The table below shows the results of their experiment.

(a) How long does it take for a watermelon to fall 5 feet from the first story window?
(b) How much longer does it take a watermelon to fall from 15 feet than from 5 feet? From 45 ft than from 35 ft?
(c) What pattern or trend do you see in these data?

Solution

Story	Distance (ft)	Time (sec)
First	5	0.60
Second	15	0.96
Third	25	1.26
Fourth	35	1.49
Fifth	45	1.68

CONCEPTS

✓ **1.1 Exercises – Concepts – Fundamentals** Complete the Fundamentals section of the exercises at the end of Section 1.1. Compare your answers to those at the back of the textbook, and make corrections as necessary.

📖 **Read the Concept Check for Section 1.1**, in the Chapter Review at the end of Chapter 1.

1.1 Solutions to ✎ Exercises

✎ **Section 1.1 Exercise 7**

(a) Since the number of data points totals 7, we find the average by adding up the data and dividing by 15: $\dfrac{113+21+16+16+19+29+21}{7} = \dfrac{235}{7} = 33.6$. So the average is 33.5.

(b) To find the median income, we first order the list of values:

$$16, 16, 19, 21, 21, 29, 113$$

Since there are 7 numbers and seven is odd, the median is the middle number 21.

(c) There is one data point greater than the average. There are two data points greater than the median.

✎ **Section 1.1 Exercise 21 Investment Seminar**

(a) Since we are finding the average income of 7 participants, we find the sum of all the incomes and divide by 7.

$$\frac{56,000+58,000+48,000+45,000+59,000+72,000+63,000}{7}=57,285.71$$

So the average income of the participants is $57,285.71.

To find the median, we first order the list of values:

$$45,000, 48,000, 56,000, 58,000, 59,000, 63,000, 72,000$$

Since there are 7 numbers and seven is odd, the median is the middle number 58,000. So the median income of the participants is $58,000.

(b) Four participants have income greater than the average.

(c) After the new participant joins the group, there are 8 participants, so we find the sum of the incomes and divide by 8.

$$\frac{401,000+500,000}{8}=\frac{901,000}{8}=112,625$$

So the average income of the 8 participants is $112,625. There is only 1 income that is greater than the average.

To find the median, we first order the list of values:

$$45,000, 48,000, 56,000, 58,000, 59,000, 63,000, 72,000, 500,000$$

Since there are 8 numbers and 8 is even, the median is the average of the middle two numbers:

$$\frac{58,000+59,000}{2}=\frac{117,000}{2}=58,500$$

So the median income of the 8 participants is $58,500.

Since the data has an outlier ($500,000), the average of $112,625 is not typical, being much greater than most of the data. However the median income of $58,500 is reasonably close to most of the participant's incomes, making it a better measure of central tendency.

✐ Section 1.1 Exercise 25 Pets per Household

(a) Since the survey includes 6 households, so we find the total number of pets and divide by 6:

$$\frac{5+11+2+0+15+1}{6} = \frac{34}{6} = 5.7$$

So the average number of pets per household is between 5 and 6 pets.

(b) No. The largest family has 5 people, but only 2 pets. Another household has 11 pets, so the largest family does not have the most pets.

(c) The household with the most pets has 1 person.

✐ Section 1.1 Exercise 27 Snowfall

(a) February is the second month, so the table shows that the snowfall in February was 12 inches.

(b) The highest snowfall was in the third month, which is March. The first month, January, had 49 inches of snow.

(c) There are 12 months, so we find the sum of the snowfalls for each month and divide by 12:

$$\frac{49+12+83+42+2+8+14}{12} = \frac{210}{12} = 17.5$$

So the average monthly snowfall in 2006 is 17.5 inches.

(d) To find the median, we first order the list of values:

$$0, 0, 0, 0, 0, 2, 8, 12, 14, 42, 49, 83$$

Since there are 12 months, and 12 is even, the median is the average of the middle two values:

$$\frac{2+8}{2} = \frac{10}{2} = 5$$

So the median monthly snowfall for 2006 is 5 inches.

✐ Section 1.1 Exercise 29 Falling Watermelons

(a) The watermelon takes 0.60 seconds to fall 5 feet from the first story window.

(b) The watermelon takes 0.96 seconds to fall 15 feet from the second story window, which is $0.96 - 0.60 = 0.36$ seconds longer than it takes for the watermelon to fall from the first story window.

(c) The distances are increasing by 10 feet each story. The times increase, but not by a constant amount. For example, it takes $0.96 - 0.60 = 0.36$ seconds longer to fall from the second story than from the first story, whereas it takes $1.26 - 0.96 = 0.30$ seconds longer to fall from the third story than from the second. In fact, the amount by which the fall-time increases each story is decreasing.

1.2 Visualizing Relationships in Data

OBJECTIVES

☑ *Check the box when you can do the exercises addressing the objective, such as those included in this study guide, which are listed here.*   **Exercises**

Relations: Input and Output

☐	Express a table of two variable data as a relation.	35
☐	Determine the domain and range of a relation.	35
☐	Explain what an ordered pair from a relation represents in context.	35(c)

Graphing Two-Variable Data in a Coordinate Plane

☐	Choose a scale for the *x*- and *y*-axis that displays the data clearly.	29(a), 39
☐	Create a scatter plot of two-variable data.	29(a), 39
☐	Describe the relationship between the inputs and outputs of a relation by describing how the outputs change as the inputs increase.	29(b), 39

Reading a Graph

☐	Express a scatter plot as a relation.	15
☐	Identify the output(s) that correspond to a given input, or the input(s) that correspond to a given output from a graph of the relation.	15

GET READY...

This section uses vocabulary, like *ordered pair*, *x-* and *y-coordinates*, and *coordinate plane*, and skills, like plotting points, and reading coordinates from a graph. Refresh your understanding of these in Algebra Toolkit D.1.

✓ **Algebra Checkpoint** Test yourself by completing the Algebra Checkpoint at the end of this section of the textbook, and comparing your answers to those in the back of the textbook. Refer to Algebra Toolkit D.1 as necessary.

SKILLS

Relations: Input and Output

These exercises ask you to turn a table into a relation, state the domain and range, and identify inputs and outputs. In Section 1.2 of the textbook, find the definition of a relation in a blue box, and the definitions of domain, range, input and output in bold typeface.

📖 **Read Section 1.2 Example 1 and the paragraph that precedes it**, in which the authors construct a relation from a table relating students' sleep time before an exam and their exam scores.

[1] *The inputs are _____, and the outputs are _____. How many students slept for 7 hours before the exam? _____ How many times does the number 7 appear in the range of the relation, when it is written as a set, as in part (e)? _____*

---

[1] The inputs are scores on the exam, and the outputs are hours of sleep. Two students slept for 7 hours before the exam. The number 7 appears one time is the range of the relation, since the range of the relation is a set.

✎ **Section 1.2 Exercise 35(a, b, d, e) Income and Home Prices** The median incomes and median home prices in several neighborhoods across the U.S. are shown in the table below. (Part (c) appears in the Contexts section.)

(a) Express the data as a relation.
(b) Draw a diagram of the relation.
(d) Find the output(s) corresponding to the input 80,000.
(e) Find the domain and range of the relation.

Solution

Median income	Median home price
80,000	400,000
30,000	250,000
70,000	300,000
55,000	250,000
80,000	450,000
60,000	300,000
55,000	300,000
150,000	1,500,000

Graphing Two-Variable Data in a Coordinate Plane

These exercises give you a table of data and ask you to construct a scatter plot, which, if you do it by hand, requires you to choose an appropriate scale for the *x*- and *y*-axes.

📖 **Read Section 1.2 Example 3**, in which the authors draw scatter plots of temperature versus time and depth versus pressure, the sets of two variable data studied in Section 1.1 Examples 5 and 6, and comment on the resulting patterns.

[2]*In both examples, one of the data points lies on the y-axis of the graph. In the time-temperature data that point is (____, ____), and in the depth-pressure data, that point is (____, ____).*

📖 **Read Section 1.2 Example 4**, in which the authors create scatter plots of different enzyme levels and discuss whether or not there is a relationship between them.

[3]*Which of samples 1 through 20 has the highest level of enzyme A? _____ Which data point in each of the two graphs corresponds to this sample?*
Answer in your own words:

―――――――――――――――――

[2] In the time-temperature data, the data point (0, 60) lies on the *y*-axis. In the depth-pressure data, the data point (0, 15) lies on the *y*-axis.

[3] Sample number 14 has the highest level of enzyme A. On each graph, sample number 14 is the data point that lies furthest to the right.

> **Hints and Tips 1.2a: Choosing scales for the x- and y-axes of a scatter plot**
> Make graphs easy to read by choosing an appropriate scale. Choose the scale for the x-axis to
> - include all of the values in the domain of the relation, and
> - not include too much above the largest or below the smallest value in the domain.
>
> Similarly choose the values on the y-axis to accommodate the range of the representation. See also Algebra Toolkit D.3, about choosing the viewing rectangle when using a graphing calculator.

These guidelines explain the author's choices of scale in Section 1.2 Examples 3 and 4. The tidy, computer-generated graphs in the textbook are easy to read, even though they are small. When creating a scatter plot by hand, use graph paper, and make the diagram large enough to convey the information clearly.

⇨ **Section 1.2 Example 4(a)** Make a scatter plot of the levels of the pairs of enzymes A and B.

Partial solution: choosing a scale for the x-axis
The inputs (and hence the domain) of this relation are the concentrations of enzyme A. The smallest value for enzyme A is 0.8 mg/dL (sample 20), and the largest value is 4.1 mg/dL (sample 14). Thus the domain of the relation lies within the interval [0.8, 4.1], so the scale for the x-axis should include these. The authors' x-axis has tick marks at the whole numbers 0, 1, 2, 3, and 4, and the axis line extends a short distance beyond 4 to include 4.1 in the diagram.

⇨ **Section 1.2 Example 3 (excerpt)** Draw a scatter plot of the time-temperature data…

Partial solution: Choosing the scale for the y-axis
The outputs of the relation (and hence the range) are the temperatures. The lowest temperature is 58°F, while the highest is 68°F, so the range of the relation lies in the interval [58, 68]. The authors chose the scale on the y-axis to extend from 0 to 70. Alternatively, to emphasize how much the temperature is changing, you could select a scale extending from 56 to 68, as shown. The graph in Section 1.2 Example 5 also includes a "broken" y-axis.

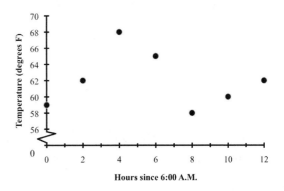

Hours since 6:00 A.M.

Reading a Graph

A scatter plot presents the same information as a table of two-variable data, so exercises that give you a scatter plot ask the same questions as exercises that give you a table, like expressing the data as a relation, finding the domain and range and discussing relationships between particular inputs and outputs.

📖 **Read Section 1.2 Example 5**, in which the authors read information about the average annual precipitation in Medford, Oregon from a scatter plot.

[4]*What year was the wettest? _____ How much rain did Medford receive in its driest year? _____ inches*

✏ **Section 1.2 Exercise 15 (extended)** The graph of a relation is given.

(a) List three ordered pairs in the relation.
(b) Find the output(s) corresponding to the input 4.
(c) Find the inputs(s) corresponding to the output 50.
(d) Find the domain and range of the relation.

Solution

CONTEXTS

Relations: Input and Output

Exercises about relations ask you to connect the algebraic notation to the context.

📖 **Read Section 1.2 Example 1(c),** in which the authors explain what an ordered pair represents in the context of sleep time before an exam.

Hints and Tips 1.2b: What does the algebra represent in context?
Here are some steps to help you connect the algebraic concepts to the context of the word problem.
Algebraically: Explain what the algebraic concept is, without the context.
In context: Rewrite your explanation, replacing the references to the x-values and y-values with their meanings in this context, including units on any numbers.
More eloquently: Fix the English so that your answer is eloquent.

---

[4] In 1996, Medford had the most rain in the ten-year period. The driest year in the ten-year period was 2001, in which Medford had 15 inches of rain.

The guidelines above explain the authors' answers for Example 1(c):

⇨ **Section 1.2 Example 1(c)** What does the pair (80,5) represent?

Solution
We write the following sentences as scratch work. Algebraically, the *x*-value is 80, and the *y*-value is 5. In context, the score on the exam is 80, and the number of hours of sleep is 5. Here is our final, more eloquent answer. The pair (80,5) represents a student who scored 80 on the exam and had 5 hours of sleep before the exam.

📖 **Read Section 1.2 Example 2**, in which the authors construct a relation from a table of bird watching data.

[5]*The inputs are* _____, *and the outputs are* _____.
What does the pair (21, 15) represent?
Answer in your own words:

✎ **Section 1.2 Exercise 35(c) Income and Home** *Find this exercise on page 8.* What does the ordered pair (55,000, 250,000) represent?

Solution

Graphing Two-Variable Data in a Coordinate Plane

📖 **Read Section 1.2 Examples 3 and 4 again**, paying particular attention to the authors' description of the trends displayed in the scatter plots.

> **Hints and Tips 1.2c: Describing trends in relations**
> When describing the relationship between the inputs and outputs of a relation, explain what happens to the output when the input increases.

The authors use this tip:

⇨ About the time-temperature graph in Example 3: "…as time increases, the temperature first increases, then decreases, then increases again."

⇨ About the depth-pressure graph is Example 3 "As the depth increases, so does the pressure."

---

[5] The inputs are days in February, and the outputs are the numbers of house finches seen. We write the following sentences as scratch work. Algebraically, the *x*-value is 21, and the *y*-value is 15. In context, the day in February is 21 and the number of house finches is 15. We write our final answer more eloquently, the pair (21, 15) tells us that 15 house finches were seen at the bird feeder on February 21.

⇨ About the Enzymes in Example 4(c), "…when the level of enzyme A goes up, the level of enzyme C tends to go down."

⇨ For more examples of verbal descriptions of trends in two-variable data, see Section 1.2 Exercises 19-22.

✎ **Section 1.2 Exercise 39 Christmas Bird Count (extended)** For the past 100 years, ornithologists have traditionally made a worldwide bird count at Christmas time. The table below shows the number of house finches observed in a Christmas Bird Count in California.

(a) Make a scatter plot of the data in the table.
(b) What trends do you detect in the house finch population of California?
(c) What does the pair (15, 86,507) represent?

Solution

Year	Years Since 1960	Bird Count
1960	0	33,621
1965	5	44,787
1970	10	53,838
1975	15	86,507
1980	20	73,767
1985	25	91,659
1990	30	87,632
1995	35	107,190
2000	40	69,733
2005	45	61,053

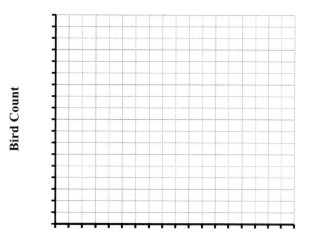

Bird Count

Years Since 1960

CONCEPTS

✓ **1.2 Exercises – Concepts – Fundamentals** Complete the Fundamentals section of the exercises at the end of Section 1.2. Compare your answers to those at the back of the textbook, and make corrections as necessary.

📖 **Read the Concept Check for Section 1.2**, in the Chapter Review at the end of Chapter 1.

1.2 Solutions to ✐ Exercises

✐ Section 1.2 Exercise 15 (extended)

(a) For example, this relation includes (1, 30), (2, 20) and (4, 50).

(b) The vertical line that passes through the input 4 also pass through only one data point at (4, 40). That data point has y-value 40, so the output corresponding to the input 4 is 40.

(c) The horizontal line passing through the output 50 also passes through (3, 50) and (6, 50), so the inputs corresponding to the output 50 are 3 and 6.

(d) The domain is the set of all inputs in the relation. If there is a data point on the vertical line passing through a value on the x-axis, then that value is in the domain. The domain is

$$\{1, 2, 3, 4, 5, 6, 7, 8, 9, 10, 11\}$$

The range is the set of all outputs in the relation. If there is a data point on the horizontal line passing through a value on the y-axis, then that value is in the range. The range is

$$\{20, 30, 40, 50, 70, 80, 90\}$$

✐ Section 1.2 Exercise 35 Income and Home Prices

(a) The set of ordered pairs that defines this relation is

{(80,000, 400,000), (30,000, 250,000), (70,000, 300,000), (55,000, 250,000), (80,000, 450,000), (60,000, 300,000), (55,000, 300,000), (150,000, 1,500,000)}

(b) A diagram of the relation is shown below.

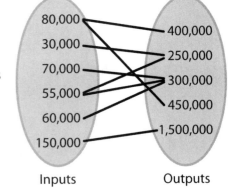

(c) Following the **Hints and Tips 1.2b**, we write the following sentences as scratch work. Algebraically, the x-value is 55,000 and the y-value is 250,000. In context, the median income is $55,000 per year and the median home price is $250,000.

Here is our final more eloquent answer. The pair (55,000, 250,000) represents a neighborhood in which the median yearly income is $55,000 and the median home price is $250,000.

(d) There are two outputs that correspond to the input 80,000; they are 400,000 and 450,000.

(e) The domain of the relation is the set {80,000, 30,000, 70,000, 55,000,60,000, 150,000}, and the range is the set {400,000, 250,000, 300,000, 450,000,1,500,000}.

✎ Section 1.2 Exercise 39 Christmas Bird Count (extended)

(a)

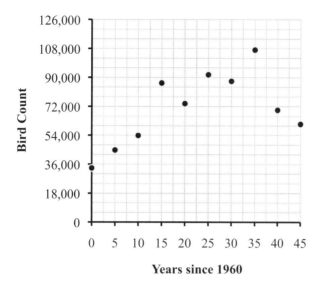

(b) Following **Hints and Tips 1.2c**, we describe the trend as the inputs increase. Since 1960, the population increased pretty steadily until 1995, when it began to decrease.

(c) Following the **Hints and Tips 1.2b**, we write the following sentences as scratch work. Algebraically, the x-coordinate is 15 and the y-coordinate is 86,507. In context, the year is 1975 (15 years after 1960), and the number of birds is 86,507.

Here is our final, more eloquent answer. The pair (15, 86,507) represents the 86,507 California house finches that bird watchers counted in the 1975 Christmas Bird count.

1.3 Equations: Describing Relationships in Data

OBJECTIVES

☑ *Check the box when you can do the exercises addressing the objective, such as those listed here, which are included in this study guide.*

Exercises

Making a Linear Model from Data

☐ Determine the initial value and the amount by which the output changes as the input increases by 1 from a table of data with evenly spaced inputs and constant first differences. **7, 11, 23(a, b)**

☐ Construct a linear equation to model data with evenly spaced inputs and constant first differences. **7, 11, 23(a, b)**

☐ Determine if a linear model is appropriate for a given data set, using first differences. **11, 13, 23(a, b)**

Getting Information from a Linear Model

☐ Use a linear model to predict outputs corresponding to inputs that are not in the domain of the data. **7, 11, 23(c)**

Get Ready...

This section uses vocabulary, like *solution*, and what it means to *satisfy an equation*, and skills, like graphing equations, and reading coordinates from a graph. Refresh your understanding of these in Algebra Toolkit D.2.

✓ **Algebra Checkpoint** Test yourself by completing the Algebra Checkpoint at the end of this section of the textbook, and comparing your answers to those in the back of the textbook. Refer to Algebra Toolkit D.2 as necessary.

Skills

Making a Linear Model from Data

Exercises in this section include determining if data can be modeled by a linear equation, and creating a linear equation to model data when appropriate. In Section 1.3, find the definition of a linear model and of first differences in blue boxes, and the definition of an initial value in bold typeface.

[1] *In the linear model $y = A + Bx$, what are A and B?*
Answer in your own words:

[2] *In the linear model $y = 3 + 6x$, the initial value is _____, and if the x-value increases by 1, then the y-value _____ (increases / decreases) by _____.   In the linear model $y = -2 - 6x$, the initial value is _____, and if the x-value increases by 1, then the y-value _____ (increases / decreases) by _____.*

📖 **Read Section 1.3 Example 1**, in which the authors construct a linear model for the cost to make chairs.

[3] *The input in this example is _____, and the output is _____.*

The variable _____ represents the input, and the variable _____ represents the output.

What x-value corresponds to the initial value 80? _____ When x increases from 2 to 3, C _____ (increases / decreases) by _____. When x increases from 3 to 4, C _____ (increases / decreases) by _____. The model tells us that to make 5 chairs, it will cost the furniture maker _____ dollars.

---

[1] The parameter *A* is the initial value, and *B* is the amount by which *y* changes when *x* increases by 1.

[2] In the linear model $y = 3 + 6x$, the initial value is 3, and if the *s*-value increases by 1, then the *y*-value increases by 6. In the linear model $y = -2 - 6x$, the initial value is –2, and if the *s*-value increases by 1, then the *y*-value decreases by 6.

[3] The input is the number of chairs produced, and the output is the total cost. The variable *x* represents the input, and the variable *C* represents the output. The input $x = 0$ always corresponds to the initial value. When *x* increases from 2 to 3, or from 3 to 4, *C* increases by 12. It costs $140 to make 5 chairs.

✐ **Section 1.3 Exercise 7** A set of data is given.

   (a) Find a linear model for the data.

   (b) Use the model to complete the table.

   (c) Draw a graph of the model.

Solution

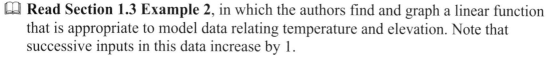

x	y
0	5
1	12
2	19
3	26
4	
5	
6	

📖 **Read Section 1.3 Example 2**, in which the authors find and graph a linear function that is appropriate to model data relating temperature and elevation. Note that successive inputs in this data increase by 1.

[4] *The inputs for the data must be _____, in order to compute first differences. Are the inputs evenly spaces in Section 1.2 Exercise 27?_____ (Yes / No) What about Exercise 28? _____ (Yes / No) If you have 6 data points, how many first differences would you calculate?*

[5] *How can you tell that a linear model is appropriate for the data in Section 1.3 Example 2? Answer in your own words:*

[6] *How do the authors find the parameters A and B in the model? Answer in your own words:*

_____

[4] The data must be evenly spaced to compute first differences. In Section 1.2 Exercise 27, the inputs are evenly spaced (skipping by 2), but in Section 1.2 Exercise 28, the inputs are not evenly spaced (some of the inputs even repeat!). With 6 data points, you calculate 5 first differences.

[5] You can tell that a linear model is appropriate, because the first differences are constant (equal to -10).

[6] The authors find the initial value A by looking at the output corresponding to an input of 0. The authors find B by finding the amount the outputs change when the input increases by 1. Since the (evenly spaced) inputs increase by 1 on the table, B turns out to be -10, from the first differences.

✎ **Section 1.3 Exercises 11 and 13** A set of data is given.

(a) Find the first differences.

(b) Is a linear model appropriate? If so, find a linear model for the data.

(c) Use the model to complete the table.

11.

x	y	First difference
0	205	–
1	218	
2	231	
3	244	
4		
5		
6		

13.

x	y	First difference
0	23	–
1	19	
2	16	
3	11	
4		
5		
6		

📖 **Read Section 1.3 Example 3**, in which the authors find a linear function that is appropriate to model data relating depth and pressure. Note that successive inputs in this data increase by 10.

[7] *What do the authors do to determine if a linear model is appropriate?*
Answer in your own words:

[8] *How do the authors find the parameter A in the model?*
Answer in your own words:

[9] *How do the authors find the parameter B in the model?*
Answer in your own words:

---

[7] The authors calculate the first differences. The authors find that the first differences are all the same (equal to 4.5), so a linear model is appropriate.

[8] The authors find the initial value A by looking at the output corresponding to an input of 0.

[9] The authors find B by finding the amount the outputs change when the inputs increase by 1. The table shows that when the input increases by 10 ft, the output increases by the first difference, 4.5 lb/in^2, so when the input increases by 1 ft, the output increases by $4.5/10 = 0.45$ lb/in^2.

📖 **Compare the data in Section 1.3 Exercises 19-21,** in which you are asked to construct linear models from data with evenly spaced inputs.

[10] In Exercise 19, as the inputs increase by _____, the outputs change by the first difference. This means that as the inputs increase by 1, the outputs change by $B = \dfrac{\text{First difference}}{\boxed{}}$.

In Exercise 20, as the inputs increase by _____, the outputs change by the first difference. This means that as the inputs increase by 1, the outputs change by $B = \dfrac{\text{First difference}}{\boxed{}}$.

In Exercise 21, as the inputs increase by _____, the outputs change by the first difference. This means that as the inputs increase by 1, the outputs change by $B = \dfrac{\text{First difference}}{\boxed{}}$.

In Exercise 22, as the inputs increase by _____, the outputs change by the first difference. This means that as the inputs increase by 1, the outputs change by $B = \dfrac{\text{First difference}}{\boxed{}}$.

✎ **Section 1.3 Exercise 23(a, b) Chocolate Powered** Car Two British entrepreneurs, Andy Pag and John Grimshaw, drove 4500 miles from England to Timbuktu, Mali, in a truck powered by chocolate. They used an ethanol that is made from old, unusable chocolate, and it took about 17 pounds of chocolate to make one gallon of ethanol. The table below gives data for the relationship between the amount of chocolate used and the number of miles driven.

(a) Use first differences to show that a linear model is appropriate for the data.

[11] How many first differences will you calculate? _____

(b) Find a linear model for the relations between the amount of chocolate used and the number of miles driven.

Miles driven	Pounds of chocolate used
0	0
20	17
40	34
60	51
80	68
100	85

---

[10] For each exercise, the amount by which each successive, evenly spaced input increases is the answer for both blanks. Exercise 19: 0.5 hours; Exercise 20: 10 chargers; Exercise 21: 1 thousand feet; and Exercise 22: 400 meters.

[11] You will calculate 5 first differences, since there are 6 data points.

CONTEXTS

Getting Information from a Linear Model

These exercises ask you to construct models in order to predict outputs for inputs that are not part of the data.

📖 **Read Section 1.3 Example 4,** in which the authors use the model for ocean pressure as a function of depth that they developed in Example 3 to predict the pressure at depths beyond the deepest measurement.

[12]*The inputs for this model are _____ (depths / pressures).*

The outputs for this model are _____ (depths / pressures).

The initial value of this model is _____ lbs/in².

If the depth increases by 1, then the pressure increases by _____ lbs/in².

✎ **Section 1.3 Exercise 23(c)** *Find this exercise on page 19.* Use the model you found in part (b) to predict how many pounds of chocolate it took to drive from England to Timbuktu.

CONCEPTS

✓ **1.3 Exercises – Concepts – Fundamentals** Complete the Fundamentals section of the exercises at the end of Section 1.3. Compare your answers to those at the back of the textbook, and make corrections as necessary.

📖 **Read the Concept Check for Section 1.3,** in the Chapter Review at the end of Chapter.

---

[12] The inputs are depths. The outputs are pressures. The initial value is 14.7 lbs/in². If the depth increases by 1, then the pressure increases by 0.45 lbs/in².

1.3 Solutions to ✐ Exercises

✐ **Section 1.3 Exercise 7**

(a) The linear model we seek is an equation of the form

$$y = A + Bx$$

When the input x is 0, the output y is 5, so the initial value A is 5. The first differences are $12 - 5 = 7$, $19 - 12 = 7$, and $26 - 19 = 7$. So for each 1-unit increase in the inputs, the amount B by which the outputs increase is 7. We now express the model as

$$y = 5 + 7x$$

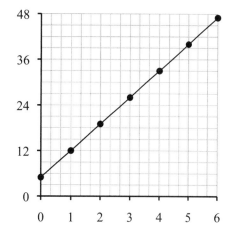

(b) When $x = 4$, $y = 5 + 7(4) = 33$. When $x = 5$, $y = 5 + 7(5) = 40$. When $x = 6$, $y = 5 + 7(6) = 47$.

x	y
0	5
1	12
2	19
3	26
4	**33**
5	**40**
6	**47**

✐ **Section 1.3 Exercises 11 and 13**

11(a) The first differences are in the table below.

(b) The inputs are evenly spaced, and the first differences are constant (each is 13), so a linear model is appropriate for this data. We are looking for a model of the form $y = A + Bx$.

When the input x is 0, the output y is 205, so the initial value A is 205. The first differences are 13, so for each 1-unit increase in the inputs, the amount B by which the outputs increase is 13. We now express the model as $y = 205 + 13x$.

x	y	First difference
0	205	–
1	218	**218 – 205 = 13**
2	231	**231 – 218 = 13**
3	244	**244 – 231 = 13**
4	**205 + 13(4) = 257**	
5	**205 + 13(5) = 270**	
6	**205 + 13(6) = 283**	

13(a) The first differences are in the table.

(b) The first differences are not constant, so a linear model is not appropriate.

x	y	First difference
0	23	–
1	19	**19 – 23 = –4**
2	16	**16 – 19 = –3**
3	11	**11 – 16 = –5**
4		
5		
6		

✐ **Section 1.3 Exercise 23 Chocolate Powered Car**

(a) The inputs are evenly spaced, and the first differences are constant, each equal to 17, so a linear model is appropriate.

(b) We are looking for a model of the form $y = A + Bx$.

When the number of miles driven x is 0, the amount of chocolate y is 0, so the initial value A is 0. The first differences are 17, so for each 20-mile increase in the distance driven, the amount of chocolate used increases by 17 lbs. Thus for each 1-mile increase in the distance driven, the amount of chocolate used increases by $17/20 = 0.85$ lbs. So B in the model is 0.85. We now express the model as $y = 0 + 0.85x$, or simply $y = 0.85x$.

(c) The distance from England to Timbuktu is 4500 miles. Applying the model, we have $y = 0.85(4500) = 3825$. So we can predict that it would take 3825 pounds of chocolate to make the 4500-mile trip.

1.4 Functions: Describing Change

OBJECTIVES

☑ *Check the box when you can do the exercises addressing the objective, such as those included in this study guide, which are listed here.*

Exercises

Definition of Function

☐ Determine if a relation written as a set of ordered pairs is a function. **7, 9**

☐ When a relation is not a function, give a specific example (with numbers) of an input that has more than one output. **9, 11, 25, 35**

Which Two-Variable Data Represent Functions?

☐ When you are given that y is a function of x, determine the input, output, independent variable and dependent variable. **11**

☐ Given a table of two-variable data, determine if one variable is a function of the other. **11, 53(b), 61**

☐ Given a function, determine the net change between two inputs. **53(b), 57**

Which Equations Represent Functions?

☐ Put an equation in function form to determine if the equation defines a function. **25, 57**

☐ When a relation given by an equation is not a function, evaluate the equation to find a numerical example of an input that gives multiple outputs. **25, 35,**

Which Graphs Represent Functions?

☐ Use the Vertical Line Test to determine if a relation given by a graph defiles a function. **35**

☐ When a relation is given by a graph, find the output(s) that correspond to a particular input. See Section 1.2 Example 5. **35**

Four Ways to Represent a Function

☐ Represent a given function verbally, numerically, symbolically, and graphically. **55**

📖 **Read Section 1.4 Example 5**, in which the authors use an equation to show that the speed that the wind blows is a function of the power that a windmill generates, and vice versa.

[6]*Suppose x is a positive real number. The relation given by $y^3 = x$ _____ (does / does not) define y as a function of x. The relation given by $y^2 = x$ _____ (does / does not) define y as a function of x. See Algebra Toolkit C.1.*

🖊 **Section 1.4 Exercise 25** These questions refer to the equation $4x^3 + 2 = y^2$.

(a) Does the equation define y as a function of x?
(b) Does the equation define x as a function of y?

Solution

Which Graphs Represent Functions?

These exercises ask you to determine if a relation represented by a graph is a function, using the Vertical Line Test. Find the Vertical Line Test in a blue box in the text.

📖 **Read Section 1.4 Example 7**, in which the authors use the Vertical Line Test to determined whether or not a relation given by a scatter plot is a function.

[7] *As in Hints and Tips 1.4, to show that y is not a function of x, the authors give a specific example; they explain that the input _____, corresponds to the outputs _____ and _____. If you know one of the people surveyed is 40 years old, do you know how in many elections he or she voted? Answer in your own words:*

---

[5] Computing, $\sqrt{4} = 2$. If $x^2 = 4$, then x = –2 or 2. Note that the square root symbol $\sqrt{n}$ always only refers to the positive square root of n.

[6] The relation given by $y^3 = x$ is a function, since x has exactly one real cube root. The relation given by $y^2 = x$ is not a function, since positive real numbers have two square roots, namely $\sqrt{x}$ and $-\sqrt{x}$.

[7] The authors show that the input x = 40 corresponds to the outputs y = 1 and y = 5. No, if you know that a surveyed person is 40 years old, then you know he or she voted in either 1 or 5 elections, but you do not know which.

📖 **Read Section 1.4 Example 8**, in which the authors use the Vertical Line Test to determine whether or not a relation given by a graph is a function.

[8] *As in Hints and Tips 1.4, to show that y is not a function of x in part (b), the authors give a specific example; they explain that the input _____, corresponds to the outputs _____ and _____. In part (b), for what do the authors use the equation, and for what do they use the Vertical Line Test? Answer in your own words:*

✏ **Section 1.4 Exercise 35** An equation and its graph are given. Use the Vertical Line Test to determine whether the equation defines y as a function of x.

CONTEXTS

Which Two-Variable Data Represent Functions?

These exercises ask you to determine if a relation given by a table of two-variable data is a function, in order to study the net change in the dependent variable as the independent variable changes. Find net change in bold typeface in Section 1.4 of the textbook.

📖 **Section 1.4 Example 3**, in which the authors find the net change in the average price of gasoline over different intervals of time in California.

[9] *When the net change is positive, there is a net _____ (increase / decrease) in the dependent variable. When the net change is negative, there is a net _____ (increase / decrease) in the dependent variable.*

✏ **Section 1.4 Exercise 53(b) (Excerpt)** The number of "big box" retail stores has increased nationwide in recent years. The following table shows the number of existing big box retail stores in the given years.

(b) Show that the number of big box retail stores is a function of the year. Find the net change in the number of big box stores from 2003 to 2005, and from 1997 to 2006.

Year	1997	1998	1999	2000	2001	2002	2003	2004	2005	2006
Number of big box retail stores	27	30	32	38	38	42	46	46	46	49

Solution

---

[8] The authors show that the input $x = 4$ corresponds to the outputs $y = -2$ and $y = 2$. The authors use the Vertical Line Test to determine that the equation (and its graph) does not define a function and to find an input that has multiple outputs, and they use the equation to find the those outputs.

[9] A positive net change indicates an increase in the dependent variable. A negative net change indicates a decrease in the dependent variable.

Which Equations Represent Functions?

These exercises ask you to determine if a relation given by an equation is a function, in order to study the net change in the dependent variable as the independent variable changes.

📖 **Read Section 1.4 Example 5 again**, noting that the authors show that the speed the wind blows is a function of the power that a windmill generates, in order to calculate the net change of wind speed as power increases.

[10] *As the wind speed changes from 7 to 10 km/h, the net change in the power is _____ (positive / negative), there is a net _____ (increase / decrease) in the power.*

✐ **Section 1.4 Exercise 57 Profit** A university music department plans to stage the opera *Carmen*. The fixed cost for the set, costumes, and lighting is $5000, and they plan to charge $15 a ticket. So if they sell x tickets, then the profit P they will make from the performance is given by the equation

$$P = 15x - 5000$$

(a) Show that P is a function of x.
(b) Find the net change in the profit P when the number of tickets sold increases from 100 to 200.
(c) Express x as a function of P.
(d) Find the net change in the number of tickets sold when the profit changes from $0 to $5000.

Solution

Which Graphs Represent Functions?

These exercises ask you to determine if a relation given by a graph is a function, in order to study the net change in the dependent variable as the independent variable changes.

📖 **Read Section 1.4 Example 6**, in which authors use the Vertical Line Test to show that the temperature of a cup of coffee is a function of time, in order to find the net change in temperature as time passes.

[11] *The net change in the temperature of the coffee over the first 30 minutes is a net _____ (increase / decrease) of _____ °F.*

---

[10] The net change in the power as the wind speed changes from 7 to 10 km/h is positive, so there is a net increase in the power.

Four Ways to Represent a Function

These exercises ask you to represent a given function verbally, numerically, symbolically and graphically.

📖 **Read Section 1.4 Example 9**, in which the authors represent the cost of a pizza as a function of the number of toppings verbally, numerically, algebraically and graphically.

[12] *In this exercise, y represents _____, and x represents _____.*

✏ **Section 1.4 Exercise 55 Deliveries** A feed store charges $25 for each bale of hay plus a $15 delivery charge. The cost C of a load of hay is a function of the number x of bales purchased. Express this function verbally, symbolically, numerically, and graphically.

Solution

Sketch your graphical representation here.

Concepts

✓ **1.4 Exercises – Concepts – Fundamentals** Complete the Fundamentals section of the exercises at the end of Section 1.4. Compare your answers to those at the back of the textbook, and make corrections as necessary.

📖 **Read the Concept Check for Section 1.4**, in the Chapter Review at the end of Chapter 1.

---

[11] There is a net decrease of 100°F in the first 30 minutes.

[12] In this exercise, y represents the cost of a pizza, and x represents the number of toppings. From the equation, the cost of a pizza is $9 plus $0.75 for each additional topping. From the table, the equation is $y = 9 + x$.

1.4 Solutions to ✎ Exercises

✎ Section 1.4 Exercise 7 and 9

7. This relation is a function because each input corresponds to exactly one output.

9. Following **Hints and Tips 1.4**, we show this relation is not a function by giving an example of an input with multiple outputs: the input 3 corresponds to three different outputs (4, 5, and 9).

✎ Section 1.4 Exercise 11

(a) Following **Hints and Tips 1.4**, we show the variable y is not a function of the variable x by giving an example of an input with multiple outputs: when the input x is 2, there are two different outputs (9 and 3). Also, when the input x is 4, there are two different outputs (2 and 8).

(b) The variable x is a function of the variable y because each value of y corresponds to exactly one value of x. Since x is a function of y, the variable y is the independent variable and the variable x is the dependent variable.

✎ Section 1.4 Exercise 25

(a) To answer this question, we try to write the equation in function form, with the dependent variable y alone on one side:

$$y = \pm\sqrt{4x^3 + 2}$$

We see from this that y is not a function of x. For example, if $x = 0$, then $y = \sqrt{2}$ or $y = -\sqrt{2}$.

(b) Again, we try to write the equation in function form, with the dependent variable alone on one side.

$$4x^3 + 2 = y^2 \qquad \text{Equation}$$
$$4x^3 = y^2 - 2 \qquad \text{Subtract 2 from each side}$$
$$x^3 = \frac{y^2 - 2}{4} \qquad \text{Divide each side by 4}$$
$$x = \sqrt[3]{\frac{y^2 - 2}{4}} \qquad \text{Take the cube root}$$

We see that x is a function of y because for each value of y we can use the equation $x = \sqrt[3]{\frac{y^2 - 2}{4}}$ to calculate exactly one corresponding value of x.

✎ Section 1.4 Exercise 35

Following **Hints and Tips 1.4**, we show the equation $4y^2 - 3x^2 = 1$ does not define y as a function of x by giving an example of an input x with multiple outputs y: The vertical line through $x = 2$ passes through the graph at two different points, so. The input 2 corresponds to two different outputs. To find these outputs, we use the equation. If $x = 2$, then $4y^2 - 3(2)^2 = 1$, so $y = \sqrt{13/4}$ or $y = -\sqrt{13/4}$.

✎ Section 1.4 Exercise 53(b) (Excerpt)

(b) The number of big box retail stores is a function of the year because for each year there is exactly one number of retail stores. Since the number of big box retail stores was 46 in 2003 and 46 in 2005, the net change from 2003 to 2005 is $46 - 46 = 0$. A net change of zero means that there was neither net increase nor decrease in the number of big box stores between 2003 and 2005.

Since the number of retail stores was 27 in 1997 and 49 in 2006, the net change from 1997 to 2006 is $49 - 27 = 22$. A positive net change means that there was a net increase in the number of big box stores from 1997 to 2006.

✎ Section 1.4 Exercise 57 Profit

(a) P is a function of x because for each value of x, the equation gives exactly one corresponding value for P.

(b) When the number of tickets sold is 100, the profit is

$$P = 15(100) - 5000 = -3500$$

dollars. (A negative profit indicates that the company is losing money.) When the number of tickets sold is 200, the profit is

$$P = 15(200) - 5000 = -2000$$

So the net change in profit is $-2000 - (-3500) = 5500$ dollars.

(c) We need to rewrite the equation with x alone on one side:

$$P = 15x - 5000 \qquad \text{Equation}$$

$$P + 5000 = 15x \qquad \text{Add 5000 to each side}$$

$$x = \frac{P + 5000}{15} \qquad \text{Divide by 15 (and switch sides)}$$

This equation defines x as a function of P.

(d) When the profit is $0, the number of tickets that the music department sells is $(0+5000)/15 = 1000/3$ tickets. When the profit is $5000, the number of tickets it sells is $(5000+5000)/15 = 2000/3$ tickets.

So the net change in the number of tickets the music department sells is $2000/3 - 1000/3 = 1000/3 \approx 333$ tickets. This shows the expected net increase in ticket sales between earning $0 profit and earning $5000 profit.

✎ Section 1.4 Exercise 55 Deliveries

Verbally We can express this function verbally by saying, "The cost of a bale of hay is \$15 for delivery plus \$25 per bale."

Symbolically Let C be the cost of a load of hay, and let x be the number of bales purchased. We can express the function symbolically by

$$C = 15 + 25x$$

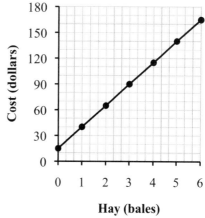

Numerically We make a table of values that gives a numerical representation of the function.

Hay (bales)	Cost (dollars)
0	15
1	40
2	65
3	90
4	115

Graphically We plot the data in the table to get the graphical representation.

1.5 Function Notation: The Concept of Function as a Rule

OBJECTIVES

☑ *Check the box when you can do the exercises addressing the objective, such as those included in this study guide, which are listed here.*

	Exercises
Function Notation	
☐ Use function notation to express a rule as a function.	9, 13, 17
☐ Apply orders of operation when translating a rule into function notation, and vice versa.	13, 17
☐ Convert a function given in function notation to an equation of two variables. Conversely, when an equation represents a function, write it in function notation.	27, 29
Evaluating Functions—Net Change	
☐ Calculate $f(a)$ for a given function f and input a.	39, 55, 61, 67
☐ Use a function that models the real world to answer questions about the context.	61, 67
☐ Given a function that models the real world, explain what expressions written using function notation represent in context.	61, 67, 69
☐ Given a function in function notation, determine the net change between two inputs. Interpret your answer in context.	61
The Domain of a Functions	
☐ Determine the domain of a function from its symbolic representation in function notation. In particular, determine the domains of functions with fractions or square roots.	49, 51
Piecewise Defined Functions	
☐ Evaluate a piecewise defined function given in function notation.	67, 69
☐ Write the function notation for a piecewise defined function that models the real world, and use it to answer questions about the context.	69

SKILLS

Function Notation

These exercises ask you to use function notation $f(x)$, and the language "f of x" or "the image of x under f." Find these expressions in Section 1.5 of the textbook. In this section, the exercises also use the vocabulary, dependent and independent variables, which was introduced in Section 1.4.

[1] *If r is a function of s, then r is the _____ (independent / dependent) variable, and s is the _____ (independent / dependent) variable.*

If r is a function of s, then r is the _____ (input / output), and s is the _____ (input / output).

In a function, the independent variable corresponds to the _____ (input / output), and the dependent variable corresponds to the _____ (input / output).

📖 **Read Section 1.5 Example 1,** in which the authors get information about a function from its symbolic representation in function notation.

[2] *Write this using function notation: "f of 2 equals 16." _____*
What would you say if you read this aloud? f(3) =24
Answer in your own words:

What would you say if you read this aloud? f(x) = 8x
Answer in your own words:

✏ **Section 1.5 Exercise 9** Let $f(x) = 2x + 1$.
   (a) What is the name of the function?
   (b) What letter represents the input? What is the output?
   (c) What rule does this function represent?
   (d) Find $f(10)$. What does $f(10)$ represent?

   Solution

📖 **Read Section 1.5 Example 2,** in which the authors express a given rule in function notation.

📖 **Read Section 1.5 Example 3,** in which the authors give a verbal description of a function expressed in function notation.

---

[1] If r is a function of s, then r is the dependent variable and s is the dependent variable. If r is a function of s, then r is the output and s is the input. In a function, the independent variable corresponds to the input, and the dependent variable corresponds to the output (since the output depends on the input).

[2] In function notation, f of 2 equals 16 is $f(2) = 16$. Aloud, $f(3) = 24$ reads, "f of 3 equals 24." Aloud, $f(x) = 8x$ reads, "f of x equals eight x," or "f of x equals eight times x."

[3]*Match the rule with the function notation.*

Function Notation	Rule	Answers
1. $2x + 5$	A. Add 5, then multiply by 2.	1. _____
2. $2(x + 5)$	B. Add 5, then square.	2. _____
3. $\dfrac{x}{2} + 5$	C. Multiply by 2, then add 5.	3. _____
4. $(x + 5)^2$	D. Divide by 2, then add 5.	4. _____
5. $\dfrac{x + 5}{2}$	E. Add 5, then divide by 2.	5. _____
6. $x^2 + 5$	F. Square, then add 5.	6. _____

✏ **Section 1.5 Exercise 13 and 17**

13. Express the given rule in function notation. "Add 2, then multiply by 5."

17. Give a verbal description of the function, $f(x) = \dfrac{x}{2} + 7$.

Solution

Solution

📖 **Read Section 1.5 Example 4**, in which the authors write a function as an equation and an equation as a function.

[4]*Does every equation define a function? See Section 1.4 Example 4.*
Answer in your own words:

Does every function given in function notation define an equation?
Answer in your own words:

✏ **Section 1.5 Exercise 27** Consider the equation, $y = 3x^2$.
(a) Does the equation define a function with x as the independent variable and y as the dependent variable? If so, express the equation in function notation with x as the independent variable.
(b) Does the equation define a function with y as the independent variable and x as the dependent variable? If so, express the equation in function notation with y as the independent variable.

Solution

---

[3] 1. C; 2. A; 3. D; 4. B; 5. E; 6. F

[4] Not every equation defines a function; in Section 1.4 Example 4, the given equation does not define w as a function of z. Every function in function notation defines an equation; to get an equation, replace the output by a variable.

✐ **Section 1.5 Exercise 29** Express the function, $f(x) = 3x^2 - 1$, as an equation. Identify the independent and dependent variables.

Solution

Evaluating Functions—Net Change

You practice interpreting function notation in these exercises.

📖 **Read Section 1.5 Example 5**, in which the authors evaluate a function written in function notation.

[5] *True or false: If f(x) = x², then f(−2) = −2² = −4.*

✐ **Section 1.5 Exercise 39** Evaluate the function, $f(x) = x^2 + 2x$, at the indicated values.

(a) $f(0) =$ (b) $f(\sqrt{2}) =$ (c) $f(-2) =$ (d) $f(a) =$

The Domain of a Function

Review Section 1.2 Example 1(e), in which the authors find the domain of a relation from its numeric representation in a table. In that example, the authors state the domain by listing the 5 inputs. In contrast, the exercises in this section ask you to find the domain of a function from its symbolic representation in function notation.

Function notation explains what to do to an input to get the corresponding output, but function notation does not explicitly specify the inputs (i.e. the domain) to use. Read the introduction to the subsection, 'The Domain of a Function,' on page 56 of the textbook, in which the authors explain the conventions for specifying the domain of a function represented in function notation. In these exercises, you will determine the domains of functions given in function notation, using these conventions.

📖 **Read Section 1.5 Example 7**, in which the authors find the domains of functions with square roots or quotients.

[6] *The domain of the quotient in part (a) is the set of values of x at which the denominator is _____.*
The domain of the radical in part (b) is the set the values of x for which the term under the radical sign is _____.

✐ **Section 1.5 Exercise 49 and 51** Find the domain of the function.

49. $h(x) = \dfrac{1}{x-5}$

Solution

51. $k(x) = \sqrt{x-5}$

Solution

---

[5] False. $f(-2) = (-2)^2 = 4$. Tip: Include the parentheses when substituting; substitute (-2), not -2, for x.

[6] The domain of a quotient is the set of values of x at which the denominator is not zero. The domain of a radical is the set of values of x such that the term under the radical is greater than or equal to zero.

Piecewise Defined Functions

In these exercises, you will use function notation to represent piecewise defined functions. Read the definition of a piecewise defined function in Section 1.5 in the textbook.

📖 **Read Section 1.5 Example 8**, in which the authors evaluate a piecewise defined function that models the cost of cell phone usage.

[7]*What is the domain of C(x)?* _____

To find C(x) for an input that is between 0 and 400, the authors use the formula C(x) = _____

To find C(x) for an input that is greater than 400, the authors use the formula C(x) = _____

✎ **Section 1.5 Exercise 55** Evaluate the piecewise defined function at the indicated values.

$$f(x) = \begin{cases} -x & \text{if } x < 0 \\ x & \text{if } x \geq 0 \end{cases}$$

(a) $f(-1) =$ (b) $f(0) =$ (c) $f(-2) =$ (d) $f(1) =$

📖 **Read Section 1.5 Example 9**, in which the authors represent the water bill for household as a piecewise defined function.

[8]*When creating the formula 0.008x, the authors multiply (fill in the blank with the appropriate units)*
0.008 _____ by x _____, to get 0.008x _____.

✎ **Section 1.5 Exercise 69(a) Internet Purchases** An Internet bookstore charges $15 shipping for orders under $100, but provides free shipping for orders of $100 or more. The cost C of an order is a function of the total price x of the books purchased.

[9]*For example, if your books cost $35, what is the total cost? C(35) =* _____ *dollars. If your books cost $145, what is the total cost? C(145) =* _____ *dollars.*

(a) Express C as a piecewise defined function.

Solution

$$C(x) = \begin{cases} \underline{\hspace{3cm}} & \text{if } x < 100 \\ \underline{\hspace{3cm}} & \text{if } x \geq 100 \end{cases}$$

---

[7] From the context or from the formula, the domain is all real numbers greater than or equal to zero. If x is between 0 and 400, then $C(x) = 39$. If x is greater than 400, the formula is $C(x) = 39+0.2(x-400)$.
[8] The authors multiplied 0.008 dollars per gallon by x gallons, to get 0.008 dollars.
[9] $C(35) = 35 + 15 = \$50$, and $C(145) = \$145$.

CONTEXTS

Evaluating Functions—Net Change

These exercises revisit the concept of net change introduced in Section 1.4, this time taking advantage of the convenience of function notation. Find the definition of net change, using function notation, in a blue box in Section 1.5 of the textbook. Compare this to the definition of net change at the bottom of page 37. In these exercises, you will calculate the net change using the concise formula given in function notation.

📖 **Read Section 1.5 Example 6**, in which the authors use function notation to express the net change in an astronaut's weight as her distance from Earth changes.

[10]*Write the first row of the table in part (b) in function notation.* _____
To find the net change in the astronaut's weight in part (c), the authors use the definition of net change, replacing f by _____, *a by* _____, *and b by* _____.

> Refer back to page 11
> **Hints and Tips 1.2b: What does the algebra represent in context?**

Hints and Tips 1.2b explains the author's answer to Example 6(a) and 6(c).

⇨ **Section 1.5 Example 6(a)** Evaluate $w(100)$. What does your answer represent?

Solution
The authors find $w(100) \approx 124$. Using Hint 1.2b, we write the following sentences as scratch work. Algebraically, 124, is the output, when the input is 100. In context, 124 pounds is the weight when the height is 100 miles. Here is our final more eloquent answer. The astronaut weighs 124 lbs when she is 100 miles from Earth's surface.

⇨ **Section 1.5 Example 6(c)** Find the net change in the weight of the astronaut from an elevation of 100 miles to an elevation of 400 miles. Interpret your results.

Solution
The authors find the net change from 100 to 400 miles is –17 pounds. We write the following sentences as scratch work. Algebraically, the net change, –17, is the amount the output changes when the input increases from 100 to 400 miles. In context, the net change, –17 pounds, is the amount the astronaut's weight changes as her height increases from 100 miles to 400 miles. Here is our final, more eloquent answer. As her height increases from 100 miles to 400 miles above Earth's surface, the astronaut's weight has a net decrease of 17 pounds.

✎ **Section 1.5 Exercise 61 How Far Can You See?** Because of the curvature of the earth, the maximum distance D that one can see from the top of a tall building or from an airplane at height h is modeled by the function

$$D(h) = \sqrt{7920h + h^2}$$

where D and h are measured in miles.
(a) Find $D(0.1)$. What does this value represent?

---

[10] The first row says $w(0) = 130$. The authors solved part (c) by replacing f by w, a by 100 and b by 400.

(b) How far can one see from the observation deck of Toronto's CN Tower, 1135 ft above the ground? (Remember that 1 mile is 5280 feet.)

(c) Commercial aircraft fly at an altitude of about 7 mi. How far can the pilot see?

(d) Find the net change of the distance D as one climbs the CN Tower from a height of 100 ft to a height of 1135 ft.

[11] *The value 1135 ft in (b) is the _____ (distance one can see / height).*
The value 7 mi in (c) is the _____ (distance one can see / height).

Solution

Piecewise Defined Functions

In these exercises, you explore models governed by different formulas on different parts of their domains.

📖 **Read Section 1.5 Example 8 again**, paying particular attention to part (c) in which the authors explain what the outputs mean in context.

[12] *What does the input x represent in context and what are its units?*
Answer in your own words:

What does the output C(x) represent in context and what are its units?
Answer in your own words:

---

[11] The 1135 ft is the height of the observation deck. The 7 mi is the height of the airplane.

[12] The input x represents the time spent on the phone, in minutes. The output $C(x)$ represents the monthly charges when x minutes are used, in dollars.

✎ **Section 1.5 Exercise 67 Income Tax** In a certain country, income tax T is assessed according to the following function of income x (in dollars):

$$T(x) = \begin{cases} 0 & \text{if } 0 \le x \le 10{,}000 \\ 0.08x & \text{if } 10{,}000 < x \le 20{,}000 \\ 1600 + 0.15(x - 20{,}000) & \text{if } 20{,}000 < x \end{cases}$$

(a) Find $T(5000)$, $T(12{,}000)$, and $T(25{,}000)$. What do these values represent?
(b) Find the tax on an income of $20,000.
(c) If a businessman pays $25,000 in taxes in this country, what is his income?

[13] *In part (c), which is bigger, the businessman's income, x, or the amount of tax he pays, $25,000? Symbolically, this means that $25,000 _____ x.*

Solution

📖 **Read Section 1.5 Example 9 again**, paying particular attention to part (b), in which the authors interpret the results verbally.

[14] *What does the input x represent in context and what are its units?*
Answer in your own words:

What does the output C(x) represent in context and what are its units?
Answer in your own words:

---

[13] The businessman's income is greater than the tax he pays (as one would hope!). So $x > \$25{,}000$.

[14] The input x represents the volume of water used, in gallons. The output $C(x)$ represents the water bill for a household using x gallons of water per month. When creating a formula for $C(x)$, the authors first determine the domains $x < 4000$, and $x \ge 4000$.

✐ **Section 1.5 Exercise 69 Internet Purchases** An Internet bookstore charges $15 shipping for orders under $100, but provides free shipping for orders of $100 or more. The cost C of an order is a function of the total price x of the books purchased.

(a) See your work on page 35, where you wrote the cost as a piecewise defined function.

(b) Find $C(75)$, $C(100)$, and $C(105)$. What do these values represent?

Solution

CONCEPTS

✓ **1.5 Exercises – Concepts – Fundamentals** Complete the Fundamentals section of the exercises at the end of Section 1.5. Compare your answers to those at the back of the textbook, and make corrections as necessary.

📖 **Read the Concept Check for Section 1.5**, in the Chapter Review at the end of Chapter 1.

1.5 Solutions to ✐ Exercises

✐ **Section 1.5 Exercise 9**

(a) The name of the function is f.

(b) The letter x represents the input. The output is $2x + 1$.

(c) The rule is, "Multiply the input by 2, and then add 1."

(d) $f(10) = 2(10) + 1 = 21$. The symbol $f(10)$ represents the value of the function at $x = 10$.

✐ **Section 1.5 Exercise 13**

First we need to choose a letter to represent this rule. Let's let h stand for the rule, "Add 2, then multiply by 5." Then for any input x, adding 2 yields $x + 2$, then multiplying by 5 gives $5(x + 2)$. Thus we can write

$$h(x) = 5(x + 2)$$

✐ **Section 1.5 Exercise 17**

The rule f is "Divide the input by 2, and then add 7."

✐ **Section 1.5 Exercise 27**

(a) The equation is in function form, with the dependent variable y alone on one side. Thus, the equation determines a function with dependent variable y and independent variable x. If we call this rule k, then we can write this function as $k(x) = 3x^2$.

(b) To answer this question, we try to write the equation in function form, with the dependent variable x alone on one side.

$$y = 3x^2 \qquad\qquad \text{Equation}$$

$$\frac{y}{3} = x^2 \qquad\qquad \text{Divide each side by 3}$$

$$x = \pm\sqrt{\frac{y}{3}} \qquad\qquad \text{Take the square root (and switch sides)}$$

This does not define a function with independent variable y and dependent variable x, since the input $y = 3$ corresponds to two different outputs (-1 and 1).

Section 1.5 Exercise 29

If we write y for the output $f(x)$, then we can write the function as

$$y = 3x^2 - 1$$

The dependent variable is y, and the independent variable is x.

Section 1.5 Exercise 39

(a) $f(0) = (0)^2 + 2(0) = 0$ (b) $f(\sqrt{2}) = (\sqrt{2})^2 + 2(\sqrt{2}) = 2 + 2\sqrt{2}$

(c) $f(-2) = (-2)^2 + 2(-2) = 4 - 4 = 0$ (d) $f(a) = a^2 + 2a$

Section 1.5 Exercise 49 and 51

49. $h(x) = \dfrac{1}{x-5}$ 51. $k(x) = \sqrt{x-5}$

The function is not defined when the denominator is 0. that is, when x is 5. So the domain of f is $\{x \mid x \neq 5\}$.

The function is defined only when $x - 5 \geq 0$. This means that $x \geq 5$. So the domain of k is $\{x \mid x \geq 5\}$.

Section 1.5 Exercise 55

(a) Since $-1 < 0$, we have $f(-1) = -(-1) = 1$.

(b) Since $0 \geq 0$, we have $f(0) = 0$.

(c) Since $-2 < 0$, we have $f(-2) = -(-2) = 2$.

(d) Since $2 \geq 0$, we have $f(2) = 2$.

Section 1.5 Exercise 61 How Far Can You See?

(a) $D(0.1) = \sqrt{7920(0.1) + (0.1)^2} \approx 28.14$.

Using **Hints and Tips 1.2b**, we write the following sentences as scratch work. Algebraically, 28.14 is the output, when the input is 0.1. In context, 28.14 miles is the distance that one can see when the height is 0.1 miles. Here is our final, more eloquent answer. One can see 28.14 miles from a vantage point a tenth of a mile high.

(b) First we convert 1135 ft to miles.

$$\frac{1135 \text{ feet}}{(5280 \text{ feet/mile})} \approx 0.215 \text{ miles}$$

Now we evaluate $D(0.215) = \sqrt{7920(0.215) + (0.215)^2} \approx 41.27$. One can see approximately 41.27 miles from atop the observation deck of Toronto's CN Tower.

(c) From a height of 7 miles, the pilot can see $D(7) = \sqrt{7920(7) + (7)^2} \approx 235.56$ miles.

(d) First we convert 100 feet to miles. We have $\dfrac{100 \text{ feet}}{(5280 \text{ feet/mile})} \approx 0.0189 \text{ miles}$.

Now

$$D(0.0189) = \sqrt{7920(0.0189) + (0.0189)^2} \approx 12.24$$

So, the net change is the distance D one can see as one climbs from 100 ft to 1135 ft is

$$D(0.215) - D(0.0189) = 235.56 - 12.24 = 223.32$$

The net change of 223.33 miles is the net increase in the distance one can see when climbing from 100 to 1135 feet.

✐ **Section 1.5 Exercise 67 Income Tax**

(a) Since $0 \le 5000 \le 10,000$, we have $T(5000) = 0$. Using **Hints and Tips 1.2b**, we rite the following sentences as scratch work. Algebraically, 0 is the output when 5000 is the input. In context, $0 of tax is assessed when the income is $5000. Here is our more eloquent final answer. The expression $T(5000) = 0$ means that an individual with an income of $5000 does not pay any taxes.

Since $10,000 < 12,000 \le 20,000$, we have $T(12,000) = 0.08(12,000) = 960$. Following the model above, $T(12,000) = 960$ means that an individual who earns $12,000 pays $960 in taxes.

Since $20,000 < 25,000$, $T(25,000) = 1600 + 0.15(25,000 - 20,000) = 2350$. So $T(25,000) = 2350$ means that an individual who earns $25,000 pays $2350 in taxes.

(b) The tax on an income of $20,000 is $T(20,000)$. Since $10,000 \le 20,000 \le 20,000$, we have $T(20,000) = 0.08(20,000) = \$1600$.

(c) We want to find x so that $T(x) = 25,000$. To do this, we first determine which of the three formulas were used to calculate the businessman's tax. The amount of tax the businessman pays is less than the amount of income he earns, forcing his income to be greater than his tax of $25,000. This implies that we use the third formula to calculate his tax.

Now we solve:

$25{,}000 = 1600 + 0.15(x - 20{,}000)$	The equation $25{,}000 = T(x)$
$25{,}000 = 0.15x - 1400$	Distribute and combine constants
$26{,}400 = 0.15x$	Add 1400 to each side
$x = 176{,}000$	Divide each side by 0.15 (and switch sides)

Thus the businessman's income is $176,000.

✎ Section 1.5 Exercise 69 Internet Purchases

(a) This piecewise defined function has a different formula for purchases under $100 and purchases over $100. Thus we will define $C(x)$ in two pieces: for $x < 100$ and for $x \geq 100$.

The total cost is the cost of the order plus the cost of shipping, which is $x + 15$ for $x < 100$, and $x + 0$ for $x \geq 100$. We have

$$C(x) = \begin{cases} x + 15 & \text{if } x < 100 \\ x & \text{if } x \geq 100 \end{cases}$$

(b) Since $0 \leq 75 \leq 100$, we have $C(75) = 75 + 15 = 90$. Using **Hints and Tips 1.2b**, we write the following sentences as scratch work. Algebraically, 90 is the output when 75 is the input. In context, $90 is the total cost when the order is $75. Here is our more eloquent final answer. The expression $C(75) = 90$ means that a $75 order costs $90 with shipping.

Since $100 \leq 100$, we have $T(100) = 100$. Following the model above, $T(100) = 100$ means that a $100 order costs $100 since there is no shipping fee.

Since $100 \leq 105$, $T(105) = 105$. Following the model above, $T(105) = 105$ means that a $105 order costs $105 since there is no shipping fee.

1.6 Working with Functions: Graphs and Graphing Calculators

OBJECTIVES

☑ *Check the box when you can do the exercises addressing the objective, such as those included in this study guide, which are listed here.*   **Exercises**

Graphing a Function from a Verbal Description

☐ Sketch the graph of a function given its verbal description.   **49**

Graphs of Basic Functions

☐ Create a table of values and sketch the graphs of basic functions. These include the constant function, the identity function, powers of x like $\sqrt{x}$, x^2, and x^3, and related functions such as $x^3 - 8$ and $\sqrt{4x}$.   **13, 15, 19, 23, 33**

Graphing with a Graphing Calculator

☐ Change the viewing rectangle on a graphing calculator to better see the features of a graph.   **35, 41**

Graphing Piecewise Defined Functions

☐ Sketch the graphs of piecewise functions, including functions whose formulas include absolute values.   **31, 45, 55(a)**

☐ Interpret the meaning of jumps in the graph of a piecewise defined function that models the real world.   **55(b)**

GET READY...

This section uses vocabulary pertaining to a graphing calculator, like viewing rectangle, Xmin, Ymin, Xmax and Ymax, and skills, like choosing an appropriate viewing rectangle for a graph and graphing two functions on the same screen. This section also warns about the pitfalls of relying on a graph from a graphing calculator or computer. Refresh your understanding of these in Algebra Toolkit D.3.

✓ **Algebra Checkpoint** Test yourself by completing the Algebra Checkpoint at the end of this section of the textbook, and comparing your answers to those in the back of the textbook. Refer to Algebra Toolkit D.3 as necessary.

SKILLS

Graphs of Basic Functions

These exercises ask you to graph basic functions, like the constant function, the identity function, and powers of x like x^2, x^3 and $\sqrt{x}$. In order to prepare for Chapter 4, in which you will study how changes in the formula for a function are reflected in its graph, these exercises also include functions like $2x^2$, $x^2 + 3$ and $(x + 4)^2$, which are made by scaling and translating basic functions. Look for the definitions of the constant function and the identity function in bold typeface in Section 1.6 of the textbook.

📖 **Read Section 1.6 Examples 2 and 3**, in which the authors use the formula for a function to make a table of values, plot the points in the table and join them by a smooth curve.

[1] The authors use the same viewing rectangle on the three graphs in Figures 4, 5 and 6, so you could compare them easily. That viewing rectangle is [_____, _____] by [_____, _____].

📖 **Read Section 1.6 Example 4**, in which the authors graph functions that involve square roots.

[2] Why do the authors fail to include any negative x-values on the table?
Answer in your own words:

Hints and Tips 1.6: Choosing the x-values to put in your table when graphing a function

When you are graphing by hand, you must choose the x-values of the points that you will include in your table and plot on your graph.

- Choose the x-values within the domain of the function.
- Choose an interval of x-values that includes any features you know about, like x-intercepts, or maxima and minima.
- Begin with integer values of x, and then add x-values between the integers where you want to see more detail. For example, to see more detail near $x = 2$, include $x = 1.5, 2$ and 2.5. To see even more detail, include $x = 1.5, 1.75, 2, 2.25$ and 2.5.

In the textbook, an interest in making the tables and graphs fit nicely on the page influences the authors' choices of x-values, however you can see evidence that they use these guidelines.

⇨ In Section 1.6 Example 3(a), the authors choose x-values from –3 to 3, including the x-intercepts at $x = 0$.

⇨ In Section 1.6 Example 3(b) and (c), the authors choose x-values from –2 to 2, and include $x = -\frac{1}{2}$ and $x = \frac{1}{2}$ to provide more detail near $x = 0$.

⇨ In Section 1.6 Example 4, the authors use values within the domain, $\{x \mid x \geq 0\}$. They choose to provide more detail near $x = 0$, by including $x = \frac{1}{4}$ (choosing $\frac{1}{4}$ instead of $\frac{1}{2}$ because the square root of 4 is an integer).

Refer back to page 10
Hints and Tips 1.2a: Choosing scales for the x- and y-axes of a scatter plot

---

[1] [–5, 5] by [–5, 5].

[2] The x-values on the tables belong to the domain of the functions, which does not include negative numbers.

✏ **Section 1.6 Exercises 13, 15, 19 and 23** Sketch the graph of the function by first making a table of values.

13. $f(x) = 8$

x	$f(x)$

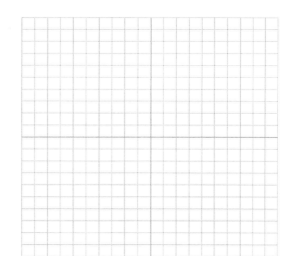

15. $g(x) = x - 4$

x	$g(x)$

19. $k(x) = -x^2$

x	$k(x)$

23. $G(x) = x^3 - 8$

x	$G(x)$

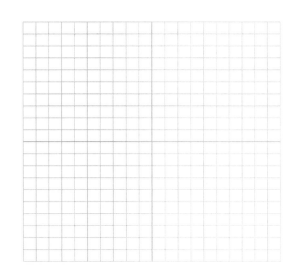

🖊 **Section 1.6 Exercise 33** Graph the functions $f(x) = \sqrt{x}$, $g(x) = \sqrt{4x}$, and $h(x) = \sqrt{x} - 4$, on the same coordinate axis.

x	$f(x)$	$g(x)$	$h(x)$

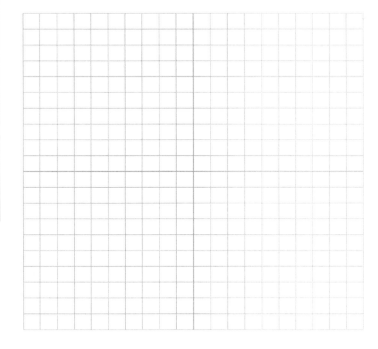

Graphing with a Graphing Calculator

These exercises ask you to choose an appropriate viewing rectangle to display graphs of functions on your graphing calculator. You also use the calculator to zoom in on the graph, in order to more accurately read the coordinates of points of interest, like where two graphs intersect.

📖 **Read Section 1.6 Example 5**, in which the authors graph a function on a graphing calculator.

[3] *What is the first thing the authors do to prepare to graph the function on the graphing calculator?*
Answer in your own words:

How do the authors know what viewing rectangle to use? Refer to Algebra Toolkit D.3 for a detailed discussion of viewing rectangles.
Answer in your own words:

✎ **Section 1.6 Exercise 35** Draw a graph of the function $f(x) = 4 + 6x - x^2$ in an appropriate viewing rectangle.

Solution

The viewing rectangle you use is [_____, _____] by [_____, _____].
Copy the graph from your calculator here.

📖 **Read Section 1.6 Example 6**, in which the authors find the coordinates at which two graphs intersect, using a graphing calculator.

[4] *The authors use the graphing calculator to estimate the coordinates (–1,0), (0, –1) and (2,3) where the graphs intersect. Check that the point (–1,0) actually does lie on the graph of both functions by showing that f(–1) = 0 and g(–1) = 0.*
Answer in your own words:

---

[3] The authors first express the function in equation form. The authors experiment with different viewing rectangles until they find one that seems to show all the main features of the function.
[4] $f(-1) = (-1)^2 - 1 = 1 - 1 = 0$, so $(-1, 0)$ lies on the graph of f. $g(-1) = (-1)^3 - 2(-1) - 1 = 1 + 2 - 1 = 0$, so $(-1, 0)$ lies on the graph of g.

✎ **Section 1.6 Exercise 41** Do the graphs of the two functions

$$f(x) = 3x^2 + 6x - \frac{1}{2}, \text{ and } g(x) = \sqrt{7 - \frac{7}{12}x^2}$$

intersect in the viewing rectangle [–4, 4] by [–1, 3]? If they do, how many points of intersection are there?

Solution

The graphs intersect _____ times in the given viewing rectangle.

Copy the graph from your calculator into the space provided.

Graphing Piecewise Defined Functions

These exercises ask you to graph piecewise defined functions, some given in the notation you learned in Section 1.5, and some expressed in terms of the absolute value function. Find the definition of a piecewise defined function in bold typeface in the textbook in Section 1.5, and the definition of the absolute value function in bold typeface in the textbook in Section 1.6. Practice evaluating absolute values by reading Example 6 and completing Exercise 59, in Algebra Toolkit A.2.

📖 **Read Section 1.6 Example 9**, in which the authors graph the absolute value function by first making a table of values.

[5]*The absolute value of a number is always positive. Why is there a minus sign in the piecewise definition of the absolute value given in the textbook?*
Answer in your own words:

_____

[5] The absolute value of, for example, –5 is equal to 5, which is what you get when you substitute $x = -5$ into the formula $-x$, since $-(-5) = 5$.

✎ **1.6 Exercise 31** Graph the functions $f(x) = |x|$, $g(x) = 2|x|$, and $h(x) = |x+2|$ on the same coordinate axes.

x	$f(x)$	$g(x)$	$h(x)$

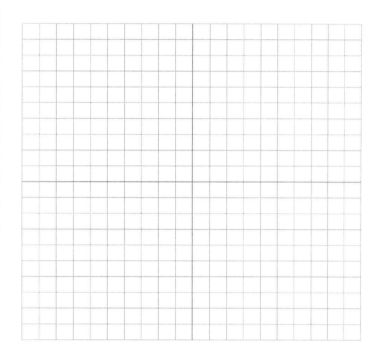

📖 **Read Section 1.6 Example 7**, in which the authors graph a piecewise defined function.

[6]*How did the authors find the y-coordinate of the open circle in the graph?*
Answer in your own words:

📖 **Read Section 1.6 Example 8**, in which the authors graph the piecewise defined water bill studied in Section 1.5 Example 9.

[7]*How did the authors find the y-coordinate of the open circle in the graph?*
Answer in your own words:

---

[6] The authors evaluate $f(x) = x + 1$ at the endpoint $x = 2$ of the domain $x > 2$ to find the coordinates (2,3) of the open circle.

[7] The authors evaluate $C(x) = 0.008x$ at the endpoint $x = 4000$ of the domain $x < 4000$ to find the coordinates (4000, 32) of the open circle.

🖉 **Section 1.6 Exercise 45** Sketch a graph of the piecewise defined function,

$$f(x) = \begin{cases} 1-x & \text{if } x < -2 \\ 5 & \text{if } x \geq -2 \end{cases}$$

Solution:

🖉 **Section 1.6 Exercise 55(a) Toll Road Rates** The toll charged for driving on a certain stretch of toll road depends on the time of day. The amount of the toll charged is given by $T(x)$, where x is the number of hours since 12:00 A.M..

(a) Graph the function T.

$$T(x) = \begin{cases} 5.00 & \text{if } 0 \leq x < 7 \\ 7.00 & \text{if } 7 \leq x \leq 10 \\ 5.00 & \text{if } 10 < x < 16 \\ 7.00 & \text{if } 16 \leq x \leq 19 \\ 5.00 & \text{if } 19 < x < 24 \end{cases}$$

Solution:

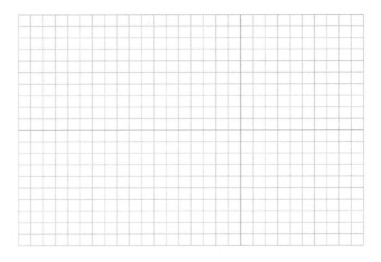

CONTEXTS

Graphing a Function from a Verbal Description

Though most of the problems in this section involve creating a graph from a formula, these exercises remind you that you can also sketch a graph from a verbal description. The desire to create formulas for functions that models these contexts motivates your study of basic functions and their graphs. Review Section 1.4, Example 9 in the textbook, in which a function is represented verbally, graphically, symbolically and numerically as a table of numbers.

📖 **Read Section 1.6 Example 1**, in which the authors draw a rough graph of a function describing the temperature of the water coming out of a faucet.

[8] *The given verbal and numerical descriptions of the function T, as well as the graphical representations the authors construct are not as precise as a formula would be. Why do the authors teach us to uses these representations?*
Answer in your own words:

✎ **Section 1.6 Exercise 49 Filling a Bathtub** A bathtub is being filled by a constant stream of water from the faucet. Sketch a rough graph of the water level in the tub as a function of time.

Solution

Graphing Piecewise Defined Functions

These exercises connect jumps in a graph of a piecewise defined function to the context of word problems, in which piecewise functions describe the billing rates for various commodities. Review the connections between a context and the symbolic representations of such functions in Section 1.5 Examples 8 and 9 in the textbook, and Exercise 67 and 69, which are included in Section 1.5 of this guide.

---

[8] In this case we do not know a formula that models the temperature of the water as a function of time. Read the paragraph in the textbook after Example 1.

📖 **Read Section 1.6 Example 8 again**, paying particular attention to part (b), in which the authors discuss what a break in the graph of a piecewise defined function represents in context.

> Refer back to page 11
> **Hints and Tips 1.2b: What does the algebra represent in context?**

Hints and Tips 1.2b explains the authors' answer to Section 1.6 Example 8(b).

⇨ **Section 1.6 Example 8(b)** What does the break in the graph represent?

Solution

Following **Hints and Tips 1.2b**, we write the following sentences as scratch work. Algebraically, the break in the graph is where the y-value jumps from 32 to 48, when $x = 4000$. In context, the break in the graph is where the water bill jumps from \$32 to \$48, when the household water use is 4000 gallons. Here is our more eloquent final answer. The break in the graph represents the increase in the water bill from \$32 to \$48, as a household's water use exceeds the 4000-gallon mark.

[9] *How can you tell from the formula that there is a jump in the graph at x = 4000?*
Answer in your own words:

🖉 **Section 1.6 Exercise 55(b) Toll Road Rates** The toll charged for driving on a certain stretch of toll road depends on the time of day. Read the formula and part (a) of this exercise on page 50.

(b) What do the breaks in the graph represent?

Solution

CONCEPTS

✓ **1.6 Exercises – Concepts – Fundamentals** Complete the Fundamentals section of the exercises at the end of Section 1.6. Compare your answers to those at the back of the textbook, and make corrections as necessary.

📖 **Read the Concept Check for Section 1.6**, in the Chapter Review at the end of Chapter 1.

---

[9] In the graphs of piecewise defined functions, jumps can happen at the endpoints of the intervals. We look for a jump in the graph of C at the endpoint $x = 4000$. To see the jump, evaluate the functions $0.008x$ and $0.012x$ at $x = 4000$; since the answers, 32 and 48 are different, the graph will jump.

1.6 Solutions to ✐ Exercises

✐ **Section 1.6 Exercises 13, 15, 19 and 23**

Following **Hints and Tips 1.6**, we choose integers for the *x*-values in our table. Since we can see that *g* has an *x*-intercept at $x = 4$, we include 4 in the domain of our graph.

13. $f(x) = 8$

x	$f(x)$
−3	8
−2	8
−1	8
0	8
1	8
2	8
3	8

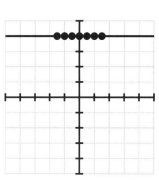

15. $g(x) = x - 4$

x	$g(x)$
−1	−5
0	−4
1	−3
2	−2
3	−1
4	0
5	1

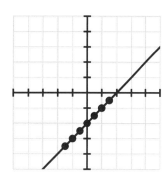

Following **Hints and Tips 1.6**, we choose to include 0.5 and −0.5, to show more detail in our graphs near $x = 0$.

19. $k(x) = -x^2$

x	$k(x)$
−2	−4
−1	−1
−0.5	−0.25
0	0
0.5	−0.25
1	−1
2	−4

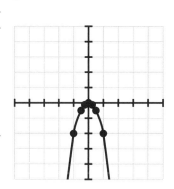

23. $G(x) = x^3 - 8$

x	$G(x)$
−2	−16
−1	−9
−0.5	−8.125
0	−8
0.5	−7.875
1	−7
2	0

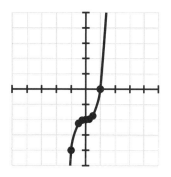

✐ **Section 1.6 Exercise 31** $f(x) = |x|$, $g(x) = 2|x|$, and $h(x) = |x + 2|$.

$x.$	$f(x)$	$g(x)$	$h(x)$
−3	3	6	1
−2	2	4	0
−1	1	2	1
0	0	0	2
1	1	2	3
2	2	4	4
3	3	6	5

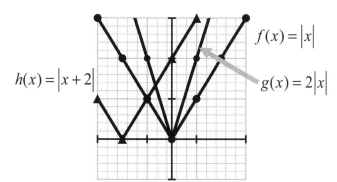

✎ **Section 1.6 Exercise 33** $f(x) = \sqrt{x}$, $g(x) = \sqrt{4x}$, and $h(x) = \sqrt{x} - 4$.

x	$f(x)$	$g(x)$	$h(x)$
0	0	0	-4
0.25	0.5	1	-3.5
1	1	2	-3
2	1.41	2.83	-2.59
3	1.73	3.46	-2.27
4	2	4	-2

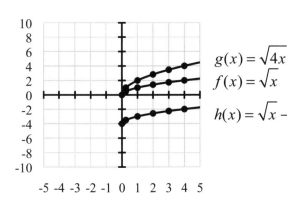

✎ **Section 1.6 Exercise 35**

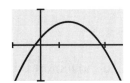

The viewing rectangle we use is $[-3, 9]$ by $[-20, 15]$.

✎ **Section 1.6 Exercise 41**

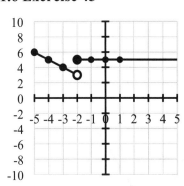

The graphs intersect 2 times in the given viewing rectangle.

✎ **Section 1.6 Exercise 45**

x	$f(x)$
-4	5
-3	4
-2	5
-1	5
0	5
1	5
2	5

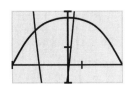

✎ **Section 1.6 Exercise 49 Filling a Bathtub**

In this sketch, we imagine that there is 2 inches of water in the tub after 1 minute, and 4 inches after 2 minutes, and so on.

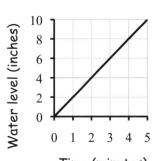

✐ Section 1.6 Exercise 55(a) Toll Road Rates

(b) Following **Hints and Tips 1.2b**, we write the following sentences as scratch work. Algebraically, the first break in the graph is where the y-value jumps from 5 to 7, when $x = 7$. In context, the break in the graph is where the toll jumps from \$5 to \$7, when the number of hours since 12 is 7. Here is our final more eloquent answer. The first break in the graph represents the increase in the toll from \$5 to \$7 at 7 A.M.. Similarly the other jumps represent changes in the toll at various times throughout the day.

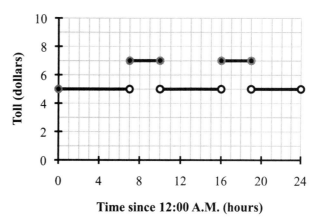

1.7 Working with Functions: Graphs and Graphing Calculators

OBJECTIVES

☑ *Check the box when you can do the exercises addressing the objective, such as those included in this study guide, which are listed here.*

	Exercises
Reading the Graph of a Function	
☐ Graph functions in appropriate viewing rectangles, using a graphing calculator.	**11, 19, 27, 37, 49**
☐ Use the relationship between the height of the graph at each input and the corresponding output to answer questions.	**7, 11, 19, 37, 41, 43, 49**
☐ Identify the output that corresponds to a given input, or the input(s) that correspond to a given output, from a function's graph.	**7(a, c), 41(a)**
☐ Determine the inputs at which two functions are equal, or one function is greater than (or less than) another, from a graph.	**7(d), 11**
☐ Find the net change in a function over a given interval by first reading the necessary information from the graph of the function.	**7, 41(d)**
Domain and Range from a Graph	
☐ Determine the domain and range of a function from its graph. Use interval notation to write your answer.	**7(b), 19**
Increasing and Decreasing Functions	
☐ Determine the intervals on which a function is increasing or decreasing from its graph. Use interval notation to write your answer.	**27(b), 37(b), 49(b)**
Local Maximum and Minimum Values	
☐ Use a graph to determine the local maximum values and local minimum values of a function, and the inputs at which each occurs.	**37(a), 41(b, c), 49(c)**
☐ Relate the endpoints of the intervals of increase and decrease to the local maximum and minimum values.	**49**

GET READY...

This section uses interval notation and graphs of intervals on the number line. Refresh your understanding of these in Algebra Toolkit A.2.

✓ **Algebra Checkpoint** Test yourself by completing the Algebra Checkpoint at the end of this section of the textbook, and comparing your answers to those in the back of the textbook. Refer to Algebra Toolkit A.2 as necessary.

This section also uses vocabulary pertaining to a graphing calculator, like *viewing rectangle,* and skills, like graphing two functions on the same screen. Refresh your understanding of these in Algebra Toolkit D.3 Example 2.

SKILLS

Reading the Graph of a Function

These exercises ask you to find the information about a function from a graph, including the information you need to calculate net change. Reread the definition of net change in a blue box in Section 1.5.

📖 **Read Section 1.2 Example 5 again**, to remind yourself how the authors begin the discussion about how to read information from a scatter plot.

[1]*The authors find that the precipitation in year 7 is 18 inches. How do they get the number 18 from the graph?*
Answer in your own words:

What do the authors add to the graph to help answer part (c)?
Answer in your own words:

📖 **Read Section 1.7 Example 1**, in which the authors answer questions about Mr. Hector's weight, using a graph showing his weight as a function of his age.

[2]*The authors find that Mr. Hector weighed 80 lbs when he was 10 years old. How do they get the number 80 from the graph?*
Answer in your own words:

_____

[1]   The number 18 is the height of the point in the graph above year 7. The authors draw a horizontal line on the graph at the average, to make it easy to see the data points with higher than average outputs, since they lie above the line.

[2]   Hector's weight at age 10 is $W(10)$, which is the height of the graph above the x-value 10.

📖 **Read Section 1.7 Example 2**, in which the authors answer questions that are written using function notation, from a graph showing the temperatures from noon to 6 P.M. at a certain weather station.

[3] *What could you add to the graph to help you find the values of x for which T(x) ≥ 25 (part (d))?*
Answer in your own words:

✎ **Section 1.7 Exercise 7(a, c, d, and e)** The graph of a function *h* is given.

(a) Find $h(-2)$, $h(0)$, $h(2)$, and $h(3)$.
(c) Find the values of *x* for which $h(x) = 3$.
(d) Find the values of *x* for which $h(x) \le 3$.
(e) Find the net change in the value of *h* when *x* changes from −2 to 4.

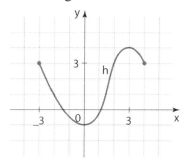

Solution

📖 **Read Section 1.6 Example 6 again**, in which the authors find the coordinates at which two graphs intersect, using a graphing calculator.

[4] *The values of x at which f(x) = g(x) are _____, _____, and _____.*

📖 **Read Section 1.7 Example 3**, in which the authors determine the *x*-values at which one function is greater than or equal to, equal to, or less than, another function.

[5] *When x lies in the interval [−1, 2], f(x) ≥ g(x), so the graph of f lies _____ (above / below / to the left of / to the right of) the graph of g.*

The authors use the viewing rectangle [_____, _____] by [_____, _____] to display the graphs. Would the viewing rectangle [0, 6] by [0,6] also be a good choice? Answer in your own words:

---

[3] You could draw a horizontal line on the graph at a height of 25. The *x*-values you are looking for correspond to the part of the graph that lies above the line.
[4] You have $f(x) = g(x)$ at the *x*-values where the graphs of *f* and *g* meet. The authors find that the two graphs meet at (−1, 0), (0,−1) and (2, 3), so the *x*-values are −1, 0, and 2.
[5] The graph of *f* lies above the graph of *g*. *The viewing rectangle is [−3, 4] by [−1, 6]. The viewing rectangle [0, 6] by [0,6]* would not be a good choice, since the two graphs intersect outside that rectangle.

✎ **Section 1.7 Exercise 11** Graph the function $f(x) = x^2 - 5x + 1$ and $g(x) = -3x + 4$ with a graphing calculator. Use the graphs to find the indicated points or intervals; state your answers correct to two decimal places.

(a) Find the value(s) of x for which $f(x) = g(x)$.

(b) Find the value(s) of x for which $f(x) \geq g(x)$.

(c) Find the value(s) of x for which $f(x) < g(x)$.

Solution

Copy the graph from your calculator in the box provided.

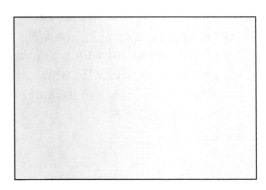

Domain and Range from a Graph

These exercises use the vocabulary, domain and range, introduced in Section 1.2 for relations in general, and used in Section 1.5 for functions in particular. Look for a summary of this material and a description of how to visualize the domain and range of a function from its graph in Section 1.7 of the textbook.

📖 **Read the paragraph before Section 1.7 Example 4**, in which the authors find the domain and range of the function from Example 1, describing Mr. Hector's weight.

📖 **Read Section 1.7 Example 4**, in which the authors read the domain and range of a function from its graph.

[6]*You find the domain by looking at the _____ (vertical / horizontal) axis, and the range by looking at the _____ (vertical / horizontal) axis.*

✎ **Section 1.7 Exercise 7(b)** Refer to the graph for this problem on page 57 of this study guide.

(b) Find the domain and range of h.

Solution

---

[6] You find the domain by looking at the horizontal axis, and the range by looking at the vertical axis.

✎ **Section 1.7 Exercise 19** Consider the function $f(x) = \sqrt{16 - x^2}$.

(a) Use a graphing calculator to draw the graph of f.
(b) Find the domain and range of f from the graph.

Solution

Copy the graph from your calculator in the box provided.

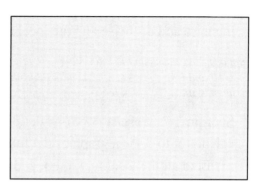

Increasing and Decreasing Functions

These exercises ask you to determine the intervals on which a function is increasing or decreasing from its graph. Find the definitions of increasing and decreasing in a blue box in Section 1.7 of the textbook.

📖 **Read Section 1.7 Example 5**, in which the authors find the intervals on which a function is increasing and those on which the function is decreasing.

[7] *The authors use the viewing rectangle [_____, _____] by [_____, _____]. However, the authors say that the intervals on which f is increasing is (_____, _____] and [_____, _____). The authors assume that the function continues up to the right and down to the left; in Chapter 5, you will study polynomial function, and learn that this assumption is correct.*

✎ **Section 1.7 Exercise 27** Consider the function $f(x) = x^2 - 5x$.

(a) Use a graphing device to draw the graph of f.
(b) State approximately the intervals on which f is increasing and on which f is decreasing.

Solution

Copy the graph from your calculator in the box provided.

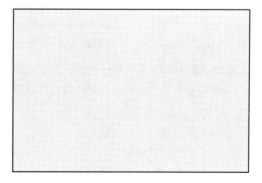

---

[7] The viewing rectangle is [–2, 4] by [–5, 5], and the intervals on which the function is increasing are $(-\infty, 0]$ and $[2, \infty)$.

Local Maximum and Minimum Values

These exercises refer to the local maxima and minima of a function. Read the definition of local maximum and minimum values in a blue box in Section 1.7 of the textbook.

📖 **Read Section 1.7 Example 6**, in which the authors find the local maximum and minimum values of a cubic polynomial. Note that the authors use the same function in Examples 5 and 6. As you read, look for a relationship between the intervals of increase and decrease and the local extrema.

[8]*Suppose f increases on the intervals (– ∞, 1] and [4, ∞), and decreases on the interval [1, 4], and that f(1) = 10 and f(4) = –3. Then f has a local minimum value _____ at the x-value _____ and a local maximum value _____ at the x-value _____. Tip: sketch a graph of f.*

✏ **Section 1.7 Exercise 37** Consider the function $f(x) = x^3 - x$. Use a graphing calculator to draw a graph of the function.

(a) Find all the local maximum and minimum values of the function and the value of x at which each occurs. State such answers correct to two decimal places.

(b) Find the intervals on which the function is increasing and on which the function is decreasing. State each answer correct to two decimal places.

Solution

Copy the graph from your calculator in the box provided.

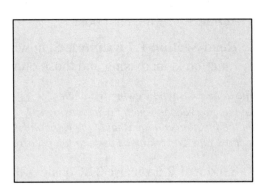

CONTEXTS

Reading the Graph of a Function

In these exercises, you connect questions about the real world to models given in graphs, and apply the skills you practiced above.

📖 **Read Section 1.7 Example 1 again**, paying attention to the connections between the context (Mr. Hector's age and weight) and the features of the graph.

[9]*When you read Mr. Hector's weight from the graph, the values are on the _____ (vertical / horizontal) axis. When you read Mr. Hector's age from the graph, the values are on the _____ (vertical / horizontal) axis.*

---

[8] Then f has a local minimum value –3 at the x-value 4, and a local maximum value 10 at the x-value 1.

[9] Mr. Hector's weight is on the vertical axis, and his age is on the horizontal axis.

✎ **Section 1.7 Exercise 41 Power Consumption** The figure shows the power consumption in San Francisco for September 19, 1996. (P is measured in megawatts; t is measured in hours starting at midnight.)

(a) What was the power consumption at 6:00 A.M.? At 6:00 P.M.?

(b) When was the power consumption a maximum?

(c) When was the power consumption a minimum?

(d) What is the net change in the values of P as the values of x changes from 0 to 12?

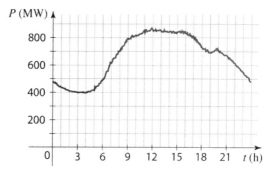

Source: Pacific Gas & Electric.

Solution

Local Maximum and Minimum Values

These exercises ask you to find maxima, minima, and intervals of increase and decrease from a graph, in order to answer questions relevant to the contexts being modeled.

📖 **Read Section 1.7 Example 7 and the paragraph before it**, in which the authors answer questions about traffic flow. **Study more about this in Chapter 6 Exploration 3.**

[10] *What the carrying capacity of a road?*
Answer in your own words:

_____

[10] The carrying capacity of a road is the maximum number of cars that can safely travel along it.

✎ **Section 1.7 Exercise 49 Migrating Fish** Suppose a fish swims at a speed v relative to the water, against a current of 5 mi/h. Using a mathematical model of energy expenditure, it can be shown that the total energy E required to swim a distance of 10 mi is given by

$$E(v) = 2.73v^3 \frac{10}{v-5}, \quad 5.1 \le v \le 10$$

Biologists believe that migrating fish try to minimize the total energy required to swim a fixed distance.

(a) Graph the function E in the viewing rectangle [5.1, 10] by [4000, 13,000].

(b) Where is the function E increasing, and where is the function E decreasing?

(c) Find the local minimum value of E. For what velocity v is the energy E minimized?

[Note: This result has been verified; migrating fish swim against a current at a speed 50% greater than the speed of the current.]

[11] *What speed is 50% greater than 5 mi/h (the speed of the current)?* _____

Solution

Copy the graph from your calculator in the box provided.

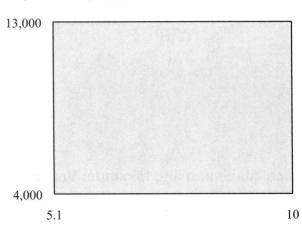

13,000

4,000

5.1   10

CONCEPTS

✓ **1.7 Exercises – Concepts – Fundamentals** Complete the Fundamentals section of the exercises at the end of Section 1.7. Compare your answers to those at the back of the textbook, and make corrections as necessary.

📖 **Read the Concept Check for Section 1.7**, in the Chapter Review at the end of Chapter 1.

---

[11] First, 50% of 5 is (0.5)(5)=2.5. So 50% greater than 5 is 5+2.5 = 7.5 mi/h.

1.7 Solutions to ✐ Exercises

✐ **Section 1.7 Exercise 7**

(a) The height of the graph above –2 is 1, so $h(-2) = 1$. Similarly, $h(0) = -1$, $h(2) = 3$, and $h(3) = 4$.

(b) The domain is the interval $[-3, 4]$. The range is the interval $[-1, 4]$.

(c) The height of the graph is 3 at $x = -3$, $x = 2$, and $x = 3$.

(d) The graph is at or below 3 between $x = -3$ and $x = 2$, and also at $x = 4$. So the domain is $[-3, 2] \cup \{4\}$.

(e) The height of the graph at $x = -2$ is 1, and when $x = 4$ is 3. So the net change is

$$h(4) - h(-2) = 4 - 1 = 3$$

✐ **Section 1.7 Exercise 11** We use the viewing rectangle $[-2, 5]$ by $[-10, 10]$.

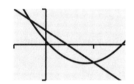

(a) Recall that the value of the function is the height of the graph. So $f(x) = g(x)$ at the x-values where the graphs of f and g meet. By zooming in, we see graphs meet at $x = -1$ and $x = 3$.

(b) We need to find the x-values at which $f(x) \geq g(x)$. These are the x-values where the graph of f is above the graph of g. From the diagram, we see that this happens for x less than –1 and for x greater than 3. So $f(x) \geq g(x)$ for x in the intervals $(-\infty, -1]$ and $[3, \infty)$.

(c) We need to find the x-values at which $f(x) < g(x)$. These are the x-values where the graph of g is above the graph of f. From the diagram, we see that this happens for x strictly between –1 and 3. So $f(x) < g(x)$ for x in the interval $(-1, 3)$. (We do not include the points –1 and 3 because of the strict inequality.)

✐ **Section 1.7 Exercise 19**

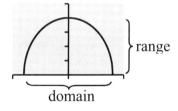

(a) We use the viewing rectangle $[-5, 5]$ by $[0, 5]$.

(b) The domain is $[-4, 4]$, and the range is $[0, 4]$.

✐ **Section 1.7 Exercise 27**

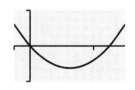

(a) We use the viewing rectangle $[-1, 6]$ by $[-8, 8]$.

(b) By zooming in, we see that the graph is decreasing on $(-\infty, -2.5]$.

✎ **Section 1.7 Exercise 37** We use the viewing rectangle [–2, 2] by [–4, 4].

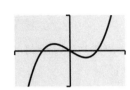

(a) From the graph, we see that f has a local minimum value of –0.38 at the x-value 0.58, and a local maximum value of 0.38 at the x-value –0.58.

(b) From the graph, we see that f is increasing on $(-\infty, -0.58]$ and $[0.58, \infty)$, and f is decreasing on $[-0.58, 0.58]$.

✎ **Section 1.7 Exercise 41 Power Consumption**

(a) The time 6:00 A.M. is 6 hours after midnight, so the power consumption at 6:00 A.M. is the height of the graph above 6, which is approximately 500 MW. Similarly, The power consumption at 6:00 P.M. is the height of the graph above 18, which is approximately 700 MW.

(b) The power consumption reaches its highest point at approximately $x = 12$, which is at 12 noon.

(c) The power consumption reaches its lowest point at approximately $x = 4$, which is 4:00 A.M..

(d) From the graph, we see that $P(0)$ is approximately 475, and $P(12)$ is approximately 850. Thus the net change between $x = 0$ and $x = 12$ is approximately

$$P(12) - P(0) = 850 - 475 = 375$$

So the net change between midnight and noon is a net increase of approximately 375 MW.

✎ **Section 1.7 Exercise 49 Migrating Fish**

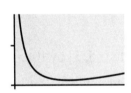

(b) From the graph, E is decreasing on [5.1, 7.5], and increasing on [7.5, 10].

(c) The graph of E reaches its lowest value of approximately 4607 at $v = 7.5$. This means that fish use 4607 Jewels of energy to swim 10 miles at 7.5 miles per hour.

1.8 Working with Functions: Modeling Real-World Problems

GET READY...

This section uses vocabulary pertaining to a graphing calculator, like *viewing rectangle,* and skills, like graphing two functions on the same screen. Refresh your understanding of these in Algebra Toolkit D.3.

SKILLS

Modeling with Functions

These exercises ask you to use ratios (conversion factors) to create models.

🖉 **Section 1.8 Exercise 1 (extended)** There are 7 days in a week.

(a) Find a function that models the number N of days in w weeks.

(b) Find a function that models the number w of weeks in N days.

[1] *How many weeks are there in 364 days? When you divide 364 days by 7 days per week, the units on the answer are _____.*

Solution

---

[1] The 364 days is 52 weeks. When you divide 364 days by 7 days per week, the units on the answer are weeks (you get 52 weeks). (364 days) / (7 days/week) = 52 weeks. Note that saying there are 52 weeks per year, and 365 days per year yields a slight inconsistency.

🖉 **Section 1.8 Exercise 2** There are 25 cents in a quarter.

(a) Find a function that models the number N of cents in q quarters.

(b) Find a function that models the number q of quarters it would take to have N cents.

[2]*How many quarters make 700 cents ($7)? The units on the answer you get when you divide 700 cents by 25 cents per quarter are _____.*

Solution

Getting Information from the Graph of a Model

These exercises ask you to use a graphing calculator to estimate the solutions to inequalities.

📖 **Read Section 1.7 Example 3 again**, paying particular attention to parts (b) and (c) in which the authors solve the inequalities $f(x) \geq g(x)$ and $f(x) < g(x)$.

[3]*Use your calculator to graph the functions f(x) =5 – x² and g(x) = 4 in the same viewing rectangle. For what values of x is f(x) > 4? _____*

Copy the graph from your calculator in the box provided.

CONTEXTS

Modeling with Functions

These exercises ask you to create linear models for situations that are described verbally. Recall that you create formulas for linear functions that model data in Section 1.3 (see for example Section 1.3 Example 3 about depth-pressure data). Find the vocabulary word model in bold typeface in Section 1.8 of the textbook.

---

[2] It takes 28 quarters to make 700 cents. When you divide 700 cents by 25 cents per quarter, the units on the answer are quarters (you get 28 quarters). (700 cents) / (25 cents/quarter) = 28 quarters.

[3] We use the viewing window [-2, 2] by [2, 6]. When $-1 < x < 1$, we have $f(x) > 4$.

📖 **Read Section 1.8 Example 1**, in which the authors create a linear model for the price of baseball caps.

[4] *In context, what is the input and what is the output of the function C(n)?*
Answer in your own words:

In part (b), the authors substitute the value 225 for _____ (C / n).

✐ **Section 1.8 Exercise 17** A tee shirt company makes tee shirts with school logos. the company charges a fixed fee of $200 to set up the machines plus $3.50 per tee shirt.

(a) Find a function C that models the cost for of purchasing x tee shirts.
(b) Use the model to find the cost of purchasing 600 tee shirts.

Solution

📖 **Read Section 1.8 Example 2**, in which the authors model the number of trees used in one year by U.S. residents.

[5] *In context, what is the input and what is the output of the function N(x)?*
Answer in your own words:

The units on the answer you get when you divide 650 pounds by 4130 pounds per tree are _____.
The authors first find the number of trees used by one resident (0.157 trees). What do they do with that information to find the number of trees used by x residents?
Answer in your own words:

_____

[4] The input n is the number of caps to be purchased, and the output $C(n)$ is the total cost of purchasing n caps. The authors substitute 225 for n since 225 is a number of caps, not a cost.

[5] The input x is a number of U.S. residents, and the output $N(x)$ is the number of trees used by x residents. When you divide 650 pounds by 4130 pounds per tree, the units on the answer are trees (you get 0.157 trees). (650 lbs) / (4130 lbs/tree) = 0.157 trees. The authors first find the number of trees used by one resident (0.157 trees), and then they multiply that number by x to get the number of trees used by x residents (0.157x trees).

✐ **Section 1.8 Exercise 19 Gas Cost** The cost of driving a car depends on the number of miles driven and the gas mileage of the car. Kristi owns a Honda Accord that gets 30 miles to the gallon.

(a) Find a function C that models the cost of driving Kristi's car x miles if the cost of gas is $3.20 per gallon.

[6] *What did the authors do first to find N(x) in Example 2? If you follow their strategy, what will you do first here?*

Answer in your own words:

(b) Use the model to find the cost of driving Kristi's car 500 miles.

(c) Kristi's budget for gas is $250 a month. Use the model to find the number of miles Kristi can drive each month without exceeding her monthly gas budget.

Solution

---

[6] In Example 2, the authors first find the number of trees used by one resident. Using their strategy in this problem, you first find the cost to drive the car one mile.

Hints and Tips 1.8a: Thinking about the problem
To help you understand the verbal description of a function, calculate some numerical examples, before tackling the task of generating the function's formula.
- Determine what the inputs and outputs of your model will be in context.
- Make up some examples of inputs, and determine their corresponding outputs from the verbal description of the function.
- To see how the given numbers in the problem and the input you choose contributes to the formula, write out your calculations, and avoid doing computations in your head.

The authors use these guidelines in the "Thinking about the Problem" blocks in Section 1.8 Examples 3, 4, and 5. They also guide you to use this strategy in Section 1.8 Exercises 21(a) and 29(a), included in this study guide.

📖 **Read Section 1.8 Example 3**, in which the authors model the amount of water in an irrigation tank with a linear function.

[7]*Before beginning the solution, the authors discuss thinking about the problem. What analysis do they do to help you understand the situation?*
Answer in your own words:

The authors construct a table to help you understand the parts of the model. What is the first quantity in the table? What is the last?
Answer in your own words:

[8]*In part (c), the model gave W(20) = –400. It is impossible for the tank to have –400 gallons of water in the tank. What happened?*
Answer in your own words:

---

[7] Following Hints and Tips 1.8a, the authors think about the problem by making up an example. In particular, they find the output of the model (the amount of water in the tank), when the input (days since the tank was filled) is equal to 10 days. The first quantity in the table is x, which is the input of the model. The last entry in the table is $120 - 80x$, which is the output of the model.

[8] The tank ran out of water somewhat before 20 days had passed.

🖉 **Section 1.8 Exercise 21 Cost of Wedding** Sherri and Jonathan are getting married. They have a budget of $5000. The are planning the reception and choose a reception hall that costs $700, a DJ that costs $300, a caterer that charges $18.50 a plate, and a wedding cake that costs $1.50 per guest.

(a) Complete the table for the cost of the reception for the given number of guests.

Number of guests	Cost of hall	Cost of DJ	Cost of caterer	Cost of wedding cake	Total cost of reception
10	$700	$300	$185	$15	$1200
20					
30					
40					
50					

(b) Find a function C that models the cost of the reception when x guests attend.
(c) Determine how much the reception would cost if 75 people attended. That is, find $C(75)$.
(d) Determine how many people can attend the reception if Sherri and Jonathan spend their total budget of $5000; that is find the value of x when $C(x) = 5000$.

Solution

Getting Information from the Graph of a Model

These exercises ask you to create a linear model from the verbal description of a real-world situation, and then use the graph of your model to answer questions about the context.

Hints and Tips 1.8b: Thinking about the Problem
To help you understand the verbal description of a geometric problem (like one involving volumes, areas or other geometric quantities), draw a diagram, and label the dimensions according to the constraints of the context.

The authors use these guidelines in the "Thinking about the Problem" blocks in Section 1.8 Examples 4, and 5. In both of these examples, the authors draw several diagrams as they calculate some examples (following Hints and Tips 1.8a). They also guide you to use this strategy in Section 1.8 Exercise 29, included in this study guide.

📖 **Read Section 1.8 Example 4**, in which the authors model the volume of a cereal box.

[9] *Before beginning the solution, the authors discuss thinking about the problem. What analysis do they do to help you understand the situation?*
Answer in your own words:

The authors graph y = _____ and y = _____, on the graphing calculator, in order to find the depths of boxes for which the volume is greater than 60 in³. They find that the volume is greater than 60 in³ when the depth is greater than _____ inches.

✎ **Section 1.8 Exercise 27 Volume of a Container** A Florida orange grower ships orange juice in rectangular plastic containers that have square ends and are one and a half times as long as they are wide (see the figure).

(a) Find a function V that models the volume of a container of width x.
(b) Use the model to find the volume of a container of width 10 in.
(c) Graph the function V. Use the graph to find the width of the plastic container that has a volume of 315 in².
(d) Use the graph from part (c) to find the width for which the container has volume greater than 450 in³. Express your answer in interval notation.

x

1.5x

x

Solution

Copy your graph for (c) here.

_____

[9] Following Hints and Tips 1.8a and b, the authors think about the problem by making up some example dimensions of cereal boxes, and drawing diagrams. In particular, they find the output of the model (the volume of the cereal box), when the input (the depth of the cereal box) is equal to 1, 2, 3 or 4 inches. The authors graph $y = V(x) = 15x^3$, and $y = 60$. The authors find that $V(x) \geq 60$ when $x \geq 1.59$. (The number 1.59 is the x-coordinate of the point of intersection of the two graphs.)

📖 **Read Section 1.8 Example 5**, in which the authors find a model for the area of a garden that can be enclosed with a given length of fencing.

[10]*Before beginning the solution, the authors discuss thinking about the problem. What analysis do they do to help you understand the situation?*
Answer in your own words:

The authors graph y = _____ and y = _____, on the graphing calculator, in order to find the widths of gardens for which the area is greater than 825 ft². They find that the area is greater than 825 ft² when the width is between _____ and _____ feet.

✎ **Section 1.8 Exercise 29 Fencing a Field** Consider the following problem: A farmer has 2400 ft of fencing and wants to fence off a rectangular field that borders a straight river. He does not need a fence along the river (see the figure). What are the dimensions of the field of largest area that he can fence?

(a) Experiment with the problem by drawing several diagrams illustrating the situation. Calculate the area of each configuration, and use your results to estimate the dimensions of the largest possible field.

(b) Find a function A that models the area of a field in terms of one of its sides x.
(c) Use a graphing calculator to find the dimensions of the field of largest area. Compare with your answer to part (a).

Solution

Copy your graph for (c) here.

_____

[10] Following Hints and Tips 1.8a and b, the authors think about the problem by making up some example dimensions of gardens, and drawing diagrams. In particular, they find the output of the model (the area of the garden), when the input (width of the garden) equals to 10, 20, 30, 40, 50 or 60 feet. The authors graph $y = A(x) = 70x-x^2$, and $y = 825$. The authors find that $A(x) \geq 825$ when $15 \geq x \geq 55$. (The numbers 15 and 55 are the x-coordinates of the points of intersection of the two graphs.)

CONCEPTS

✓ **1.8 Exercises – Concepts – Fundamentals** Complete the Fundamentals section of the exercises at the end of Section 1.8. Compare your answers to those at the back of the textbook, and make corrections as necessary.

📖 **Read the Concept Check for Section 1.8**, in the Chapter Review at the end of Chapter 1.

1.8 Solutions to 🖊 Exercises

🖊 Section 1.8 Exercise 1 (extended)

(a) Since there are 7 days in one week, there are $7w$ days in w weeks. Our model is $N = 7w$.

(b) We have

$$\text{weeks in 1 day} = \frac{1 \text{ day}}{7 \text{ days/week}} = \frac{1}{7}\text{week}$$

Since there is $1/7$ week in one day, there are $(1/7)N$ weeks in N days. Our model is $w = N/7$.

🖊 Section 1.8 Exercise 2

(a) Since there are 25 cents in a quarter, there are $25q$ cents in q quarter. Our model is $N = 25q$.

(b) We have

$$\text{quarters in 1 cent} = \frac{1 \text{ cent}}{25 \text{ cents/quarters}} = \frac{1}{25}\text{quarter}$$

Since there is $1/25$ quarter in a cent, there are $(1/25)N$ quarters in N cents. Our model is $q = N/25$.

🖊 Section 1.8 Exercise 17

Following **Hints and Tips 1.8b**, we think about the problem by calculating an example. If we want to purchase 2 tee shirts, it will cost us $200 + \$3.50*2 = \$207$.

(a) The cost to make x tee shirts $200 plus $3.50x$, so our model is $C(x) = 200 + 3.5x$.

(b) To find the cost of purchasing 600 tee shirts, we calculate

$$C(600) = 200 + 3.5(600) = 2{,}300$$

Thus the cost to purchase 600 tee shirts is $2,300.

✐ **Section 1.8 Exercise 19 Gas Cost**

(a) Since cost of gas is \$3.20 per gallon, and Kristi's car gets 30 miles per gallon, the cost to drive the car each mile is

$$\text{cost to drive each mile} = \frac{\text{price for each gallon (\$/gal)}}{\text{distance Kristi drives for each gallon of gas (mi/gal)}}$$

$$= \frac{\$3.20/\text{gal}}{30 \text{ mi/gal}} \approx \$0.11/\text{mi}$$

So each mile costs Kristi about \$0.11. Thus x miles will cost her $0.11x$ dollars. We can express our model as $N(x) = 0.11x$.

(b) The cost of driving Kristi's car 500 miles is $C(500) = 0.11(500) = \$55$.

(c) To find the distance Kristi can drive for \$250, we solve $C(x) = 250$ for x.

$C(x) = 250$	Equation
$0.11x = 250$	Substitute $C(x) = 0.11x$
$x \approx 2273$	Divide each side by 0.11

So Kristi can drive approximately 2,273 miles on her monthly budget of \$250.

✐ **Section 1.8 Exercise 21 Cost of Wedding**

(a)

Number of guests	Cost of hall	Cost of DJ	Cost of caterer	Cost of wedding cake	Total cost of reception
10	\$700	\$300	\$185	\$15	\$1,200
20	\$700	\$300	18.50*20 = \$370	1.50*20 = \$30	\$1,400
30	\$700	\$300	18.50*30 = \$555	1.50*30 = \$45	\$1,600
40	\$700	\$300	18.50*40 = \$740	1.50*40 = \$60	\$1,800
50	\$700	\$300	18.50*50 = \$925	1.50*50 = \$75	\$2,000

(b) Following the pattern we see in the table, we have

Total Cost	=	cost of hall	+	cost of DJ	+	cost of caterer	+	cost of cake
$C(x)$	=	700	+	300	+	$18.50x$	+	$1.50x$

Simplifying, we have $C(x) = 1000 + 20x$.

(c) If 75 guests attend, the total cost would be $C(75) = 1000 + 20(75) = \$2,500$.

(d) To find the number of guests Sherri and Jonathan can host for \$5,000, we solve $C(x) = 5000$ for x.

$C(x) = 5000$	Equation
$1000 + 20x = 5000$	Substitute $C(x) = 1000 + 20x$
$20x = 4000$	Subtract 1000 from each side
$x = 200$	Divide each side by 20

So Sherri and Jonathan can host 200 people with a budget of \$5000.

✎ Section 1.8 Exercise 27 Volume of a Container

(a) To find the function that models the volume of the container, we first recall the formula for the volume of a rectangular box.

$$\text{volume} = \text{depth} \times \text{width} \times \text{height}$$

We have width = x, height = x, and depth = $1.5x$. Thus, our model is

$$V(x) = x \, x \, 1.5x = 1.5x^3$$

(b) The volume of a container of width 10 in is $V(10) = 1.5(10)^3 = 1{,}500 \text{ in}^3$.

(c) We use the viewing rectangle [0, 8] by [0,800]. By zooming in on the graph, we see that $V(x) = 450$ when x is approximately 6.69. Thus such a container with volume 450 in^3 has width approximately 6.69 in.

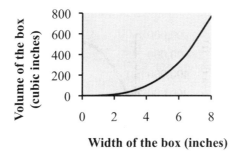

(d) The graph is higher than 450 when x is greater than 6.69. The volume $V(x)$ is greater than 450 when x is in the interval $[6.69, \infty)$.

✎ Section 1.8 Exercise 29 Fencing a Field

(a) Following **Hints and Tips 1.8b**, we draw several fences with length 2400 ft, and with one side along a river. If the farmer fences a plot with width 600 ft perpendicular to the river, then the length must be 1200 ft, because $2(600) + 1200 = 2400$, which is the total length of the fence. So the area is:

$$\text{Area} = \text{width} \times \text{length} = 600 \, 1200 = 720{,}000 \text{ ft}^2$$

The table shows various choices for fencing the field. We see that as the width increases, the area of the garden increases then decreases. From the table, we estimate that the maximum area is 720,000 ft^2, achieved when the width perpendicular to the river is 600 ft.

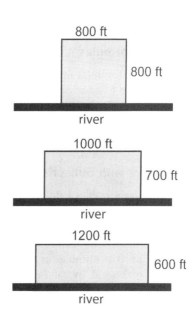

(b) Let x be the width of the field. The total amount of fencing is 2400 ft. From our experiments in part (a), we see that if we let l be the length of the field along the river, we have

$$2x + l = 2400$$

So the length is given by $l = 2400 - 2x$. Thus the model we want is

$$\text{Area} = \text{width} \times \text{length}$$
$$= x(2400 - 2x) = 2400x - 2x^2$$

Width (ft)	Length (ft)	Area (ft^2)
300	1800	540,000
400	1600	640,000
500	1400	700,000
600	1200	720,000
700	1000	700,000
800	800	640,000
900	600	540,000

The area of the field can be modeled by the function $A(x) = 2400x - 2x^2$.

(c) We use the viewing rectangle [0, 1200] by [0, 800,000]. From the graph we see that the maximum area is achieved when the width is 600 ft, as we predicted in part (b).

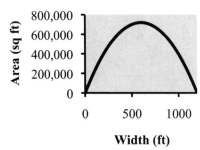

Width (ft)

1.9 Making and Using Formulas

OBJECTIVES

☑ *Check the box when you can do the exercises addressing the objective, such as those included in this study guide, which are listed here.*

	Exercises
Finding Formulas	
☐ Use formulas for the areas of a rectangle and a circle to find formulas for the areas of more complicated regions.	**17**
☐ Use ratios to find formulas.	**21**
☐ Think about a model before writing its formula by making up examples of inputs for your model and figuring out the corresponding outputs.	**21, 23**
Variables with Subscripts	
☐ Use variables with subscripts to make it easier to remember what the variables in a formula represent in context.	**23, 27**
Reading and Using Formulas	
☐ Rewrite formulas in a form that is convenient to answer questions about the real world, by solving for one of the variables.	**7, 27**
☐ Use formulas to answer questions about the real world.	**17, 21, 23, 27, 33**

GET READY...

Review Algebra Toolkit C.1 Examples 9 and 10, in which the authors solve equations for one variable in terms of others.

SKILLS

What is a Formula?

Read a discussion of formulas and their uses in this subsection of the textbook.

[1] *Find two examples of formulas in Algebra Toolkit D.1. For each example, what quantities does the formula relate?*
Answer in your own words:

Finding Formulas

These exercises ask you to use your understanding of areas and volumes to create formulas.

📖 **Read Section 1.9 Example 2**, in which the authors find a formula for the area of plywood needed to build a box.

[2] *Before beginning the solution, the authors discuss thinking about the problem. What analysis do they do to help you understand the situation?*
Answer in your own words:

In the formula for surface area that the authors find in part (a), where do the 2's come from?
Answer in your own words:

✏ **Section 1.9 Exercise 17** Find a formula that models the surface area A of a box with an open top of dimensions l, w, and h.

Solution

---

[1] The Distance Formula relates the distance between two points in the coordinate plane to the points' coordinates. The Midpoint Formula relates the coordinates of the midpoint of a line segment to the coordinates of the segment's endpoints.

[2] Following Hints and Tips 1.8b, the authors think about the problem by drawing diagrams that make it easier to see how to calculate the surface area of the box. The front and back of the box have identical areas, so twice that area, $2lh$, appears in the formula. Similarly, the left and right sides have the same area, and the top and bottom have the same area, hence the terms $2wh$ and $2lw$ in the formula.

Reading and Using Formulas

These exercises ask you to rewrite formulas so that it is convenient to calculate particular variables.

📖 **Read the introduction to the subsection, 'Reading and Using Formulas,'** in which the authors discuss a formula relating temperature, pressure and volume of a gas.

[3] *You might you choose to write the formula in the form $P = k\dfrac{T}{V}$ if you want to calculate _____ .*

You might choose to write the formula in the form $T = \dfrac{1}{k}PV$ if you want to calculate _____ .

📖 **Read Section 1.9 Example 4**, in which the authors use the formula for average to relate a student's scores on each exam to her exam average.

[4] *Why do the authors solve for s_3?*
Answer in your own words:

✐ **Section 1.9 Exercise 27 (extended) Electrical Resistance** When two resistors with resistances R_1 and R_2 are connected in parallel, their combined resistance R is given by the formula

$$R = \frac{R_1 R_2}{R_1 + R_2}$$

(a) Suppose that an 8-Ω resistor is connected in parallel with a 10-Ω resistor. Use the formula to find their combined resistance R.

(b) Suppose instead that you want to decrease the resistance of an 8-Ω resistor by connecting it in parallel to another resistor. Solve the equation for R_2. Use the resulting formula to find the resistance needed to make the combined resistance 5-Ω.

Solution

---

[3] The first form is most useful for calculating pressure (P), while the second is most useful for calculating temperature (T).

[4] The authors solve for s_3 to make it easy to find the score on exam 3.

📖 **Read Section 1.9 Example 5**, in which the authors rewrite the formula from Example 2 for the surface area of a box to make it easy to find the box's height.

[5]*To find the height of the box using this formula, you need to know the _____, the _____, and the _____.*

✏ **Section 1.9 Exercise 7** Solve the equation $S = 2lw + 2wh + 2lh$ for w.

Solution

CONTEXTS

Finding Formulas

These exercises ask you to use ratios, like miles per gallon or pounds per gallon, to create formulas.

📖 **Read Section 1.9 Example 1**, in which the authors find a formula for the gas mileage of a car.

[6]*Before beginning the solution, the authors discuss thinking about the problem. What analysis do they do to help you understand the situation?*
Answer in your own words:

✏ **Section 1.9 Exercise 21 CO₂ Emissions** Many scientists believe that the increase of carbon dioxide (CO_2) in the atmosphere is a major contributor to global warming. The

---

[5] To find the height, you need to know the other variables in the formula: surface area, length and width.

[6] Following Hints and Tips 1.8a, the authors think about the problem by making up an example. In particular, they find the output of the model (the gas mileage), when the inputs (the number of miles driven and the number of gallons of gasoline used) are equal to 100 miles and 2 gallons, respectively.

Environmental Protection Agency estimates that one gallon of gasoline produces on average about 19 pounds of CO_2 when it is combusted in a car engine.

(a) Find a formula for the amount A of CO_2 a car produces in terms of the number n of miles driven and the gas mileage G of the car.

[7]*Following Hints and Tips 1.8a, what analysis can you do to help you understand the situation? Answer in your own words:*

(b) Debbie owns an SUV that has a gas mileage of 21 mi/gal. Debbie drives 15,000 miles in one year. Use the formula you found in part (a) to find out how much CO_2 Debbie's care produces in one year.

(c) Debbie's friend Lisa owns a hybrid car that has a gas mileage of 55 mi/gal Lisa also drives 15,000 miles in one year. Use the formula you found in part (a) to find how much CO_2 Lisa's car produces in one year.

Solution

---

[7] To better understand the situation, make up an example. Make up example values for the inputs (distance driven and the gas mileage) and use them to determine the output (amount of CO_2 produced).

Variables with Subscripts

In these exercises, you use variables with subscripts in order to help your reader recall what the formula represent in context.

📖 **Read the introduction to the subsection, 'Variables with Subscripts,'** in which the authors discuss Newton's formula for the gravitational force between two objects.

[8]*Why do the authors choose to use m_1 and m_2 instead of two different variables in this formula? Answer in your own words:*

📖 **Read Section 1.9 Example 3**, in which the authors find a formula for a student's average score for three exams.

[9]*Before beginning the solution, the authors discuss thinking about the problem. What analysis do they do to help you understand the situation? Answer in your own words:*

📖 **Read Section 1.9 Exercise 23,** which you will complete below.

[10]*The authors use subscripts for these variables to make it easier to recall the context from the formula. In Section 1.9 Exercise 23, what do the p, 0 and s stand for in the variables p_0 and p_s? Answer in your own words:*

[11]*Following Hints and Tips 1.8a, what analysis can you do to help you understand the situation? Answer in your own words:*

---

[8] The two quantities represented by m_1 and m_2 are both masses of objects; using the same letter to represent them makes the formula easier to understand.

[9] Following Hints and Tips 1.8a, the authors think about the problem by making up an example. In particular, they first average 10 and 30, to recall the definition of average, and then they find the output of the model (the average), when the inputs (the exam scores) are equal to 10, 30 and 80, respectively.

[10] The p stands for price, the 0 indicates the original price, and the s stands for selling.

[11] To better understand the situation, make up an example. Make up example values for the inputs (original price, selling price and number of shares) and use them to determine the output (profit).

✎ **Section 1.9 Exercise 23 Investing in Stocks** Some investors buy shares of individual stocks and hope to make money by selling the stock when the price increases.

(a) Find a formula for the profit P an investor makes in terms of the number of shares n she buys, the original price p_0 of a share and the selling price p_s of a share.
(b) Silvia plans to make money by buying and selling shares of stock in her favorite retail store. She buys 1000 shares for $21.50 and waits patiently for many months until the price finally increases to $25.10. Use the formula found in part (a) to find the profit Silvia will make on her investment if she sells at $25.10.

Solution

Reading and Using Formulas

These exercises ask you to use formulas to reveal information about the contexts they represent.

📖 **Read Section 1.9 Example 6**, in which the authors use Newton's formulas for force to find the weight of the world.

[12]*To find the mass of the earth using the formula in part (a), you need to know the* _____ *, the* _____ *, and the* _____ *.*

_____

[12] To find the height, you need to know the other variables in the formula: surface area, length and width.

✎ **Section 1.9 Exercise 33 Flying Speeds of Migrating Birds** Many birds migrate thousands of miles each year between their winter and feeding grounds and their summer nesting sites. For instance, the 15-gram blackpoll warbler travels 12,000 miles from the western Alaska to South America. The air speed v at which a migrating bird flies depends on its weight w and the surface area S of its wings; these quantities satisfy the equation

$$Sv^2 = 94,700w$$

where w is measured in pounds, S in square inches, and v in miles per hour.
(a) Find a formula for v in terms of w and S.
(b) Complete the table to find the ratio w/S and the migrating air speed v for the indicated sea birds.
(c) The ratio w/S is called the wing loading; the greatest a bird's wing loading, the faster it must fly. A certain bird has a wing loading twice that of the sooty albatross. What is its migrating air speed?

Solution

Bird	w (lb)	S(in^2)	w/S	v (mi/h)
Common tern	0.26	76		
Black-headed gull	0.52	120		
Common gull	0.82	180		
Royal tern	1.1	170		
Herring gull	2.1	280		
Great skua	3.0	330		
Sooty albatross	6.3	530		
Wandering albatross	19.6	960		

CONCEPTS

✓ **1.9 Exercises – Concepts – Fundamentals** Complete the Fundamentals section of the exercises at the end of Section 1.9. Compare your answers to those at the back of the textbook, and make corrections as necessary.

📖 **Read the Concept Check for Section 1.9**, in the Chapter Review at the end of Chapter 1.

1.9 Solutions to ✐ Exercises

✐ **Section 1.9 Exercise 17**

The surface area of a box with an open top is

$$\text{surface area} = 2 \times (\text{area of front}) + 2 \times (\text{area of side}) + \text{area of bottom}$$

Now the area of the front is lh, the area of a side is wh and the area of the bottom is lw. So we can express the formula as

$$S = 2lh + 2wh + lw$$

✐ **Section 1.9 Exercise 27 (extended) Electrical Resistance**

(a) We have $R_1 = 8$ and $R_2 = 10$. Then

$$R = \frac{8 \cdot 10}{8 + 10} \approx 4.44 \; \Omega$$

so the combined resistance is 4.44 Ω.

(b) We solve the given formula for R_2:

$$R = \frac{R_1 R_2}{R_1 + R_2} \qquad \text{Formula}$$

$$R(R_1 + R_2) = R_1 R_2 \qquad \text{Multiply on each side by } R_1 + R_2$$

$$RR_1 + RR_2 = R_1 R_2 \qquad \text{Distribute } R \text{ on the left hand side}$$

$$RR_1 = R_1 R_2 - RR_2 \qquad \text{Subtract } RR_1 \text{ from each side}$$

$$RR_1 = (R_1 - R)R_2 \qquad \text{Factor } R_2 \text{ from the right hand side}$$

$$\frac{RR_1}{R_1 - R} = R_2 \qquad \text{Divide each side by } R_1 - R$$

So we can write the formula as $R_2 = \dfrac{RR_1}{R_1 - R}$. We use this formula when $R_1 = 8$ and

$R = 5$. We have $R_2 = \dfrac{8 \cdot 5}{8 - 5} \approx 13.3 \; \Omega$. If we connect an 8-$\Omega$ resistor in parallel to a 13.3-Ω resistor, the combined resistance is 5-Ω.

✏ Section 1.9 Exercise 7

To solve for w we need to put w on one side of the equation. We proceed as follows:

$$S = 2lw + 2wh + 2lh \qquad \text{Formula}$$

$$S - 2lh = 2lw + 2wh \qquad \text{Subtract } 2lh \text{ from each side}$$

$$S - 2lh = w(2l + 2h) \qquad \text{Factor } w \text{ from the right hand side}$$

$$\frac{S - 2lh}{2l + 2h} = w \qquad \text{Divide each side by } 2l + 2h$$

So we can write the formula as $w = \dfrac{S - 2lh}{2l + 2h}$. This formula allows us to find w if we know l and h.

✏ Section 1.9 Exercise 21 CO_2 Emissions

(a) The carbon emissions each mile we drive are

$$\text{emissions produced each mile} = \frac{19\ \text{lb/gal}}{G\ \text{mi/gal}} \approx \frac{19}{G}\ \text{lb/mi}$$

So if we drive one mile, we produce $19/G$ pounds of CO_2. When we drive n miles, we produce $(19/G)n$ pounds of CO_2. Our formula is

$$A = \frac{19n}{G}$$

(b) We use the formula we found in part (a), with $n = 15{,}000$ and $G = 21$.

$$A = \frac{19n}{G} = \frac{19(15{,}000)}{21} \approx 13{,}571$$

So Debbie's carbon emissions is 13,571 pounds of CO_2 per year.

(c) We use the formula we found in part (a), with $n = 15{,}000$ and $G = 55$.

$$A = \frac{19n}{G} = \frac{19(15{,}000)}{55} \approx 5182$$

So Lisa's carbon emissions are 5182 pounds of CO_2 per year.

✏ Section 1.9 Exercise 23 Investing in Stocks

(a) Following **Hints and Tips 1.8a,** we create a simple example. The inputs of our model are the number of shares n, the original price p_0 and the selling price p_s, and the output is the profit. We make up values for the inputs, and calculate the output.

Say we buy $n = 10$ shares at the original price of $p_0 = 5$ dollars per share, and then sell them at $p_s = 12$ dollars per share. We pay out (5 dollars/share)(10 shares) = \$50. When we sell the shares, we take in (12 dollars/share)(10 shares) = \$120. So our profit is

(12 dollars/share)(10 shares) – (5 dollars/share)(10 shares) = $120 – $50 = $70.

From our example, we see that the profit is

$$\text{Profit} = \begin{array}{c}\text{selling price/share} \\ \text{times the} \\ \text{number of shares}\end{array} - \begin{array}{c}\text{original price/share} \\ \text{times the} \\ \text{number of shares}\end{array}$$

$$P = p_s n - p_0 n$$

Thus the formula is $P = p_s n - p_0 n$.

(b) We have $p_0 = 21.50$, $p_s = 25.10$, and $n = 1000$. We get

$$P = (25.10)(1000) - (21.50)(1000) = 3600$$

so Silvia will make $3,600 if she sells her shares at $25.10 per share.

Section 1.9 Exercise 33 Flying Speeds of Migrating Birds

(a) We solve the formula for v.

$$Sv^2 = 94,700w \qquad \text{Formula}$$

$$v^2 = \frac{94,700w}{S} \qquad \text{Divide each side by } S$$

$$v = \pm\sqrt{\frac{94,700w}{S}} \qquad \text{Take a square root}$$

Since the velocity is not negative (the bird does not fly backwards), we have the formula $v = \sqrt{\frac{94,700w}{S}}$. Note that we can consider the input to be the wing loading (w/S), since the formula can be written $v = \sqrt{94,700(w/S)}$.

(b) Bird	w (lb)	S (in^2)	w/S (lb/in^3)	v (mi/h)
Common tern	0.26	76	0.0034	17.94
Black-headed gull	0.52	120	0.0043	20.18
Common gull	0.82	180	0.0046	20.87
Royal tern	1.1	170	0.0065	24.81
Herring gull	2.1	280	0.0075	26.65
Great skua	3.0	330	0.0091	29.36
Sooty albatross	6.3	530	0.0119	33.57
Wandering albatross	19.6	960	0.0204	43.95

(c) The bird has wing-loading twice the sooty albatross, which is 0.0119 lb/in^3. So w/S for the bird is equal to 2(0.0119) = 0.0238. Using the formula from (c), we have

$$v = \sqrt{94,700(w/S)} = \sqrt{94,700(0.0238)} \approx 47.47$$

So the bird's migrating airspeed is 47.47 mi/h.

2.1 Working with Functions: Average Rates of Change

OBJECTIVES

☑ *Check the box when you can do the exercises addressing the objective, such as those included in this study guide, which are listed here.*

Exercises

Average Rate of Change of a Function

☐ Determine the average rate of change of a function, from a table, graph or formula.   **9, 11(a(i)), 13**

☐ Use correct units when expressing the average rate of change of a real world quantity.   **25, 27, 29**

☐ Determine the relative size of the average rate of change of a function by looking at its graph. For example, determine if the average rate of increase over one interval is greater or less than the average rate of increase over another interval.   **25**

Average Speed of a Moving Object

☐ Find the average speed of a moving object by finding the average rate of change of the distance that object has moved as a function of time.   **27, 29**

Functions Defined by Algebraic Expressions

☐ Determine average rates of change of a function using its formula.   **13, 29**

GET READY...

Review the definition of net change in Section 1.5.

SKILLS

Average Rate of Change of a Function

These exercises ask you to compute the average rate of change of functions, given in tables, graphs, or formulas. Find the definition of average rate of change in a blue box in Section 2.1 of the textbook.

📖 **Read the introduction to the subsection, 'Average Rate of Change of a Function,'** in which the authors explain the definition of the average rate of change by considering the rate at which a library purchases books.

[1]*What is the relationship between the net change of a function over an interval and the average rate of change of a function over the same interval?*
Answer in your own words:

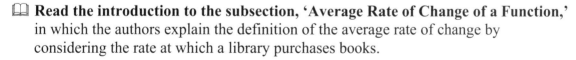

---

[1] The net change is the numerator of the formula for the average rate of change; that is, to calculate the average rate of change, you calculate the net change in the function on the interval, and divide by the length of the interval.

✎ **Section 2.1 Exercise 9** The graph of a function is given. Determine the average rate of change of the function between the indicated points.

Solution

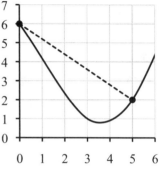

✎ **Section 2.1 Exercise 11(a(i))** A function is given by a table.

(a(i)) Determine the average rate of change of the function between $x = 2$ and $x = 4$.

Solution

x	0	1	2	3	4
$f(x)$	10	20	50	70	90

Functions Defined by Algebraic Expressions

These exercises ask you to calculate the average rate of change of a function using its formula.

📖 **Read Section 2.1 Example 4 and the paragraph that follows it**, in which the authors calculate the average rate of change of a function defined by an algebraic expression.

[2] *On the interval [–3, 1], the values of the function f _____ (increase / decrease) at an average rate of _____ units for each unit change in x.*

✎ **Section 2.1 Exercise 13 (extended)** Determine the average rate of change of the function $f(x) = 3x + 2$ between $x = 2$ and $x = 3$ and between $x = 4$ and $x = 7$.

Solution

[3] *On the intervals [2, 3] and [4, 7], the values of the function f _____ (increase / decrease) at an average rate of _____ units for each unit change in x.*

---

[2] The average rate of change on [–3, 1] is –2, so the values of f decrease at a rate of 2 units for each unit increase in x.

CONTEXTS

Average Rate of Change of a Function

These exercises ask you to calculate average rates of change of real world quantities, express them with the correct units, and find intervals with the greatest average rate of increase or decrease by looking at the function's graph.

📖 **Read the introduction to the subsection, 'Average Rate of Change of a Function,' again**, paying particular attention to the units of the quantities that you use to compute the average rate of change.

[4]*The units on the average rate of change of the number of books purchased are _____.*

📖 **Read Section 2.1 Example 1**, in which the authors calculate the average rate at which Sima installs floor tiles.

[5]*The units on f(x) are _____. The units on x are _____. The units on the average rate of change of f(x) are _____.*

What do the authors look for on the graph to find the hour with the fastest average rate of installation? Answer in your own words:

📖 **Read Section 2.1 Example 2 and the introduction preceding it**, in which the authors calculate the average rate at which the number of farms in the U.S. changes over various periods in history.

[6]*Why did the number of farms decline in the latter part of the 20th century? Answer in your own words:*

The units the output of the function are _____. The units on the input of the function are _____. The units on the average rate of change of are _____.
You can tell the number of farms declines between 1950 and 1970 since the average rate of change is _____ (positive / negative).

---

[3] The average rate of change on [2, 3] and on [4, 7] is 3, so the values of f increase at a rate of 3 units for each unit increase in x.

[4] The units of the average rate of change are books per month (or books/month).

[5] The units on $f(x)$ are tiles. The units on x are hours. The units on the average rate of change of f are tiles per hour. The authors look for the steepest rise in the graph to find the hour with the greatest installation rate.

[6] Many small family farms gave way to fewer large corporate farms. The units on the output are thousands of farms. The units on the input are years. The units on the average rate of change are thousands of farms per year. You can tell the number of farms declines between 1950 and 1970 since the average rate of change is negative.

✏ **Section 2.1 Exercise 25 Currency Exchange Rates (extended)** the Euro was introduced in 1990 as a common currency for twelve member countries of the European Union. The table in the margin shows the value of the Euro in U.S. dollars on the first business day of each year from 1999 to 2008.

(a) Draw a scatter plot of the data.

(b) Find the average rate of change of the value of the Euro in U.S. dollars between the following years: (i) 1999 and 2008; (ii) 2002 and 2006; (iii) 2005 and 2008; and (iv) 2000 to 2002.

(c) Use the scatter plot to find the year in which the value of the Euro experienced the largest average rate of increase in terms of the U.S. dollar.

(d) Use the scatter plot to find the year in which the value of the Euro experienced the largest average rate of decrease in terms of the U.S. dollar.

Year	Value (U.S. $)
1999	0.86
2000	1.01
2001	0.94
2002	0.89
2003	1.05
2004	1.26
2005	1.35
2006	1.18
2007	1.32
2008	1.46

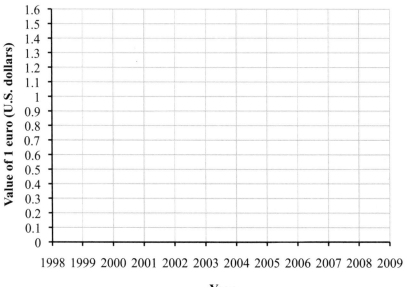

Year

Average Speed of a Moving Object

These exercises use the words, average speed, for the average rate of change of distance traveled as a function of time. Look for the definition of the average speed of a moving object in a blue box in Section 2.1 of the textbook.

📖 **Read Section 2.1 Example 3**, in which the authors find the average speed of cyclists in a race.

[7]*In this example, the units on distance are* _____. *The units on time are* _____.
The units on the average speed are _____.
How do the authors determine that the cyclists finished the race at the same time by looking at the graph (assume the race is for 60 miles)?
Answer in your own words:

✎ **Section 2.1 Exercise 27 Speed Skating** At the 2006 Winter Olympics in Turin, Italy, the United States won three gold metals in men's speed skating. The graph shows distance as a function of time for two speed skaters racing in a 500-meter event.

(a) Who won the race?
(b) Find the average speed during the first 10 seconds for each skater.
(c) Find the average speed during the last 15 seconds for each skater.

Solution

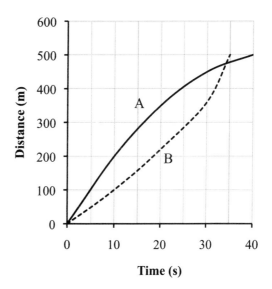

_____

[7] The units on distance are miles, the units on time are hours and the units on the average speed are miles per hour (or mi/h). The race ends after 60 miles; Jodi and James both reach 60 miles after 5 hours. On the graphs, you can see this since the ending point (5, 60) lies on both graphs.

Functions Defined by Algebraic Expressions

These exercises ask you to find the average rates of change of models of the real world that are given by formulas.

📖 **Read Section 2.1 Example 4**, in which the authors compute the average speed of a bungee jumper on his way down.

[8] *In this example, the units on distance are* _____. *The units on time are* _____ *The units on the average speed are* _____.

✏ **Section 2.1 Exercise 29 Phoenix Mars Lander** On August 4, 2007, NASA's Phoenix Mars Lander was launched into space to search for life in the icy northern region of the planet Mars; it touched down on Mars on May 25, 2008. As the ship raced into space, its jet fuel tanks dropped off when they were used up. The distance one of the tanks falls in t seconds is modeled by the function

$$s(t) = 16t^2$$

where s is measured in feet per second and $t = 0$ is the instant the tank left the ship. Use the function s to find the average speed of a tank during the following time intervals.
(a) The first 10 seconds after separation from the ship.
(b) The first 15 seconds after separation from the ship.

Solution

CONCEPTS

✓ **2.1 Exercises – Concepts – Fundamentals** Complete the Fundamentals section of the exercises at the end of Section 2.1. Compare your answers to those at the back of the textbook, and make corrections as necessary.

📖 **Read the Concept Check for Section 2.1**, in the Chapter Review at the end of Chapter 2.

---

[8] The units on distance are miles, the units on time are hours and the units on the average speed are miles per hour (or mi/h).

2.1 Solutions to ✐ Exercises

✐ Section 2.1 Exercise 9

We find the average rate of change of y between $x = 0$ and $x = 5$:

$$\text{average rate of change} = \frac{\text{net change in } y}{\text{net change in } x} = \frac{2 - 6}{5 - 0} = -\frac{4}{5}$$

✐ Section 2.1 Exercise 11(a(i))

We find the average rate of change of F between $x = 2$ and $x = 4$:

$$\text{average rate of change} = \frac{F(4) - F(2)}{4 - 2} = \frac{90 - 50}{4 - 2} = \frac{40}{2} = 20$$

✐ Section 2.1 Exercise 13 (extended)

We find the average rate of change of f between $x = 2$ and $x = 3$:

$$\text{average rate of change} = \frac{f(3) - f(2)}{3 - 2} = \frac{11 - 8}{3 - 2} = 3$$

and between $x = 4$ and $x = 7$:

$$\text{average rate of change} = \frac{f(7) - f(4)}{7 - 4} = \frac{23 - 14}{7 - 4} = \frac{9}{3} = 3$$

✐ Section 2.1 Exercise 25 Currency Exchange Rates (extended)

(b) The average rate of change
(i) between 1999 and 2008 is

$$\begin{aligned}\text{average rate of change} &= \frac{\text{net change in } y}{\text{net change in } x} \\ &= \frac{1.46 - 0.86}{2008 - 1999} \\ &= \frac{0.6}{9} \\ &\approx 0.0667 \text{ dollars per year}\end{aligned}$$

(ii) between 2002 and 2006 is

$$\text{average rate of change} = \frac{\text{net change in } y}{\text{net change in } x} = \frac{1.18 - 0.89}{2006 - 2002} = \frac{0.29}{4} = 0.0725 \text{ dollars per year}$$

(iii) between 2005 and 2008 is

$$\text{average rate of change} = \frac{\text{net change in } y}{\text{net change in } x} = \frac{1.46 - 1.35}{2008 - 2005} = \frac{0.11}{3}$$

$$\approx 0.0367 \text{ dollars per year}$$

and (iv) between 2000 and 2002 is

$$\text{average rate of change} = \frac{\text{net change in } y}{\text{net change in } x} = \frac{0.89 - 1.01}{2002 - 2000} = \frac{-0.12}{2}$$

$$= -0.06 \text{ dollars per year}$$

(c) From the graph, we see that the steepest increase in a single year happened in 2003, between the first business day of 2003 and the first business day of 2004. From the table, we see that the average rate of change over that year is 0.21 dollars per year.

(d) From the graph, we see that the steepest decrease in a single year happened in 2005, between the first business day of 2005 and the first business day of 2006. From the table, we see that the net change over that year is –0.17 dollars per year.

Section 2.1 Exercise 27 Speed Skating

(a) Skater B won the race because he reached 500 meters in 35 seconds, while skater A reached 500 meters in 40 seconds.

(b) The average speeds during the first 10 seconds are

$$\text{average speed of skater A} = \frac{\text{net change in distance}}{\text{change in time}} = \frac{200}{10} = 20 \text{ m/s}$$

$$\text{average speed of skater B} = \frac{\text{net change in distance}}{\text{change in time}} = \frac{100}{10} = 10 \text{ m/s}$$

(c) The average speeds during the last 15 seconds are

$$\text{average speed of skater A} = \frac{\text{net change in distance}}{\text{net change in time}} = \frac{d(40) - d(25)}{15} = \frac{100}{15} \approx 6.67 \text{ m/s}$$

$$\text{average speed of skater B} = \frac{\text{net change in distance}}{\text{net change in time}} = \frac{d(35) - d(20)}{15} = \frac{300}{15} = 20 \text{ m/s}$$

Section 2.1 Exercise 29 Phoenix Mars Lander

(a) The average speed of the tank in the first 10 seconds of the fall is

$$\text{average speed} = \frac{\text{net change in distance}}{\text{change in time}} = \frac{s(10) - s(0)}{10 - 0} = \frac{16(10)^2 - 16(0)^2}{10} = 160 \text{ ft/s}$$

(b) The average speed of the tank in the first 15 seconds of the fall is

$$\text{average speed} = \frac{\text{net change in distance}}{\text{change in time}} = \frac{s(15) - s(0)}{15 - 0} = \frac{16(15)^2 - 16(0)^2}{15} = 240 \text{ ft/s}$$

2.2 Linear Functions: Constant Rates of Change

OBJECTIVES

☑ *Check the box when you can do the exercises addressing the objective, such as those included in this study guide, which are listed here.*

Exercises

Linear Functions

☐ Determine if a function is linear from its formula.	**9, 11, 13, 53**
☐ Graph a line, by first finding the coordinates of two points.	**17, 25, 53**
☐ Calculate the average rate of change of a function from its formula	**21, 53**

Linear Functions and Rates of Change

☐ Determine the rate of change and initial value of a model of the real world from its formula.	**53**
☐ Write the formula for a linear function that models the real world, given a rate of change and an initial value.	**55, 59**

Linear Functions and Slope

☐ Find the slope of the line passing through two given points.	**45, 59**
☐ Find the slope of the graph of a linear function from its formula.	**25, 53**

Using Slope and Rate of Change

☐ Find the rate of change of a linear function, by finding the slope of its graph.	**59**

GET READY...

Read the definition of a linear model in a blue box in Section 1.3.

SKILLS

Linear Functions

In these exercises you decide if a given formula represents a linear function. Find the definition of a linear model in a blue box in Section 1.3 of the textbook, and of a linear function in a blue box in Section 2.2 of the textbook, and compare them. Recall that in Section 1.3, you determine whether or not data with evenly spaced inputs are linear (see Section 1.3 Example 2 in the textbook, and review Section 1.3 Exercise 21(a), which is referred in this study guide on page 19).

Since the graph of a linear function is a line, you graph linear functions in these exercises plotting two points and sketching the line through them. Recall that in Section 1.6, you graph basic functions, including linear functions (see Section 1.6 Example 3(a)), by making a table of values, plotting points, and connecting them. In these exercises, you also calculate average rates of change for a linear function. Find the definition of average rate of change in a blue box in Section 2.1.

📖 **Read Section 2.2 Example 1**, in which the authors determine whether or not a given function is linear.

[9]*How did the authors justify the assumptions that f, g and k are linear functions?*
Answer in your own words:

✎ **Section 2.2 Exercises 9, 11, and 15** Determine whether the given function is linear.

9. $f(x) = 3 + \dfrac{1}{2}x$ 11. $f(x) = 4 - x^2$

Solution   **Solution**

15. $f(t) = \dfrac{2}{3}(t-2)$

Solution

📖 **Read Section 2.2 Example 2**, in which the authors graph a line.

[10]*In Section 1.6, you graph functions by making a table of values, plotting the points and connecting the dots. What do the authors do here?*
Answer in your own words:

---

[9] To show that the functions f, g, and k are linear functions, the authors give the values of m and b in each case.

[10] Since the graph of a linear function is a line, the authors plot only two of the points from the table, and draw a straight line through them.

✎ **Section 2.2 Exercise 17** Make a table of values for $f(x) = 6x + 5$ and sketch its graph.

Solution

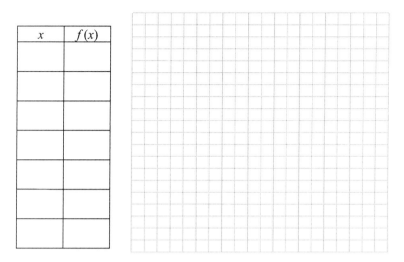

x	$f(x)$

📖 **Read Section 2.2 Example 3**, in which the authors calculate the average rate of change of a linear function.

[11] *The average rate of change of f between any two values of x is _____. Where does the average rate of change of f appear in the formula for f?*
Answer in your own words:

✎ **Section 2.2 Exercise 21** Find the average rate of change of $f(x) = 4 + 2x$ on the following intervals.

(a) Between $x = -1$ and $x = 1$
(b) Between $x = 1$ and $x = 2$
(c) Between $x = a$ and $x = h + a$

Solution

---

[11] The average rate of change between any two values is 3, which is the coefficient of x in the formula for f.

Linear Functions and Slope

In this subsection, the authors summarize the connection between the rate of change of a linear function and its formula that you explore in Section 2.2 Example 3 and Exercise 21 above. Find the definition of the slope of a line and a description of the parts of a linear function in terms of its slope and y-intercept in blue boxes in Section 2.2 of the textbook.

📖 **Read Section 2.2 Example 6**, in which the authors find the slope of a staircase.

[12]*If you move along the edge of the trim board of the staircase, and increase the x coordinate by 12 inches, what happens to the y-coordinate? What about if you increase the x coordinate by 1 inch?*
Answer in your own words:

✎ **Section 2.2 Exercise 45** Find the slope of the line passing through $(2, 2)$ and $(-10, 0)$.

Solution

📖 **Read Section 2.2 Example 7**, in which the authors graph a linear function, and calculate its slope.

[13]*The slope of the line is _____. The authors find the slope in two different ways; what are they?*
Answer in your own words:

The rate of change of f is _____. The authors find the rate of change in two different ways; what are they?
Answer in your own words:

---

[12] When you increase the x-coordinate by 12 inches, the y-coordinate increases by 8 inches. When you increase the x-coordinate by 1 inch, the y-coordinate changes by 2/3 inch.

[13] The slope of the line is 1. The authors find the slope by reading the value of m from the formula for f, and also by calculating rise over run. The rate of change is the same as the slope, which is 1. The authors find it by reading the value of m from the formula for f, and also by calculating the average rate of change between $x = 0$ and $x = 1$.

✎ **Section 2.2 Exercise 25** For the linear function $f(x) = 4 + 2x$,

(a) Sketch the graph
(b) Find the slope of the graph.
(c) Find the rate of change of the function.

Solution

Using Slope and Rate of Change

The authors sum up the relationships between slopes and rates of change that you explore in Section 2.2 Example 7 and Exercise 25 above, in this subsection.

📖 **Read the introduction to the subsection, 'Using Slope and Rate of Change,'** in which the authors relate rates of change to slopes.

[14] *In Section 2.2 Exercise 53, would you describe m = 4 in the model as a slope or as a rate of change? Answer in your own words:*

CONTEXTS

Linear Functions and Rate of Change

In these exercises, you use the vocabulary, initial value, and rate of change to describe the coefficients of a linear function. Find these words in bold typeface and in a blue box in Section 2.2 of the textbook.

📖 **Read the blue box entitled, 'Linear Models,' in Section 1.3 of the textbook,** in which the authors describe the meaning of the coefficients of a linear model.

[15] *The coefficient m from Section 2.2 is the same as the coefficient ____ (A / B) from Section 1.3. Comparing the descriptions of the coefficients from Sections 1.3 and 2.2, we find that the rate of change is the amount by which f(x) changes for each _____.*

---

[14] The 4 is the rate at which the landfill is receiving trash, in thousands of tons per day; the 4 is also the slope of the graph.

📖 **Read Section 2.2 Example 4**, in which the authors use a linear function to model a toddler's growth.

[16] *The rate of change of h is (include units with your answer) _____. How much does the height of the toddler change each month _____.*

✎ **Section 2.2 Exercise 53** The amount of trash in a country landfill is modeled by the function $T(x) = 32,400 + 4x$, where x is the number of days since January 1, 1996, and $T(x)$ is measured in thousands of tons.

(a) Is T a linear function?
(b) What is the initial amount of trash in the landfill in 1996?
(c) At what rate is the landfill receiving trash?
(d) Sketch a graph of T.

Solution

📖 **Section 2.2 Example 5**, in which the authors find a linear function that models the volume of water in a swimming pool as it is being filled.

[17] *The units on the rate of change of the volume of water in the pool are _____.*

✎ **Section 2.2 Exercise 55 Weather Balloon** Weather balloons are filled with hydrogen and released at various sites to measure and transmit data such as air pressure and temperature. A weather balloon is filled with hydrogen at the rate of 0.5 ft³/s. Initially, the balloon has 2 ft³ of hydrogen. Find a linear function that models the volume of hydrogen in the balloon after t seconds.

Solution

---

[15] The m in Section 2.2 corresponds to the B in Section 1.3. The rate of change is the amount by which f changes for each unit increase in x.

[16] The rate of change of h is 0.4 in/mo. The toddler's height increases by 0.4 inches each month.

[17] The units on the rate of change of volume of water in the pool are gallons per minute.

Using Slope and Rate of Change

📖 **Read Section 2.2 Example 8**, in which the authors compare the graphs of the distance two drivers travel along I-76 as a function of time.

[18] *Mary travels _____ miles in 1 hour. The slope of the line representing the distance Mary travels is _____ miles per hour. At what speed is Mary traveling? _____ miles per hour.*

✏ **Section 2.2 Exercise 59 Commute to work** Jade and her roommate Jari live in a suburb of San Antonio, Texas, and both work at an elementary school in the city. Each morning they commute to work traveling west on I-10. One morning Jade left for work at 6:50 am, but Jari left 10 minutes later. Both drove at a constant speed. The graphs below show the distance (in miles) each of them has traveled on I-10 at time t (in minutes), where $t = 0$ is 7:00 A.M.

(a) Use the graph to decide which of them is traveling faster.
(b) Find the speed at which each of them is driving.
(c) Find linear functions that model the distances that Jade and Jari travel as functions of t.

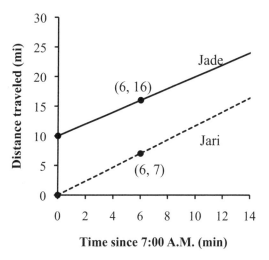

CONCEPTS

✓ **2.2 Exercises – Concepts – Fundamentals** Complete the Fundamentals section of the exercises at the end of Section 2.2. Compare your answers to those at the back of the textbook, and make corrections as necessary.

📖 **Read the Concept Check for Section 2.2**, in the Chapter Review at the end of Chapter 2.

---

[18] Mary travels 75 miles in 1 hour; the slope and Mary's speed are 75 miles per hour.

2.2 Solutions to ✎ Exercises

✎ Section 2.2 Exercises 9, 11, and 15

9. This function is linear with slope $m = 1/2$ and y-intercept $b = 3$.

11. This function is not linear since the x is squared.

15. Distributing, we have $f(t) = \frac{2}{3}(t - 2) = \frac{2}{3}t - \frac{4}{3}$. Thus, this function is linear with slope $m = 2/3$ and y-intercept $b = -4/3$.

✎ Section 2.2 Exercise 17

Since the graph of $f(x) = 6x + 5$ is a line, to make the graph, we can plot any two points from the table and connect them. Here, we plot $(0, 5)$ and $(2, 17)$.

x	$f(x)$
−3	−13
−2	−7
−1	−1
0	5
1	11
2	17
3	23

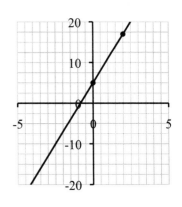

✎ Section 2.2 Exercise 21

(a) average rate of change $= \dfrac{f(1) - f(-1)}{1 - (-1)} = \dfrac{(4 + 2 \cdot 1) - (4 + 2(-1))}{2} = 2$

(b) average rate of change $= \dfrac{f(2) - f(1)}{2 - 1} = \dfrac{(4 + 2 \cdot 2) - (4 + 2 \cdot 1)}{1} = 2$

(c) average rate of change $= \dfrac{f(h + a) - f(a)}{h + a - a} = \dfrac{(4 + 2 \cdot (h + a)) - (4 + 2 \cdot a)}{h}$

$= \dfrac{4 + 2h + 2a - 4 - 2a}{h} = \dfrac{2h}{h} = 2$

✎ Section 2.2 Exercise 25

(a) The graph of f is a line. To sketch the graph, we need only find two points on the line. Since $f(0) = 4$ and $f(1) = 6$, two points on the graph are $(0, 4)$ and $(1, 6)$. Connect these points to get the graph of f.

(b) Since $f(x) = 4 + 2x$, we see that $m = 2$, so the graph of f is a line with slope 2. We can also find the slope using the definition of slope and the two points from part (a):

$$\text{slope} = \frac{\text{rise}}{\text{run}} = \frac{6 - 4}{1 - 0} = 2$$

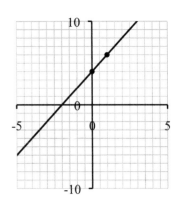

(c) Since f is a linear function in which m is 2, the rate of change of f is 2. We can also calculate the rate of change directly, using the two values from part (a):

$$\text{rate of change} = \frac{f(1) - f(0)}{1 - 0} = \frac{6 - 4}{1 - 0} = 2$$

✎ Section 2.2 Exercise 45

The slope of the line passing through (2, 2) and (−10, 0) is

$$\text{slope} = \frac{\text{rise}}{\text{run}} = \frac{2 - 0}{2 - (-10)} = \frac{2}{12} = \frac{1}{6}$$

✎ Section 2.2 Exercise 53

(a) Yes, $T(x) = 32,400 + 4x$ is a linear function, where $m = 4$ and $b = 32,400$.

(b) The initial amount of trash in the landfill is the initial value of T, which is the constant b in the formula. So the initial amount of trash is 32,400 thousand tons.

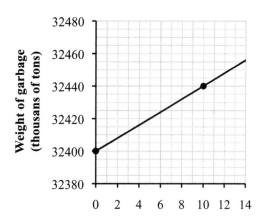

(c) Since T is a linear function, its rate of change is the constant m, which in this case is 4. This means the rate at which the landfill is receiving trash is 4 thousand tons per day.

(d) Since T is a linear function, its graph is a line. So to graph T we need to find two points on the line. Since $T(0) = 32,400$ and $T(10) = 32,440$, the points (0, 32,400) and (10, 32,440) are on the graph.

✎ Section 2.2 Exercise 55 Weather Balloon

We need to find a linear function $V(t) = b + mx$ that models the volume $V(t)$ of hydrogen in the weather balloon at time t. The rate of change of volume is 0.5 ft³/s, so $m = 0.5$. Since the balloon has 2 ft³ of hydrogen initially, we have $V(0) = 2$. So the initial value $b = 2$. Now that we know m and b, we can write $V(t) = 2 + 0.5t$.

✎ Section 2.2 Exercise 59 Commute to work

(a) Jari is traveling faster than Jade. You can see this on the graph, since Jari's line appears slightly steeper than Jade's.

(b) From the graph, we see that Jade traveled 10 miles at 7 A.M. and 16 miles at 6 minutes after 7 A.M.. The speed is the rate of change, which is the steepness of the graph, so Jade's speed is

$$\text{slope} = \frac{16 - 10}{6 - 0} = 1 \text{ mi/min}$$

Jari started at 7 A.M. and traveled 7 miles 6 minutes later. So Jari's speed is

$$\text{slope} = \frac{7-0}{6-0} \approx 1.17 \text{ mi/min}$$

So indeed, Jari is traveling faster than Jade.

(c) From the graph, the *y*-intercept for Jade's line is $b = 10$ mi. We find that the slope of her line is 1 mi/min in part (b). So an equation of her line is $y = x + 10$.

Similarly, from the graph, the *y*-intercept for Jari's line is $y = 0$ mi. We find the slope of her line is 1.17 mi/min in part (b). So an equation of her line is $y = 1.17x$.

2.3 Equations of Lines: Making Linear Models

OBJECTIVES

☑ *Check the box when you can do the exercises addressing the objective, such as those included in this study guide, which are listed here.*

Exercises

☐ Use the vocabulary "slope" and "rate of change" interchangeably; use the vocabulary "*y*-intercept" and "initial value" interchangeably.   **57, 61, 63**

☐ Recognize the slope and initial value when they are described in words.   **57, 61, 63**

☐ When constructing a linear model of the real world, determine what information you have, and what you should do to create a model.   **57, 61, 63**

Slope-Intercept Form

☐ Write the equation for a line in slope-intercept form.   **19, 23, 31, 53, 61, 63**

☐ Find an equation for a line, given the slope and the *y*-intercept.   **11**

☐ Find an equation for a line, given the *y*-intercept and another point on the graph.   **57**

☐ Find the slope of a line given two points on the line.   **31, 57, 63**

Point-Slope Form

☐ Find an equation of a line using point-slope form, given the slope and a point on the graph.   **23, 61**

☐ Find an equation of a line using point-slope form, given two points on the graph.   **31, 57, 63**

☐ Think about a model before writing its formula by making up examples of inputs for your model and figuring out the corresponding outputs.   **63**

☐ Explain what the slope, *x*- and *y*-intercepts represent in context.   **63**

Horizontal and Vertical Lines

☐ Construct or recognize a horizontal or vertical line through a given point.   **See page 108**

☐ Identify the slope, and *x*- and *y*-intercepts of a horizontal or vertical line, if it is defined.   **See page 108**

When Is the Graph of an Equation a Line

☐ Find the *x*- and *y*-intercepts of a line, using its representation in general linear form.   **53**

Get Ready...

You have encountered three interpretations of the parameters in a linear function. In the textbook, find and compare the blue boxes entitled,

- 'Linear Models,' in Section 1.3,
- 'Linear Functions and Rate of Change,' in Section 2.2, and
- 'Linear Functions and Slope,' in Section 2.2.

[19] *For the linear function f(x) = mx + b, write three different interpretations of the number m, and two different interpretations of the number b.*

Answer in your own words:

Skills

Slope-Intercept Form

In these exercises, you write equations in a slope-intercept form to determine if the equations are linear, and if so, to find the slopes and y-intercepts. Find slope-intercept form in a blue box in Section 2.3 of the textbook.

📖 **Read Section 2.3 Example 1**, in which the authors write equations in slope-intercept form.

[20] *In part (a), if the x-value increases by 1, then the y-value changes by _____. In part (b), if the x-value increases by 1, then the y-value changes by _____.*

✎ **Section 2.3 Exercise 11** Find an equation of the line with slope 5 and y-intercept 2.

Solution

✎ **Section 2.3 Exercise 19** Express the equation $9x - 3y - 4 = 0$ in slope-intercept form.

Solution

---

[19] The number m is (1) the amount by which $f(x)$ changes (increases or decreases) for each unit increase in x, (2) the slope of the graph of $f(x)$, and (3) the rate of change of $f(x)$. The number b is (1) the initial value of $f(x)$ and (2) the y-intercept of the graph of $f(x)$.

[20] Connecting concepts from 1.3 and 2.2, if the x-value increases by 1, the y-value changes by the slope, which is 3 in part (a) and 2/3 in part (b).

Point-Slope Form

In these exercises you create models for lines using the point-slope form of the equation. Find a derivation of the point-slope formula, including a summary in a blue box, in Section 2.3 of the textbook.

📖 **Read Section 2.3 Example 3**, in which the authors find an equation of the line with a given slope that passes through a given point.

[21] *How do the authors use the slope to graph the line?*
Answer in your own words:

✎ **Section 2.3 Exercise 23**

(a) Find an equation of the line with slope 2 that passes through (0, 4).
(b) Simplify the equation by putting it into slope-intercept form.
(c) Sketch a graph of the line.

Solution

---

[21] The authors explain that a slope of –1/2 indicates that when we move two units to the right, the graph drops by 1. The authors begin at the given point (1, –3) and plot a second point by moving to the right by 2 and down by 1.

📖 **Read Section 2.3 Example 5**, in which the authors find an equation of the line that passes through two given points.

[22] *To use the point-slope formula, you need to know* _____ *and* _____.
Here the slope is not given in the exercise; the authors calculate the slope using the formula m = _____.

✏ **Section 2.3 Exercise 31**

(a) Find an equation of the line that passes through (–2, 1) and (4, 7).
(b) Simplify the equation by putting it into slope-intercept form.
(c) Sketch a graph of the line.

Solution

_____

[22] To use the point-slope formula you need to know a point on the line and the slope of the line. The

authors calculate the slope using the formula, $m = \dfrac{y_2 - y_1}{x_2 - x_1}$.

Horizontal and Vertical Lines

Read the introduction to the section entitled, 'Horizontal and Vertical Lines,' in Section 2.3 of the textbook, explaining the equations for vertical and horizontal lines, and find a blue box summarizing the discussion.

📖 **Read Section 2.3 Example 7**, about horizontal and vertical lines.

For each line described in the following table, fill in the blanks with a number or "undefined."

23 *The horizontal line through (3,–2)* *An equation of this line is* _____.	*slope* _____ *y-intercept* _____ *x-intercept* _____
24 *The vertical line through (3,–2)* *An equation of this line is* _____.	*slope* _____ *y-intercept* _____ *x-intercept* _____
25 *The line x = –5* *This line is* _____ *(horizontal / vertical).*	*slope* _____ *y-intercept* _____ *x-intercept* _____
26 *The line y = 8* *This line is* _____ *(horizontal / vertical).*	*slope* _____ *y-intercept* _____ *x-intercept* _____

When Is the Graph of an Equation a Line?

These exercises ask you to use the general form of the equation of a line. You use this general form in Chapter 7, when you study systems of linear equations. Find the general form of the equation of a line in a blue box in Section 2.3 of the textbook.

📖 **Read Section 2.3 Example 8**, in which the authors use the general form of a linear equation to answer questions.

27 *What two points do the authors use to sketch the graph? (_____, _____) and (_____, _____). Every line has an equation in general form, but not every line has an equation in slope intercept form. Give an example of a line that does not have an equation in slope intercept form.*

---

23 An equation of this line is $y = -2$. Slope $m = 0$; y-intercept $b = -2$; x-intercept is undefined.

24 An equation of this line is $x = 3$. Slope is undefined; y-intercept is undefined; x-intercept is 3.

25 The line is vertical; slope is undefined; y-intercept is undefined; x-intercept is –5.

26 The line is horizontal; slope $m = 0$; y-intercept $b = 8$; x-intercept is undefined.

27 The authors use the x-intercept and the y-intercept, (6, 0) and (0, 4), to sketch the graph. The slope of a vertical line is undefined, so vertical lines, for example the line $x = 4$, does not have an equation in slope intercept form.

✐ **Section 2.3 Exercise 53** Use the general linear equation $3x + 5y - 15 = 0$ to answer the following questions.

(a) Find the slope of the line.
(b) Find the x- and y-intercepts.
(c) Use the intercepts to sketch the graph.

Solution

CONTEXTS

In these exercises you construct more linear models of the real world. Some of these exercises give you the slope, and your task is to recognize it in the context of the word problem. In other exercises, you calculate the slope, and interpret it in the context of the problem.

Slope-Intercept Form

Exercises in Section 2.2 give you the rate of change and the initial value of a real world quantity that you model with a linear function. In contrast, these exercises give you the initial value and an additional point; you use the formula for slope to calculate the rate of change.

📖 **Reread Section 2.2 Example 5**, paying particular attention to what information is given in the exercise.

[28] *Is the rate of change of the linear function given in the problem? _____ (Yes / No) Is the initial value of the linear function given in the problem? _____ (Yes / No)*

✐ **Section 2.2 Exercise 55 Weather Balloon** Find this exercise on page 100 of this study guide.

📖 **Read Section 2.3 Example 2**, in which the authors construct a linear model for the distance a car travels, in order to find its speed.

[29] *Is the rate of change of the linear function given in the problem? _____ (Yes / No) Is the initial value of the linear function given in the problem? _____ (Yes / No)*

---

[28] The rate of change, 5 gal/min, and the initial value, 200 gal, are given in the exercise.

✐ **Section 2.3 Exercise 57 Air Traffic Control** Air traffic controllers at most airports use a radar system to identify the speed, position, and other information about approaching aircraft. Using radar, an air traffic controller identifies an approaching aircraft and determines that it is 45 miles from the radar tower. Five minutes later, she determines that the aircraft is 25 miles from the radar tower. Assume that the aircraft is approaching the radar tower directly at a constant speed.

(a) Find a linear equation that models the distance y of the aircraft from the radar tower x minutes after it was first observed.

(b) What is the speed of the approaching aircraft?

Solution

Point-Slope Form

These exercises give you the rate of change of the function, and a point on the graph that is not necessarily the initial value; you use the point-slope form to write the linear model, from which you determine the initial value. Some of these exercises instead give you two points on the graph; you first find the slope, and then use the point slope form to write the model. Find the point-slope form in a blue box in Section 2.3 of the textbook.

Additionally, in these exercises, you use Hints and Tips 1.2b to interpret the slope, y-intercept and x-intercept of a linear model in context. Reread the blue box in Section 1.3 entitled linear models, in which the authors interpret the meaning of slope.

📖 **Read Section 2.3 Example 4**, in which the authors find a linear function that models the volume of water in a swimming pool as it is being filled.

[30] *Is the rate of change of the linear function given in the problem? _____ (Yes / No) Is the initial value of the linear function given in the problem? _____ (Yes / No)*

---

[29] The rate of change is not given in the problem; the authors compute it using the formula for slope. The initial value, 350 ft, is given in the problem.

[30] The rate of change, 5 gallons per minute, is given in the problem. The initial value is not given in the problem; the authors put the equation in slope intercept form to find the y-intercept.

✎ **Section 2.3 Exercise 61 Weather Balloon** A weather balloon is filled with hydrogen at the rate of 0.5 ft³/s. After 2 seconds the balloon contains 5 ft³ of hydrogen. Find a linear equation that models the volume V of hydrogen in the balloon at any time t.

Solution

📖 **Read Section 2.3 Example 6**, in which the authors model the demand for soda with a linear function.

[31] Is the rate of change of the linear function given in the problem? _____ (Yes / No) Is the initial value of the linear function given in the problem? _____ (Yes / No)

> Refer back to page 69
> **Hints and Tips 1.8a: Thinking about the problem**

[32] In this model, the input x represents _____, and the output y represents _____.
Think about the problem by using this sentence from the exercise to fill out the table of examples.

At a price of $1.00 per can of soda he sells 600 cans, but for every 25-cent increase in the price, he sells 75 fewer cans each week.

x	y
1	
1.25	

x	y
1.50	
1.75	

The authors find the slope using the two points (_____, _____) and (_____, _____). Any pair of points from your table would also allow you to find the slope.

> Refer back to page 11
> **Hints and Tips 1.2b: What does the algebra represent in context?**

[33] What does (1.25, 525) represent in context?
Answer in your own words:

---

[31] Neither the rate of change nor the initial value is given in the problem.

[32] The input x represents the price per can; the output y represents the number of cans the operator sells. The rows of the table are (1, 600), (1.25, 525), (1.50, 450), (1.75, 375). The authors find the slope using (1, 600) and (1.25, 525).

[33] Following Hints and Tips 1.2b, we write the following sentences as scratch work: Algebraically, the x-value is 1.25, when the y-value is 525. In context, the cost of the can of soda is $1.25 when the number of cans the operator sells each week is 525. We construct our final answer more eloquently: The operator sells 525 cans of soda when he sets the price at $1.25 per can.

As in the question above, you have used Hints and Tips 1.2b to describe what ordered pairs in a relation represent. Here Hints and Tips 1.2b also explains parts (c), (d) and (e) of Example 6, in which the authors determine what the slope, x-intercept and y-intercept represent in context.

⇨ **Section 2.3 Example 6(c)** What does the slope represent?

Solution

The authors find that the slope is −300. Following Hints and Tips 1.2b, we write the following sentences as scratch work.

> Algebraically, −300 is the amount by which y changes for each unit increase in x.

Now we rewrite that sentence, adding units to the numbers and replacing x and y by their meanings.

> In context, −300 cans is the amount by which weekly sales changes for each $1 increase in the price per can.

The authors state the final answer more eloquently:

> The slope is −300. That means that sales drop by 300 cans for every $1 increase in price.

⇨ **Section 2.3 Example 6(d)** What does the y-intercept represent?

Solution

The authors find the y-intercept is 900. Following Hints and Tips 1.2b, we write the following sentences as scratch work.

> Algebraically, 900 is the y-value when the x-value is 0. In context, 900 cans is the number of cans the operator would sell if the price per can were $0.

We state our final answer more eloquently:

> The y-intercept of 900 cans represents the number of cans the operator could give away for free.

⇨ **Section 2.3 Example 6(e)** What does the x-intercept represent?

Solution

The authors find the x-intercept is 3. Following Hints and Tips 1.2b, we write the following sentences as scratch work.

> Algebraically, 3 is the x-value when the y-value is 0. In context, $3.00 is the price when the number of cans the operator would sell is 0.

The authors state the final answer more eloquently:

> The x intercept is 3. This means if he charges $3.00 per can, he will sell no soda at all.

✎ **Section 2.3 Exercise 63 Demand for Bird Feeders** A community bird-watching society makes and sells simple bird feeders to raise money for its conservation activities. They sell 20 per week at a price of $10 each. They are considering raising the price, and they find that for every dollar increase, they lose two sales per week.

(a) Find a linear equation that models the number y of feeders that they sell each week at price x by a linear equation, and sketch a graph of the equation.

(b) How many feeders would they sell if they charged $14 per feeder? If they charged $6?

(c) Find the slope of the line. What does the slope represent?

(d) Find the y-intercept of the line. What does it represent?

(e) Find the x-intercept of the line. What does it represent?

[34] Following Hints and Tips 1.8a, think about the problem by making up some examples. The input x represents _____, and the output y represents _____. Complete the table of examples.

Solution

x	y
10	
11	
12	
13	

---

[34] The input is the price per feeder, and the output is the number of feeders they sell per week. The points in the table are (10, 20), (11, 18), (12, 16), and (13, 14).

2.3 Solutions to ✐ Exercises

✐ Section 2.3 Exercise 11

Using the slope-intercept form with slope 5 and y-intercept 2, we get

$$y = mx + b \qquad \text{Slope-intercept form}$$

$$y = 5x + 2 \qquad \text{Replace } m \text{ by 5 and } b \text{ by 2.}$$

So the slope-intercept form of the equation of the line is $y = 5x + 2$.

✐ Section 2.3 Exercise 19

To put the equation into slope-intercept form, we solve for y.

$$9x - 3y - 4 = 0 \qquad \text{Given equation}$$

$$-3y = -9x + 4 \qquad \text{Subtract } 9x - 4 \text{ from each side}$$

$$y = \frac{-9}{-3}x + \frac{4}{-3} \qquad \text{Divide each side by } -3$$

$$y = 3x - \frac{4}{3} \qquad \text{Simplify}$$

So the slope-intercept form of the equation of the line is $y = 3x - \frac{4}{3}$.

✐ Section 2.3 Exercise 23

(a) and (b) We know the slope m is 2, and a point (x_1, y_1) on the line is $(0, 4)$. So we use the point-slope form with m replaced by 2, x_1 by 0, and y_1 by 4:

$$y - y_1 = m(x - x_1) \qquad \text{Point-slope form}$$

$$y - 4 = 2(x - 0) \qquad \text{Replace } m \text{ by 2, } x_1 \text{ by 0, and } y_1 \text{ by 4}$$

$$y - 4 = 2x \qquad \text{Distributive Property}$$

$$y = 4 + 2x \qquad \text{Add 4 to each side}$$

So an equation for the line in slope intercept form is $y = 4 + 2x$.

(c) The fact that the slope is 2 tells us that when the run is 1 the rise is 2, so when we move 1 unit to the right, the line rises by 2 units. This enables us to sketch the line.

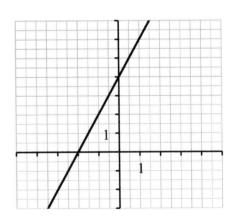

✏ **Section 2.3 Exercise 31**

(a) and (b) We first use the two given points to find the slope:

$$m = \frac{7-1}{4-(-2)} = \frac{6}{6} = 1$$

We know the slope m is 1, and a point (x_1, y_1) on the line is $(-2, 1)$. So we use the point-slope form with m replaced by 1, x_1 by -2, and y_1 by 1:

$y - y_1 = m(x - x_1)$	Point-slope form
$y - 1 = 1(x - (-2))$	Replace m by 1, x_1 by -2, and y_1 by 1
$y - 1 = x + 2$	Simplify
$y = x + 3$	Add 1 to each side

So an equation of the line in slope-intercept form is $y = x + 3$.

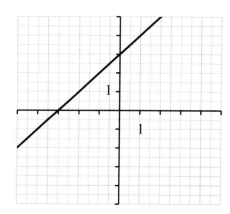

(c) The fact that the slope is 1 tells us that when the run is 1 the rise is 1, so when we move 1 unit to the right, the line rises by 1 unit. This enables us to sketch the line.

✏ **Section 2.3 Exercise 53**

(a) To find the slope, we put the equation into slope-intercept form.

$3x + 5y - 15 = 0$	General form of the equation of the line
$5y = -3x + 15$	Subtract $3x - 15$ from each side
$y = \dfrac{-3}{5}x + 3$	Divide each side by 5

Comparing this last equation with the slope-intercept form $y = b + mx$, we see that the slope m is $-\dfrac{3}{5}$.

(b) To find the x-intercept, we replace y by 0 in the general form of the equation, and solve for x.

$3x + 5(0) - 15 = 0$	Replace y by 0
$3x = 15$	Add 15 to each side
$x = 5$	Divide each side by 3

So the x-intercept is $x = 5$.

To find the y-intercept, we replace x by 0 in the general form of the equation, and solve for y.

$$3(0) + 5y - 15 = 0 \qquad \text{Replace } x \text{ by } 0$$

$$5y = 15 \qquad \text{Add 15 to each side}$$

$$y = 3 \qquad \text{Divide each side by 5}$$

So the y-intercept is $y = 3$.

(c) We plot the intercepts are $(0, 3)$ and $(5, 0)$, and sketch the line that contains them.

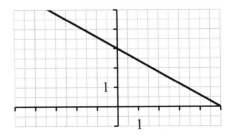

✎ Section 2.3 Exercise 57 Air Traffic Control

(a) Let's choose the time of the first measurement to be time 0, that is $x = 0$. So the next measurement is 5 minutes later at time $x = 5$. Then from the given data, we get the following two points: $(0, 45)$ and $(5, 25)$.

The first of these two points tells us that the y-intercept is $b = 45$. We use the two points together to get the slope:

$$m = \frac{25 - 45}{5 - 0} = \frac{-20}{5} = -4 \text{ mi/min}$$

So the linear equation that models the distance y of the aircraft from the radar tower after x minutes is $y = mx + b = -4x + 45$.

(b) The slope of the line we found in part (a) is the same as the rate of change of the distance with respect to time; that is , the slope is the speed of the aircraft. So the speed of the aircraft is -4 mi/min. (The speed is negative because the distance to the radar tower is decreasing.)

✎ Section 2.3 Exercise 61 Weather Balloon

We need to find a linear equation, $V = b + mt$, which models the volume V of hydrogen in the balloon at time t. The rate of change of volume is 0.5 cubic feet per second, so $m = 0.5$. Since the balloon has 5 cubic feet of hydrogen after 2 seconds, the point $(t_1, V_1) = (2, 5)$ is on the desired line. We find the equation by using the point-slope form of a line with $(t_1, V_1) = (2, 5)$, and $m = 0.5$.

$$V - V_1 = m(t - t_1) \qquad \text{Point-slope form}$$

$$V - 5 = 0.5(t - 2) \qquad \text{Replace } m \text{ by 0.5, } t_1 \text{ by 2, and } V_1 \text{ by 5}$$

$$V - 5 = 0.5t - 1 \qquad \text{Distributive Property}$$

$$V = 0.5t + 4 \qquad \text{Add 5 to each side}$$

So an equation of the line in slope-intercept form is $V = 0.5t + 4$.

✐ Section 2.3 Exercise 63 Demand for Bird Feeders

(a) Since the number of birdfeeders y that the society sells depends linearly on the price x, the model we want is a linear equation: $y = b + mx$. From the given information, we know the point $(10, 20)$ lies on the line. Since for every dollar they increase the price, they lose 2 sales, we know $(11, 18)$ is another point on the line. From these two points, we can find the slope of the line:

$$m = \frac{18 - 20}{11 - 10} = \frac{-2}{1} = -2$$

So we have the slope $m = -2$, and a point $(x_1, y_1) = (10, 20)$. Now using the point-slope form for the equation of a line, we get

$y - y_1 = m(x - x_1)$ Point-slope form

$y - 20 = -2(x - 10)$ Replace m by -2, x_1 by 10, and y_1 by 20

$y - 20 = -2x + 20$ Distributive Property

$y = -2x + 40$ Add 20 to each side

So when the price is x dollars per feeder, the society sells $y = -2x + 40$ birdfeeders per week.

(b) At a price of \$14 per feeder, the society sells $y = -2(14) + 40 = 12$ birdfeeders per week. At \$6 per feeder, they sell $y = -2(6) + 40 = 28$ feeders per week.

(c) The slope is -2. Following **Hints and Tips 1.2b**, we write the following sentences as scratch work. Algebraically, -2 is the amount by which y changes for each unit increase in x. In context, -2 birdfeeders is the amount by which sales changes for each increase in price by \$1. We write our final answer more eloquently: The slope is -2. This means that sales decreases by 2 birdhouses for each \$1-increase in price.

(d) The y-intercept is 40. Following **Hints and Tips 1.2b**, we write the following sentences as scratch work. Algebraically, 40 is the y-value when the x-value is 0. In context, 40 birdfeeders is the number of birdfeeders the society would sell if the price is \$0. We write our final answer more eloquently: The y-intercept of 40 birdfeeders is the number of birdfeeders the society could give away for free.

(e) To find the x-intercept, set $y = 0$ and solve for x.

$0 = -2x + 40$ Replace y by 0

$2x = 40$ Add $2x$ to each side

$x = 20$ Divide each side by 2

So the x-intercept is 20. Following **Hints and Tips 1.2b**, we write the following sentences as scratch work. Algebraically, 20 is the x-value when the y-value is 0. In context, \$20 is the price when the number of birdfeeders they sell is 0. We write our final answer more eloquently: The x-intercept is 20. This means that if they charge \$20 per birdhouse, they will not sell any birdhouses at all.

2.4 Varying the Coefficients: Direct Proportionality

OBJECTIVES

☑ *Check the box when you can do the exercises addressing the objective, such as those included in this study guide, which are listed here.*

Exercises

☐ Use the term, coefficient of *x*, interchangeably with slope, and the term, constant coefficient, interchangeably with *y*-intercept.   **9**

Varying the Constant Coefficient: Parallel Lines

☐ Explain what changing the constant coefficient (i.e. the *y*-intercept) changes in the graph of a line.   **9**

☐ Construct lines parallel to a given line.   **13, 19**

☐ Recognize when a line is parallel to a given line from its description in context, its equation or its graph.   **9, 51**

Varying the Coefficient of *x*: Perpendicular Lines

☐ Explain what changing the coefficient of *x* (i.e. the slope) changes in the graph of a line.   **27**

☐ Construct lines perpendicular to a given line.   **19**

Modeling Direct Proportionality

☐ Find equations of proportionality for relationships in the real world.   **47, 49, 53**

☐ Use equations of proportionality to change units.   **51, 53**

GET READY...

Review the point-slope form of an equation of a line from Section 2.3.

SKILLS

Varying the Constant Coefficient: Parallel Lines

These exercises relate the constant coefficient in a model to the graph of the model, and discuss when lines are parallel. Find the words constant coefficient and coefficient of *x* in bold typeface and the definition of parallel lines in a blue box in Section 2.4 of the textbook.

^{35}*The constant coefficient is the _____ (slope / y-intercept). The coefficient of x is the _____ (slope / y-intercept). In the equation y = mx + b, m is the _____ (constant coefficient / coefficient of x), and b is the _____ (constant coefficient / coefficient of x).*

📖 **Read Section 2.4 Example 1,** in which the authors sketch the graphs of linear functions with the same slope, but different *y*-intercepts.

^{36}*If two lines are parallel, then they have the same _____ (constant coefficient / x-coefficient). Read the paragraph after Example 1.*

---

^{35} The constant coefficient is the *y*-intercept, and the coefficient of *x* is the slope. The parameter *m* is the coefficient of *x* and the parameter *b* is the constant coefficient.
^{36} Parallel lines have the same *x*-coefficient.

✎ **Section 2.4 Exercise 9** Use a graphing device to graph the given family of linear equations in the same viewing rectangle. What do the lines have in common?

(a) $y = 2(x + b)$ for $b = 0, 1, 2, 3$
(b) $y = 2(x + b)$ for $b = 0, -1, -2, -3$

Solution

Copy the graphs from your calculator into the spaces provided.

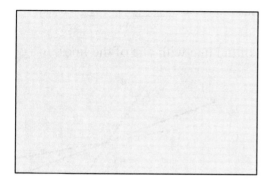
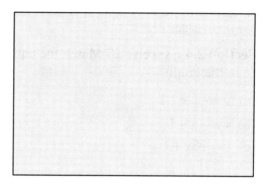

📖 **Read Section 2.4 Example 2**, in which the authors find and graph an equation of a line that is parallel to a given line.

[37] *The authors use the _____ form to write an equation for the line parallel to $y = 4 - 3x$ through (2,0). How do they find the slope?*
Answer in your own words:

✎ **Section 2.4 Exercise 13**

(a) Find an equation of the line that is parallel to $y = 3x - 4$, and passes through (2, 4).
(b) Sketch a graph of both lines on the same coordinate axes.

Solution

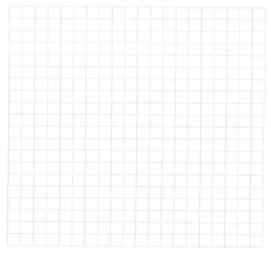

----

[37] The authors use the point-slope form. Sine the line the authors construct in part (a) is parallel to $y = 4 - 3x$, and since parallel lines have no points in common, the authors conclude that the slope of the parallel line must also be -3.

Varying the Coefficient of *x*: Perpendicular Lines

These exercises relate the coefficient of *x* in a model to the graph of the model, and discuss when lines are perpendicular. Find the definition of perpendicular lines in a blue box in Section 2.4 of the textbook.

📖 **Read Section 2.4 Example 4**, in which the authors graph lines with a variety of slopes, but a single *y*-intercept.

[38]*If two lines cross the y-axis at the same point, then they have the same* _____ *(constant coefficient / x-coefficient).*

✎ **Section 2.4 Exercise 27** Match the equation of the line with one of the lines, l_1, l_2, or l_3, in the graph.

(a) $y = -\frac{1}{2}x + 1$

(b) $y = -x + 1$

(c) $y = -3x + 1$

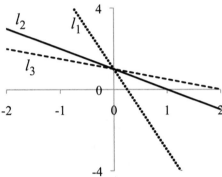

📖 **Read pages 181-182 between Example 4 and Example 5**, in which the authors discuss the slopes of perpendicular lines.

[39]*Figure 7 shows l_1 and l_2, two perpendicular lines. On l_1, when x changes by b, y changes by* _____ *, so the slope of l_1 is* _____ . *On l_2, when x changes by a, y changes by* _____ *, so the slope of l_2 is* _____ .

The slopes are negative reciprocals of each other. The negative reciprocal of 2/3 is _____ . *The negative reciprocal of 3 is* _____ .

📖 **Read Section 2.4 Example 5**, in which the authors construct parallel and perpendicular lines.

[40]*The slope of the line l is* _____ . *The slope of the line parallel to l is* _____ . *The slope of the line perpendicular to l is* _____ .

---

[38] Lines crossing the *y*-axis at the same point have the same constant coefficient.

[39] On l_1, when *x* changes by *b*, *y* changes by *a*, so the slope is *a/b*. On l_2, when *x* changes by *a*, *y* changes by –*b*, so the slope is –*b/a*. The negative reciprocal of 2/3 is –3/2, and the negative reciprocal of 3 is –1/3.

[40] The slope of *l* is 2. The slope of the line parallel to *l* is 2. The slope of the line perpendicular to *l* is –1/2.

✏ **Section 2.4 Exercise 19** The equation of the line l is $y = -\frac{1}{2}x + 1$, and the coordinates of the point P are $(-1, 4)$.

(a) Find an equation of the line that is parallel to l and passes through P.
(b) Find an equation of the line that is perpendicular to l and passes through P.
(c) Sketch a graph of all three lines on the same coordinate axes.

Solution

CONTEXTS

Modeling Direct Proportionality

These exercises ask you to write models for variables that are directly proportional. Find the definition of directly proportional in a blue box in Section 2.4 of the textbook.

📖 **Read the introduction to the subsection, 'Modeling Direct Proportionality,'** in which the authors discuss the electrical capacity of solar panels.

[41] *When y is directly proportional to x, an equation relating them has the form* _____. *The number k is called the* _____.

✏ **Section 2.4 Exercises 47 and 49 (modified)** Find the equation of proportionality that relates y to x.

47. The number of miles per hour y in x feet per second. (One mile is equal to 5280 feet.)

Solution

49. The number of barrels y in x fluid ounces of crude oil. (One barrel of crude oil contains 42 gallons of oil, and each gallon contains 128 fluid ounces.)

Solution

---

[41] When y is directly proportional to x, the variables are related by an equation of the form, $y = kx$. The number k is called the constant of proportionality.

    📖 **Read Section 2.4 Example 6**, in which the authors continue to discuss the electrical capacity of solar panels.

[42]*Since 16 solar panels produce _____ kilowatts of electricity, 2×16 = 32 solar panels produce _____ kilowatts of electricity, and 3×16 = 48 solar panels produce _____ kilowatts of electricity.*

    ✎ **Section 2.4 Exercise 53 Crude Oil in Plastic** A small amount of crude oil is used for manufacturing plastic. Scientists estimate that about 3 fluid ounces of crude oil is used to manufacture 1 million plastic bottles. So the number of fluid ounces of crude oil y used to manufacture plastic bottles is directly proportional to the number of bottles x that are manufactured.

(a) Find the equation of proportionality that relates y to x.

(b) In 2006, about 29 billion plastic bottles were used in the United States. Use the equation found in part (a) to determine the number of barrels of crude oil that were used to manufacture these plastic bottles. (Use the equation you found in Section 2.4 Exercise 49 (modified) in this study guide page 121 to convert fluid ounces to barrels.)

(c) Search the Internet to confirm your answer to part (b).

Solution

Varying the Constant Coefficient: Parallel Lines

These exercises ask you to use parallel lines to model the real world.

    📖 **Read Section 2.4 Example 3**, in which the authors model the distance two trains travel along the California coast.

[43]*How do the authors determine that the trains will not collide? What would the graphs look like if the trains were to collide?*

Answer in your own words:

---

[42] Since 16 panels produce 3.2 kW, 32 panels produce 6.4 kW and 48 panels produce 9.6 kW.

[43] The models have the same slope, so their graphs are parallel, which means that the lines do not cross, and the trains do not collide. If the trains were to collide, they would be at the same place at the same time, which means their graphs would cross.

🖉 **Section 2.4 Exercise 51 Kayaking** Mauricio and Thanh are kayaking south down a fast-flowing river, heading toward some rapids. Mauricio leaves 45 miles north of the rapids at 6:00 A.M., and Thanh leaves 24 miles north of the rapids at 8:00 A.M. Both boys maintain a constant speed of 3 ft/s. (Use the equation you found in Section 2.4 Exercise 47 (modified) in this study guide page 121 to convert ft/s to mi/h.)

(a) For each boy, find a linear equation that relates his distance y from the rapids at time x.
(b) Sketch a graph of the linear equations you found in part (a).
(c) Will Mauricio ever pass Thanh?

Solution

CONCEPTS

✓ **2.4 Exercises – Concepts – Fundamentals** Complete the Fundamentals section of the exercises at the end of Section 2.4. Compare your answers to those at the back of the textbook, and make corrections as necessary.

📖 **Read the Concept Check for Section 2.4**, in the Chapter Review at the end of Chapter 2.

2.4 Solutions to ✎ Exercises

✎ Section 2.4 Exercise 9

(a) We need to sketch graphs of the following four equations:

$$y = 2(x+0) \quad y = 2(x+1)$$
$$y = 2(x+2) \quad y = 2(x+3)$$

We simplify these using the Distributive Property:

$$y = 2x \qquad y = 2x + 2$$
$$y = 2x + 4 \qquad y = 2x + 6$$

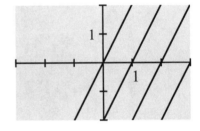

Each of these lines has the same slope (but not the same y-intercept). These lines are parallel.

(b) We need to sketch graphs of the following four equations:

$$y = 2(x+0) \quad y = 2(x-1)$$
$$y = 2(x-2) \quad y = 2(x-3)$$

We simplify these using the Distributive Property:

$$y = 2x \qquad y = 2x - 2$$
$$y = 2x - 4 \qquad y = 2x - 6$$

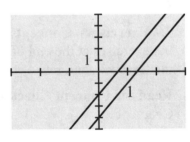

Each of these lines has the same slope (but not the same y-intercept). These lines are parallel.

✎ Section 2.4 Exercise 13

(a) The line $y = 3x - 4$ has slope 3. So the parallel line also has slope 3. We use the point-slope form, with $m = 3$ and $(x_1, y_1) = (2, 4)$ to find the equation of the parallel line.

$y - y_1 = m(x - x_1)$	Point-slope form
$y - 4 = 3(x - 2)$	Replace m by 3, x_1 by 2, and y_1 by 4
$y - 4 = 3x - 6$	Distributive Property
$y = -2 + 3x$	Add 4 to each side

So an equation for the line in slope intercept form is $y = -2 + 3x$.

(b)

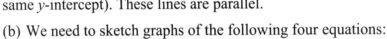

✎ Section 2.4 Exercise 27

(a) l_1, (b) l_2, (c) l_3. These lines all have the same y-intercept ($y = 1$), but not the same slope. The slopes are negative, so their graphs fall as we move from left to right. The line in part (a) has the smallest negative slope, so it must correspond to the line l_1 in the graph, which has the shallowest decline. The line in part (b) has the next smallest negative slope, so it must correspond to the line l_2, in the middle. The graph of line l_3, with the steepest decline, corresponds to the line in part (c) with the largest negative slope.

✎ Section 2.4 Exercise 19

(a) The line $y = -\frac{1}{2}x + 1$ has slope $-\frac{1}{2}$, so the parallel line also has slope $-\frac{1}{2}$. We use the point-slope form, with $m = -\frac{1}{2}$ and $(x_1, y_1) = (-1, 4)$, to find the equation of the parallel line.

$$y - y_1 = m(x - x_1) \qquad \text{Point-slope form}$$

$$y - 4 = -\frac{1}{2}(x + 1) \qquad \text{Replace } m \text{ by } -\frac{1}{2}, x_1 \text{ by } -1, \text{ and } y_1 \text{ by } 4$$

$$y - 4 = -\frac{1}{2}x - \frac{1}{2} \qquad \text{Distributive Property}$$

$$y = \frac{7}{2} - \frac{1}{2}x \qquad \text{Add 4 to each side}$$

So an equation of the parallel line in slope intercept form is $y = \frac{7}{2} - \frac{1}{2}x$.

(b) The slope of the perpendicular line is 2, which is the negative reciprocal of $-\frac{1}{2}$, since $-\frac{1}{(-1/2)} = \frac{2}{1} = 2$. We use the point-slope form, with $m = 2$ and $(x_1, y_1) = (-1, 4)$ to find the equation of the perpendicular line.

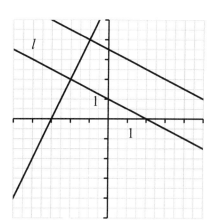

$y - y_1 = m(x - x_1)$ Point-slope form

$y - 4 = 2(x + 1)$ Replace m by 2, x_1 by -1, and y_1 by 4

$y - 4 = 2x + 2$ Distributive Property

$y = 6 + 2x$ Add 4 to each side

So an equation of the perpendicular line in slope intercept form is $y = 6 + 2x$.

✎ **Section 2.4 Exercises 47 and 49 (modified)**

47. The equation of proportionality is an equation of the form $y = kx$, where x represents the speed in feet per second, and y represents the speed in miles per hour.

To find the constant of proportionality k, we need to know a speed in feet per second and in miles per hour.

$$1 \text{ ft/sec} = \frac{1 \text{ ft}}{1 \text{ sec}} \times \frac{1 \text{ mi}}{5280 \text{ ft}} \times \frac{60 \text{ sec}}{1 \text{ min}} \times \frac{60 \text{ min}}{1 \text{ hr}} = \frac{3600 \text{ mi}}{5280 \text{ hr}} \approx 0.68 \text{ mi/hr}$$

So when x is 1 ft/sec, y is 0.68 ft/sec. We replace x by 1 and y by 0.68.

$y = kx$	Equation of proportionality
$0.68 = k(1)$	Replace x by 1 and y by 0.68
$k = 0.68$	Switch sides

Now that we know k is 0.68, we can write the equation of proportionality: $y = 0.68x$.

49. The equation of proportionality is an equation of the form $y = kx$, where x represents the volume of oil in fluid ounces, and y represents the corresponding number of barrels of oil.

To find the constant of proportionality k, we need to know a volume in fluid ounces and in barrels.

$$1 \text{ barrel} = 1 \text{ barrel} \times \frac{42 \text{ gal}}{1 \text{ barrel}} \times \frac{128 \text{ fl. oz.}}{1 \text{ gal}} = \frac{5376 \text{ fl. oz.}}{1 \text{ gal}} \approx 5376 \text{ fl. oz.}$$

So when x is 5376 fl. oz, y is 1 barrel. We replace x by 5376 and y by 1.

$y = kx$	Equation of proportionality
$1 = k(5376)$	Replace x by 5376 and y by 1
$k = \dfrac{1}{5376}$	Divide by 5376, and switch sides

$k \approx 0.000186 = 1.86 \times 10^{-4}$

Now that we know k is 1.86×10^{-4}, we can write the equation of proportionality:

$$y = (1.86 \times 10^{-4})x.$$

✎ **Section 2.4 Exercise 53 Crude Oil in Plastic**

(a) The equation of proportionality is an equation of the form $y = kx$, where x represents the number of bottles, and y represents the volume of crude oil in fluid ounces. When x is 1 million plastic bottles, y is 3 fl. oz. of crude oil. We replace x by 1,000,000 and y by 3.

$$y = kx \qquad\qquad\qquad \text{Equation of proportionality}$$

$$3 = k(1{,}000{,}000) \qquad\quad \text{Replace } x \text{ by 1,000,000 and } y \text{ by 3}$$

$$k = \frac{3}{1{,}000{,}000} = 3 \times 10^{-6} \quad \text{Divide by 1,000,000, and switch sides}$$

Now that we know k is 3, we can write the equation of proportionality:

$$y = (3 \times 10^{-6})x.$$

(b) To find how much crude oil was used to produce 29 billion plastic bottles, we replace x by $29{,}000{,}000{,}000 = 2.9 \times 10^{10}$ in the equation

$$y = (3 \times 10^{-6})x \qquad\qquad\qquad \text{Equation of proportionality}$$

$$y = (3 \times 10^{-6})(2.9 \times 10^{10}) \qquad \text{Replace } x \text{ by } 2.9 \times 10^{10}$$

$$y = 87{,}000 \qquad\qquad\qquad\qquad \text{Calculate}$$

So it takes 87,000 fl. oz. of crude oil to produce 2.9×10^{10} plastic bottles. We use the results of Exercise 49 to convert this to barrels. The equation of proportionality from Exercise 49 is $y = (1.86 \times 10^{-4})x$, where x represents the volume of oil in fluid ounces, and y represents the corresponding number of barrels of oil. We have $y = (1.86 \times 10^{-4})(87{,}000) = 16.182$. So it takes 16.182 barrels of crude oil to make the 29 billion plastic bottles used in the United States in 2006.

✎ **Section 2.4 Exercise 51 Kayaking**

(a) We convert the speed 3 ft/sec to miles per hour using our solution to Exercise 47. The equation of proportionality from Exercise 47 is $y = 0.68x$, where x represents the speed in feet per second, and y represents the speed in miles per hour. Replacing x by 3 ft/sec, we have $y = 0.68(3) = 2.04$ mi/hr, so each boy travels at 2.04 mi/hr towards the rapids.

For each boy, we need to find an equation of the form $y = b + mt$. Let's take $t = 0$ to be 6:00 A.M, and y to be the distance to the rapids. Since each boy travels at 2.04 mi/hr towards the rapids, $m = -2.04$ for each boy. The slope is negative because the distance to the rapids is decreasing.

For Mauricio, $y = 45$ when $t = 0$, so the point $(0, 45)$ is on the desired line. Using the point-slope formula for the equation of a line, we get

$$y - y_1 = m(t - t_1) \qquad\qquad \text{Point-slope form}$$

$$y - 45 = -2.04(t - 0) \qquad\quad \text{Replace } m \text{ by } -2.04,\ t_1 \text{ by 0, and } y_1 \text{ by 45}$$

$$y = -2.04t + 45 \qquad\qquad \text{Add 45 to each side}$$

For Thanh, $y = 24$ when $t = 2$, so the point $(2, 24)$ is on the desired line. Using the point-slope formula for the equation of a line, we get

$$y - y_1 = m(t - t_1)$$ Point-slope form

$$y - 24 = -2.04(t - 2)$$ Replace m by 2.04, t_1 by 2, and y_1 by 24

$$y - 24 = -2.04t + 4.08$$ Distributive Property

$$y = -2.04t + 28.08$$ Add 24 to each side

(b) We know the two lines are parallel since they have the same slope.

(c) Mauricio will never pass Than. Since the lines are parallel, they have no points in common. So the boys will never be at the same place at the same time.

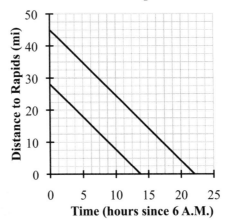

2.5 Linear Regression: Fitting Lines to Data

OBJECTIVES

☑ *Check the box when you can do the exercises addressing the objective, such as those included in this study guide, which are listed here.*

Exercises

The Line That Best Fits the Data

☐ Approximate data from the real world with linear models.   **17, 18**

☐ Explain what the slope and *x*- and *y*-intercepts represent in context.   **17**

Using the Line of Best Fit for Prediction

☐ Make predictions about inputs that are outside of the domain of the data.   **17, 18**

☐ Compare predictions to information from other sources.   **17, 18**

☐ Look for reasons your predictions may not be accurate, and for ways to refine your model to improve the results.   **18**

How Good is the Fit? The Correlation Coefficient

☐ Use the correlation coefficient to determine the degree to which data lie along a line.   **17, 18 (extended)**

GET READY...

This section teaches you how to use your graphing calculator to construct linear models for data. Review Section 1.6 and Algebra Toolkit D.3 in which you use the calculator to create graphs. This section asks you to connect the parameters in your models to the real world. Review how you describe what the slope and *y*-intercept of a line represent in context with Hints and Tips 1.2b, discussed in Section 2.3 on pages 111-112 of this study guide.

SKILLS

📖 **Read the introduction to Section 2.5,** in which the authors discuss the relationship between the amount of time an adolescent watches television and the adolescent's body mass index.

[44]*The data in Figure 1(b) do not lie on a line. Why do we model it with a linear function?*
Answer in your own words:

The Line That Best Fits the Data

In these exercises, you use technology to construct regression lines. Find the term regression line in Section 2.5 of the textbook.

📖 **Read the introduction to the subsection, 'The Line That Best Fits the Data,'** in which the authors discuss the infant mortality rate as a function of time.

[45]*What is a regression line?*
Answer in your own words:

📖 **Read Section 2.5 Example 1,** in which the authors model the infant mortality rate over the past 50 years.

[46]*The variable y represents _____. Its units are _____. The slope of the regression line that the authors create in this example is _____. The y-intercept for this regression line is _____.*

---

[44] When the data are close to linear, a linear model can be used to make approximate predictions.

[45] A regression line for a set of data is the line that best approximates the data. It is the line for which the sum of the square of the vertical distance between the data and the line is as small as possible.

[46] The variable *y* represents the infant mortality rate, which is the number of infants per thousand who were born alive, but died before their first birthdays. The units could be abbreviated as deaths per thousand. The slope of the model is approximately –0.48, and the *y*-intercept is approximately 29.41 (as shown in Figure 6(a)).

Following Hints and Tips 1.2b (see Section 2.3), we write the following sentences to figure out what the slope represents. Algebraically, –0.48 is the change in the *y*-value for each unit increase in the *x*-value. In context, –0.48 deaths per thousand is the change in the infant mortality rate when the year is increased by 1. We write our final answer more eloquently: The slope –0.48 tells us that the infant mortality rate decreases by 0.48 deaths per thousand each year.

✐ **Section 2.5 Exercise 17(a, b) Life Expectancy** The average life expectancy in the United States has been rising steadily over the past few decades, as shown in the table.

(a) Find the regression line for the data.

(b) Make a scatter plot of the data, and graph the regression line. Does the regression line appear to be a suitable model for the data?

Solution

Year	Life expectancy
1920	54.1
1930	59.7
1940	62.9
1950	68.2
1960	69.7
1970	70.8
1980	73.7
1990	75.4
2000	76.9

CONTEXTS

Using the Line of Best Fit for Prediction

In these exercises you use regression lines to interpolate and extrapolate information not included in the data. Find the terms interpolate and extrapolate in bold typeface in Section 2.5 of the textbook.

📖 **Read Section 2.5 Example 2**, in which the authors make predictions about infant mortality rate using the regression line from Example 1.

[47] *The authors find the infant mortality rate in 1995 by _____ (interpolation / extrapolation). The authors find the infant mortality rate in 2006 by _____ (interpolation / extrapolation).*

---

[47] They find the infant mortality rate in 1995 by interpolation, and the infant mortality rate in 2006 by extrapolation.

📖 **Read the paragraph after Section 2.5 Example 2**, in which the authors compare the mortality rate in 2006 the model predicted with the actual mortality rate in 2006.

[48]*Why do we have to be careful about extrapolating linear models outside the domain over which the data are spread?*
Answer in your own words:

✐ **Section 2.5 Exercise 17(d, e)** The data and context for this exercise can be found with 17(a, b) on page 130 of this study guide.

(d) Use the linear model you found in part (b) to predict the life expectancy in the year 2006.
(e) Search the Internet to find the actual 2006 average life expectancy. Compare it to your answer in part (d).

[49]*You find the life expectancy in part (d) by _____ (interpolation / extrapolation).*

Solution

📖 **Read pages 192-194, including Section 2.5 Example 3**, in which the authors study how the height of the Olympic gold metal pole vault varies with time. Read also Section 2.5 Exercise 18.

[50]*In part (d) of Example 3, the authors predict that in 2008, the winning pole vault record will be _____ meters. However, in the paragraph after the example, the authors explain that the actual record in 2008 was _____ meters. How do the authors suggest we attempt to improve the prediction?*
Answer in your own words:

Look at the graph in Figure 7. A regression line only for the data from 1975 onward appears to have a slope that is _____ (less than / greater than) the slope of the regression line that the authors found in part (a). So, extrapolating with only recent data would lead to predictions of future pole-vaulting records that are _____ (higher than / lower than) predictions made using the data since 1900.

---

[48] If you consider inputs that are far away from the domain over which the data are spread, the model may no longer be relevant. The linear trend in the data represented in the regression line may no longer exist outside the domain over which the data are spread.
[49] In part (d), you find the life expectancy by extrapolation.
[50] In part (d), the authors predict that the record will be 6.27 m; however the actual record is 5.96 m. The authors suggest using only recent data might lead to a more accurate prediction. A regression line for the data from 1975 onward appears to have a slope that is less than 0.0266 m/yr, the slope found in part (a). So, using only recent data would lead to lower predictions.

✎ **Section 2.5 Exercise 18** Read the introduction to this exercise in the textbook.

(a) Complete the second column in the table, where x is the number of years since 1972.

(b) Find the regression line for the data in part (a).

(c) Plot the data and the regression line on the same axes. Does the regression line seem to provide a good model for the data?

(d) What does the regression line predict as the winning pole vault height for the 2008 Olympics? Has this new regression line provided a better prediction than the line in Example 3? Compare to the actual result given on page 194 of the textbook.

Solution

Year	x	Height (m)
1972	0	5.64
1976	4	5.64
1980		5.78
1984		5.75
1988		5.90
1992		5.87
1996		5.92
2000		5.90
2004		5.95

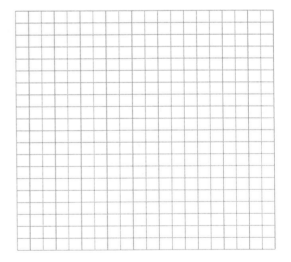

> Refer back to page 11 (see also pages 111-112)
> **Hints and Tips 1.2b: What does the algebra represent in context?**

As in Section 2.3 Example 6(c), this advice applies to interpreting the meaning of slope, and explains the authors' answer to Section 2.5 Example 3(c).

⇨ **Section 2.5 Example 3(c)** What does the slope of the regression line represent?
Solution
Following Hints and Tips 1.2b, we write the following sentences as scratch work:
  Algebraically, 0.0266 is the amount by which y changes for each unit increase in x. In context, 0.0266 meters is the amount by which the height changes each year.

The authors write the final answer more eloquently:
  The pole vault record increases by 0.0266 meters per year.

[51] *In Section 2.5 Example 1 what does the slope of the regression line represent?*
Answer in your own words:

✎ **Section 2.5 Exercise 17(c)** The data and context for this exercise can be found with 17(a, b) on page 130 of this study guide.
(c) What does the slope of the regression line represent?

Solution

📖 **Read Section 2.5 Example 4**, in which the authors model the relationship between asbestos exposure and lung tumors in mice.

[52] *What does the slope of the regression line represent?*
Answer in your own words:

---

[51] Following Hints and Tips 1.2b (see Section 2.3), we write the following sentences as scratch work, to figure out what the slope represents. Algebraically, –0.48 is the change in the y-value for each unit increase in the x-value. In context, –0.48 deaths per thousand is the change in the infant mortality rate when the year is increased by 1. We write our final answer more eloquently: The slope –0.48 tells us that the infant mortality rate decreases by 0.48 deaths per thousand each year.

[52] Following Hints and Tips 1.2b (see Section 2.3), we write the following sentences as scratch work, to figure out what the slope represents. Algebraically, 0.0177 is the change in the y-value for each unit increase in the x-value. In context, 0.0177% is the change in the percentage of mice that develop lung tumors when the exposure is increased by 1 fiber per milliliter. We write our final answer more eloquently: The slope 0.0177 tells us that increasing the asbestos exposure by 1 fiber per milliliter increases the percentage of mice that develop lung tumors by 0.0177%..

How Good Is the Fit? The Correlation Coefficient

These exercises ask you to use the correlation coefficient to describe the degree to which the data are linear. Find the term correlation coefficient in italics in Section 2.5 of the textbook.

📖 **Read the subsection, 'How Good Is the Fit? The Correlation Coefficient,'** in which the authors discuss how close the infant mortality data from Example 1 and the cancer data from Example 4 are to being linear.

[53]*The data are close to being linear if the r-value is close to _____ for increasing functions, or to _____ for decreasing functions.*

✎ **Section 2.5 Exercise 17(f) (extended)** The data and context for this exercise can be found with 17(a, b) on page 130 of this study guide. What is the correlation coefficient for this data?

Solution

✎ **Section 2.5 Exercise 18(e) (extended)** The data and context for this exercise can be found on page 132 of this study guide. What is the correlation coefficient for this data?

Solution

[54]*The correlation coefficients indicate the data in Exercise _____ (17 / 18) is closer to being linear.*

CONCEPTS

✓ **2.5 Exercises – Concepts – Fundamentals** Complete the Fundamentals section of the exercises at the end of Section 2.5. Compare your answers to those at the back of the textbook, and make corrections as necessary.

📖 **Read the Concept Check for Section 2.5**, in the Chapter Review at the end of Chapter 2.

---

[53] The data are close to linear if the correlation coefficient is close to 1 for increasing functions and close to −1 for decreasing functions.

[54] The correlation coefficient for Exercise 17 is closer to 1, so the data in Exercise 17 are closer to being linear.

2.5 Solutions to ✎ Exercises

✎ Section 2.5 Exercise 17 Life Expectancy

(a) Using a graphing calculator, we find the regression line is $y = 0.2708x - 462.9$.

(b) The line appears to fit the data.

(c) The slope of the regression line is 0.2708. Following **Hints and Tips 1.2b**, we write the following sentences as scratch work:

Algebraically, 0.2708 is change in the y-value for each unit change in the x-value. In context, 0.2708 years is the amount that the life expectancy increases for each 1-year increase in time.

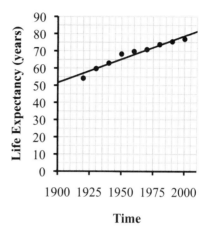

We express our final answer more eloquently:
Each year, the life expectancy increases by 0.2708 years.

(d) To predict the life expectancy in 2006, we replace x by 2006 in the regression equation to get $y = 0.2708(2006) - 462.9 \approx 80$. So we predict that the life expectancy in 2006 is about 80 years.

(f) The correlation coefficient for this data is $r = 0.9725$.

✎ Section 2.5 Exercise 18

(b) Using a graphing calculator, we find the regression line is $y = 0.0101x + 5.6553$.

(c) The regression line appears to fit the data.

(d) To predict the winning pole vault height for the 2008 Olympics, we replace x by 36 in the regression equation to get $y = 0.0101(36) + 5.6553 \approx 6.02$. So we predict that the winning pole vault height for the 2008 Olympics to be 6.02 meters.

(e) The correlation coefficient is $r = 0.9278$.

Year	x	Height (m)
1972	0	5.64
1976	4	5.64
1980	8	5.78
1984	12	5.75
1988	16	5.90
1992	20	5.87
1996	24	5.92
2000	28	5.90
2004	32	5.95

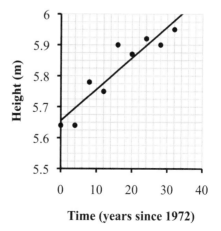

Time (years since 1972)

2.6 Linear Equations: Getting Information from a Model

OBJECTIVES

☑ *Check the box when you can do the exercises addressing the objective, such as those included in this study guide, which are listed here.*

	Exercises
Getting Information from a Linear Model	
☐ Estimate the input of a linear model that corresponds to a given output from a graph of the model.	21
☐ Calculate the input of a linear model that corresponds to a given output, using the formula.	21, 27, 29, 31
Models that Lead to Linear Equations	
☐ Think about a model before writing its formula by making up examples of inputs for your model and figuring out the corresponding outputs.	27, 29, 31
☐ Think about a model before writing its formula by drawing a diagram, and labeling it appropriately.	29, 31
☐ Use the simple interest formula to construct models.	27
☐ Use ratios, such as concentration, to model the composition of a mixture.	29
☐ Use geometry, such as perimeter, to create models of the real world.	31

GET READY...

This section involves creating and using linear models. Reread the paragraph at the top of page 89 in Section 1.8 of the textbook, summarizing the connections between a model and the real world. Using the models involves solving linear equations graphically and algebraically. Refresh your understanding of algebraic solutions in Algebra Toolkit C.1. Review how to read information from the graph of a function in Section 1.7.

✓ **Algebra Checkpoint** Test yourself by completing the Algebra Checkpoint at the end of this section of the textbook, and comparing your answers to those in the back of the textbook. Refer to Algebra Toolkit C.1 as necessary.

CONTEXTS

Getting Information from a Linear Model

These exercises give you the output of a linear model and ask you to find the corresponding input.

📖 **Read Section 2.2 Example 5 again**, paying particular attention to the meaning of the variables in context.

[55] *Here, t represents* _____, *and V(t) represents* _____.

---

[55] Here t represents the length of time the pump has been pumping water into the pool, and $V(t)$ represents the volume of water in the pool.

📖 **Read Section 2.6 Example 1**, in which the authors use the model from Section 2.2 Example 5 to find the length of time a pump needs to work to fill a swimming pool.

[56]*Here, x represents _____, and y represents _____.*

The authors add the line _____ to the graph of the model. In Solution 1, the authors estimate the coordinates at which the two lines cross to be (_____, _____). In Solution 2, the authors find that the coordinates are actually (_____, _____).

✏ **Section 2.2 Exercise 55 Weather Balloon** Find this exercise on page 100 of this study guide.

✏ **Section 2.6 Exercise 21 Weather Balloon (extended)** A weather balloon is being filled. The linear equation

$$V = 2 + 0.5t$$

models the volume V (in cubic feet) of hydrogen in the balloon at any time t (in seconds). How many minutes will it take until the balloon contains 55 ft³ of hydrogen? Answer this question in two ways:

(a) Graphically (by graphing the equation and estimating the time from the graph)

[57]*What line should you add to the graph of the model, to help you answer the question graphically? _____*

(b) Algebraically (by solving an appropriate equation)

Convert your answer for part (b) to hours.

Solution

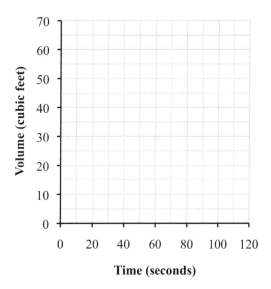

---

[56] Here x represents the length of time the pump has been pumping water into the pool, and y represents the volume of water in the pool. The authors graph the line $y = 10,000$, since they want to find the value of x when $y = 10,000$. From the graph, they estimate the coordinates to be (1950, 10,000). Using the equation, they find the coordinates are actually (1960, 10,000).

[57] Add the line $V = 55$.

Models That Lead to Linear Equations

In these exercises, you construct models of the real world. Following Hints and Tips 1.8a, you think about the problems by making up examples before constructing your formulas. Find a summary of how to construct a model in a blue box in Section 2.6 of the textbook.

📖 **Read the paragraph before Section 2.6 Example 2**, in which the authors introduce the simple interest formula.

[58]*In the simple interest formula, I represents _____, P represents _____, r represents _____, and t represents _____.*

📖 **Read Section 2.6 Example 2**, in which the authors construct a model for the interest Mary earns when she divides her inheritance between two investments.

> Refer back to page 69 (see also page 111)
> **Hints and Tips 1.8a: Thinking about the problem**

[59]*Following Hints and Tips 1.8a, think about the problem by computing an example. Say Mary invests $20,000 in the certificate that pays 6%, and the rest in the certificate that pays 4½%.*

Complete the following table.

In words	In numbers	In words	In numbers
Amount invested at 6%	**$20,000**	Interest earned at 6%	
Amount invested at 4½%		Interest earned at 4½%	
		Total interest earned	

The authors find the model _____. To find the total interest Mary earns when she invests $20,000 at 6% using the model, replace ___(x / y) by 20,000. Does your answer match your results above?

---

[58] In the formula, *I* represents the interest earned, *P* represents the principal invested, *r* represents the interest rate (in decimal form), and *t* represents the duration of the investment.

[59] If Mary invests $20,000 at 6% interest, then she invests $80,000 at 4½% interest. She earns $1,200 of interest from the certificate that pays 6%, and $3,600 from the certificate that pays 4½%, for a total of $4,800. The authors find the model $y = 4500 + 0.015x$. To find the total interest Mary when she invests $20,000, replace x by 20,000 in the model. When $x = 20,000$, we get $y = 4800$, which matches with our result above.

✏ **Section 2.6 Exercise 27 Investment** Lili invested $12,000 in two one-year certificates of deposit. One certificate pays 4%, and the other pays 4½% simple interest annually.

(a) Construct a model for the total interest Lili earns in one year on her investments. (Let x represent the amount invested at 4%.)

(b) If Lili's total interest is $526, how much money did she invest in each certificate?

[60]*Following Hints and Tips 1.8a, think about the problem by computing an example. Say Lili invests $2,000 in the certificate that pays 4%, and the rest in the certificate that pays 4½%. Complete the following table.*

In words	In numbers	In words	In numbers
Amount invested at 4%	**$2,000**	Interest earned at 4%	
Amount invested at 4½%		Interest earned at 4½%	
		Total interest earned	

Solution

---

[60] If Lili invests $2,000 at 4% interest, then she invests $10,000 at 4½% interest. She earns $80 of interest from the certificate that pays 4%, and $450 from the certificate that pays 4½%, for a total of $530.

📖 **Read the paragraph before Section 2.6 Example 3**, in which the authors introduce the concentration of a substance in a mixture.

[61] *In the formula for concentration, C represents _____, x represents _____, and V represents _____.*

📖 **Read Section 2.6 Example 3**, in which the authors construct a model for the fraction of orange soda that is pure juice.

> Refer back to page 73
> **Hints and Tips 1.8b: Thinking about the problem**

[62] *Following Hints and Tips 1.8b, we draw a diagram showing the soda and juice as if they do not mix.*

Diagram

Initially the volume of juice is _____ gallons, and the volume of soda, not including the juice is _____ gallons. Write these numbers in the appropriate boxes in the diagram.

Now following Hints and Tips 1.8a, think about the problem by computing an example. Say the manufacturer adds 100 gallons of orange juice. Write this number in the appropriate box in the diagram, and complete the following table.

In words	*In numbers*
Amount of orange juice added	**100 gal**
Amount (volume) of the mixture	
Amount of orange juice in the mixture	

In words	*In numbers*
Fraction of the mixture made up of orange juice	

The authors find the model _____. Use the model to compute fraction of soda made up of orange juice, if the manufacturer adds 100 gallons of orange juice to the vat. Does the answer match your results above?

_____

[61] In the formula for concentration, *C* represents the concentration of a substance in a mixture, *x* represents the amount of the substance in a mixture, and *V* represents the volume of the mixture.

[62] Initially the volume of juice is 45 gal, and the volume of soda, not including juice is $900 - 45 = 855$ gal. In the diagram, you write 100 gal in the top box, 855 gal in the second box, and 45 gal in the bottom box. In the table, the amount of orange juice added is 100 gal, the volume of the mixture is 1000 gal, the amount of orange juice in the mixture is 145 gal, and the fraction of the mixture made up of orange juice is

$145/1000 = 0.145$, which is 14.5%. The authors find the model $y = \dfrac{45+x}{900+x}$. Evaluating this at $x = 100$ yields

$y = 0.145$, which agrees with our results above.

✎ **Section 2.6 Exercise 29 Mixture** A jeweler has five rings, each weighing 18 grams, made of an alloy of 10% silver and 90% gold. He decides to melt down the rings and add enough silver to reduce the gold content to 75%.

(a) Construct a model that gives the fraction of the new alloy that is pure gold. (Let x denote the number of grams of silver added.)

(b) How much pure silver must be added for the mixture to have a gold content of 75%?

[63]*Following Hints and Tips 1.8b, we draw a diagram showing the gold and silver as if they do not mix.*

Silver	} Added
Gold	} Original
Silver	
Diagram	

Initially, the total mass is _____ grams. There are _____ grams of gold and _____ grams of silver. Write these numbers in the appropriate boxes in the diagram.

Now following Hints and Tips 1.8a, think about the problem by computing an example. Say the jeweler adds 10 grams of silver. Write this number in the appropriate box in the diagram, and complete the following table.

In words	In numbers		In words	In numbers
Amount of silver added	**10 g**			
Amount (weight) of the mixture			Fraction of the mixture that is pure gold	
Amount of silver in the mixture				
Amount of gold in the mixture				

The jeweler will have to add _____ (more / less) than 10 g of silver to get the fraction of the alloy that is pure gold to 75%.

Solution

_____

[63] Initially, the total mass is 5(18) = 90 g. There are 0.90(90) = 81 g of gold and 0.10(90) = 9 g of silver. In the diagram, you write 10 g in the top box, 81 g in the second box, and 9 g in the bottom box. In the table, the amount of silver added is 10 g, the mass of the mixture is 100 g, the amount of silver in the mixture is 19 g, and the amount of gold in the mixture is 81 g. The fraction of the mixture that is pure gold is 81/100 = 0.81, which is 81%. The jeweler will have to add more than 10 g of silver to get the fraction of the alloy that is pure gold to 75%.

📖 **Read Section 2.6 Example 4**, in which the authors construct a model for the perimeter of a picture frame.

[64]*Following Hints and Tips 1.8a and b, think about the problem by drawing diagrams and computing examples.*

The width of the painting without the mat is _____ inches. The length of the painting without the mat is _____ inches. Label the diagram appropriately. Suppose that the width of the strip of mat that shows is 5 inches; add labels to the diagram accordingly, and complete the following table.

In words	In numbers	In words	In numbers
Width of the strip	5 in		
Width of the mat		Perimeter of the mat	
Length of the mat			

The authors find the model _____. To find the perimeter of the picture if the strip is 5 inches wide, replace ___ (x / y) by 5. Does the answer match your results above?

✏ **Section 2.6 Exercise 31 Geometry** A graphic artist needs to construct a design that uses a rectangle whose length is 5 cm longer than its width x.

(a) Construct a model that gives the perimeter of the rectangle.

(b) If the perimeter of the rectangle is 26 cm, what are the dimensions of the rectangle?

[65]*Following Hints and Tips 1.8a and b, think about the problem by drawing diagrams and computing examples.*

Suppose the width of the rectangle is 8 cm. The length of the rectangle is _____ cm. Add labels to the diagram accordingly, and complete the following table.

In words	In numbers	In words	In numbers
Width	8 cm		
Length		Perimeter	

The artist will have to make the width _____ (more / less) than 8 cm to get the perimeter to be 26 cm.

Solution

_____

[64] Without the mat, the width is 20 inches, and the length is 15 inches. In the table, the width of the strip is 5 inches, the width of the mat is 30 inches, the length of the mat is 25 inches, and the perimeter of the mat is 110 inches. The authors find the model $y = 70 + 8x$. Evaluating this at $x = 5$ yields $y = 110$ inches, which agrees with our result above.

[65] When the width of the rectangle is 8 cm, the length of the rectangle is $8 + 5 = 13$ cm. In this case the perimeter is $2(8) + 2(13) = 42$ cm. The artist will have to make the width less than 8 cm to get the perimeter to be 26 cm (as in part (b)).

CONCEPTS

✓ **2.6 Exercises – Concepts – Fundamentals** Complete the Fundamentals section of the exercises at the end of Section 2.6. Compare your answers to those at the back of the textbook, and make corrections as necessary.

📖 **Read the Concept Check for Section 2.6**, in the Chapter Review at the end of Chapter 2.

2.6 Solutions to ✐ Exercises

✐ **Section 2.6 Exercise 21 Weather Balloon (extended)**

(a) We graph $V = 2 + 0.5t$ and $V = 55$. From the graph we see that the volume reaches 55 ft^3 after approximately 105 seconds.

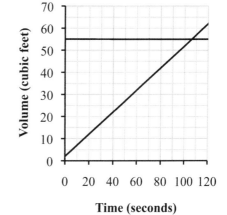

To convert 105 seconds to minutes we compute $105\ \text{sec} = 105\ \text{sec} \dfrac{1\ \text{min}}{60\ \text{sec}} = 1.75\ \text{min}$.

(b) We solve algebraically by replacing V by 55 in the equation $V = 2 + 0.5t$.

$$V = 2 + 0.5t \qquad \text{Equation}$$
$$55 = 2 + 0.5t \qquad \text{Replace } V \text{ by } 55$$
$$53 = 0.5t \qquad \text{Subtract 2 from each side}$$
$$t = \frac{53}{0.5} = 106 \qquad \text{Divide each side by 0.5 and switch sides}$$

It takes 106 seconds to fill the balloon to 55 ft^3. To convert 106 seconds to hours, we compute $106\ \text{sec} = 106\ \text{sec} \dfrac{1\ \text{min}}{60\ \text{sec}}\dfrac{1\ \text{h}}{60\ \text{min}} \approx 0.029\ \text{h}$. So it takes 0.029 hours to fill the balloon to 55 ft^3. Notice that the algebraic solution gives and exact answer, while the graphical solution is approximate.

✐ **Section 2.6 Exercise 27 Investment**

(a) We let x represent the amount invested at 4%, as directed. Since Lili's total investment is $12,000, and she invests x dollars at 4%, she must invest $12,000 - x$ at 4½%.

In words	In Algebra
Amount invested at 4%	x
Amount invested at 4½%	$12,000 - x$
Interest earned at 4%	$0.04x$
Interest earned at 4½%	$0.045(12,000 - x)$

The function we want gives the total interest Lili earns (the interest she earns at 4%
plus the interest she earns at 4½%).

$$y = 0.04x + 0.045(12{,}000 - x) \quad \text{Lili's total interest}$$
$$= 0.04x + 540 - 0.045x \quad \text{Distributive Property}$$
$$= 540 - 0.005x \quad \text{Simplify}$$

So the model we want is the linear equation $y = 540 - 0.005x$.

(b) Since Lili's interest is $526, we replace y by 526 in the model and solve the
equation for x.

$$y = 540 - 0.005x \quad \text{Model}$$
$$526 = 540 - 0.005x \quad \text{Replace } y \text{ by 526}$$
$$-14 = -0.005x \quad \text{Subtract 540 from each side}$$
$$x = \frac{-14}{-0.005} = 2{,}800 \quad \text{Divide each side by } -0.005 \text{ and}$$

switch sides

So Lili invested $2,800 at 4% and $12{,}000 - 2{,}800 = \$9{,}200$ at 4½%.

✎ Section 2.6 Exercise 29 Mixture

(a) We let x be the number of grams of silver added, as directed. First note that before
adding silver, the 5 rings weigh $18 \times 5 = 90$ g. Of the 90 g, 10% is silver, so the total
amount of silver is $0.10 \times 90 = 9$ g.

In words	In Algebra
Amount of silver added	x
Amount of the mixture	$90 + x$
Amount of silver in the mixture	$9 + x$

To find the fraction of the mixture that is silver, we divide the amount of silver in the
mixture by the amount of the mixture.

$$y = \frac{9 + x}{90 + x} \quad \begin{array}{l}\leftarrow \text{Amount of silver in the mixture} \\ \leftarrow \text{Amount of the mixture}\end{array}$$

So the model we want is $y = (9 + x)/(90 + x)$.

(b) We want the mixture to have 75% gold, and hence 25% silver. This means that the fraction of the mixture that is silver is 0.25. So we replace y by 0.25 in the model, and solve the equation for x.

$$y = \frac{9+x}{90+x} \qquad \text{Model}$$

$$0.25 = \frac{9+x}{90+x} \qquad \text{Replace } y \text{ by } 0.25$$

$$0.25(90+x) = 9+x \qquad \text{Cross multiply}$$

$$22.5 + 0.25x = 9+x \qquad \text{Distributive property}$$

$$13.5 = 0.75x \qquad \text{Subtract 9 and } 0.25x \text{ from each side}$$

$$x = \frac{13.5}{0.75} = 18 \qquad \text{Divide each side by 0.75 and switch sides}$$

So the jeweler should add 18 g of silver to the mixture.

✎ **Section 2.6 Exercise 31 Geometry**

(a) Let x be the width of the rectangle, as directed. Then the length of the rectangle is $5 + x$.

In words	In Algebra
Width of the rectangle	x
Length of the rectangle	$5 + x$

To find the perimeter of the rectangle, we add twice the width and twice the length.

$$y = 2x + 2(5+x) \qquad \text{Perimeter}$$

$$= 2x + 10 + 2x \qquad \text{Distributive Property}$$

$$= 10 + 4x \qquad \text{Simplify}$$

So the model we want is the equation $y = 10 + 4x$.

(b) Since the perimeter is 26 cm, we replace y by 26 in our model and solve for x.

$$y = 10 + 4x \qquad \text{Model}$$

$$26 = 10 + 4x \qquad \text{Replace } y \text{ by } 26$$

$$16 = 4x \qquad \text{Subtract 10 from each side}$$

$$x = 4 \qquad \text{Divide each side by 4 and switch sides}$$

So if the perimeter is 26 cm, then the width of the rectangle is 4 cm, and thus the length of the rectangle is $4 + 5 = 9$ cm.

2.7 Linear Equations: Where Lines Meet

OBJECTIVES

☑ *Check the box when you can do the exercises addressing the objective, such as those included in this study guide, which are listed here.*

Exercises

Where Lines Meet

☐ Estimate the coordinates at which lines intersect from a graph.   **5**

☐ Find the coordinates at which lines intersect using equations.   **5, 27, 33, 35**

☐ Compare the sizes of two linear models, and find the inputs for which the outputs are equal.   **27, 33, 35**

Modeling Supply and Demand

☐ Answer questions about a commodity using the supply and demand equations.   **35**

☐ Find the equilibrium point for a commodity.   **35**

GET READY...

Review 1.7 Example 3, in which the authors find the inputs at which the graphs of two functions intersect, using a graphing calculator.

SKILLS

Where Lines Meet

These exercises ask you to find the coordinates at which lines intersect both graphically and algebraically. Find instructions for finding the coordinates algebraically in a blue box in Section 2.7 of the textbook.

📖 **Read Section 2.7 Example 1**, in which the authors find the point of intersection of two linear functions.

[66] *When finding the coordinates at which the two lines intersect, the graphical solution yields an* _____ *(estimate / exact value), while the algebraic solution yields an* _____ *(estimate / exact value).*

---

[66] Reading information from a graph yields an estimate; finding the answer algebraically yields an exact solution.

✎ **Section 2.7 Exercise 5** A graph of two lines is given.

(a) Use the graph to estimate the coordinates of the point of intersection.
(b) Find an equation for each line.
(c) Use the equations from part (b) to find the coordinates of the point of intersection. Compare with your answer to part (a).

Solution

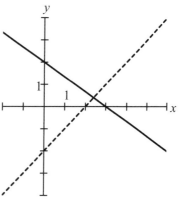

CONTEXTS

Where Lines Meet

In these exercises you study the distance traveled as a function of time. In Section 2.1 Example 3, and Section 2.2 Example 8, the authors discussed this, emphasizing the connections between the graphical representation of the distance function and the context. Read these examples again, and complete Section 2.1 Exercise 27 and Section 2.2 Exercise 59, included in this study guide on pages 91 and 101 respectively.

Also in these exercises, you use ratios to create linear functions. Read Section 1.8 Example 2 for another example in which the authors create functions using ratios. Then complete Section 1.8 Exercise 19, included in this study guide on page 68.

📖 **Read the introduction to the subsection, 'Where Lines Meet,'** in which the authors race Aesop's tortoise and hare; the tortoise has a head start, but the hare is much quicker.

[67]*How can you tell by looking at the graph in Figure 1 that the tortoise wins the race?*
Answer in your own words:

How can you tell by looking at the graph at what time the hare catches up to the tortoise?
Answer in your own words:

📖 **Read Section 2.7 Example 3,** in which two cyclists ride to the beach, one leaving early but traveling slowly, and the other leaving late but traveling quickly.

[68]*In Petra's equation, t represents _____. How do the authors find how long it takes Petra to catch up with Jordanna?*
Answer in your own words:

---

[67] The race is for 2 miles. The graph of the tortoise's distance is at $y = 2$ miles when $t = 2.5$, but the graph of the hare's distance is at $y = 2$ when $t = 2.75$. So the tortoise reached 2 miles first. Visually, we see that the point at which the tortoise's graph crosses the line $y = 2$ is to the left (and hence earlier) than the point at which the hare's graph crosses $y = 2$. The time at which the hare catches up to the tortoise is the t-coordinate of the point at which the tortoise's graph and the hare's graph intersect.

[68] In Petra's (and Jordanna's) equation, t represents the time since Jordanna left home. To find the length of time it takes Petra, take the time that Jordanna has traveled when they meet (13.3 minutes) and subtract 5 minutes, since Petra left 5 minutes after Jordanna.

✏ **Section 2.7 Exercise 33 Commute to Work** (See Exercise 59 in Section 2.2, on page 101 of this study guide). Jade and her roommate Jari live in a suburb of San Antonio, Texas, and both work at an elementary school in the city. Each morning they commute to work traveling west on I-10. One morning, Jade leaves for work at 6:50 A.M., but Jari leaves 10 minutes later. On this trip, Jade drives at an average speed of 65 mi/h, and Jari drives at an average speed of 72 mi/h.

(a) Find a linear equation that models the distance y Jari has traveled x hours after she leaves home.
(b) Find a linear equation that models the distance y Jade has traveled x hours after Jari leaves home.
(c) Determine how long it takes Jari to catch up with Jade. How far have they traveled at the time they meet?

Solution

📖 **Read Section 2.7 Example 2**, in which the authors compare the cost effectiveness of two cars.

[69] *Before Kevin drives 53,333 miles, the* _____ *(gas-powered / hybrid) costs him less money, but after Kevin drives 53,333 miles, the* _____ *(gas-powered / hybrid) costs him less money.*

If you graph the functions that the authors created in parts (a) and (b) on the same coordinate axes, then the graph for the gas-powered car lies _____ *(above / below) the graph for the hybrid car when x is less than 53,333, and the graph of the gas-powered car lies* _____ *(above / below) the graph for the hybrid car when x is greater than 53,333.*

---

[69] Before 53,333 miles, the gas-powered car costs Kevin less, but after 53,333 miles, the hybrid car costs less. For $x < 53,333$, the graph for the gas-powered car lies below the graph for the hybrid car, but for $x > 53,333$, the graph for the gas-powered car lies above the graph for the hybrid car.

✎ **Section 2.7 Exercise 27 Cell Phone Plan Comparison** Dietmar is in the process of choosing a cell phone, and a cell phone plan. The first plan charges 20¢ per minute plus a monthly fee of $10, and the second plan offers unlimited minutes for a monthly fee of $100.

(a) Find a linear function f that models the monthly cost $f(x)$ of the first plan in terms of the number x of minutes used.

(b) Find a linear function g that models the monthly cost $g(x)$ of the second plan in terms of the number x of minutes used.

(c) Determine the number of minutes for which the two plans have the same monthly cost.

Solution

Modeling Supply and Demand

These exercises use the vocabulary of supply and demand from economics. Find the definitions of the variables in the supply and demand equations in italics and the definition of the equilibrium point in bold typeface in Section 2.7 of the textbook.

📖 **Read Section 2.7 Example 4**, in which the authors discuss the supply and demand equations used in economics.

[70] *The variable y in the supply equation represents* _____, *and the variable y in the demand equation represents* _____. *Review the blue box about linear models on page 26 of the textbook. If the price increases by $1, then the amount suppliers produce* _____ *(increases / decreases) by* _____ *bushels, and the amount consumers buy (increases / decreases) by* _____ *bushels.*

---

[70] The variable y in the supply equation represents the quantity of the commodity that the suppliers produce; the variable y in the demand equation represents the quantity of the commodity that the consumers buy. If the price increases by $1, the amount the suppliers produce increases by 3.48 bushels, and the amount consumers buy decreases by 0.99 bushels.

✎ **Section 2.7 Exercise 35 Supply and Demand for Corn** An economist models the market for corn by the following equations:

$$\text{Supply:} \qquad y = 4.18p - 11.5$$
$$\text{Demand:} \qquad y = -1.06p + 19.3$$

Here, p is the price per bushel (in dollars), and y is the number of bushels produced and sold (in billions).

(a) Use the model for supply to determine at what point the price is so low that no corn is produced.

(b) Use the model for demand to determine at what point the price is so high that no corn is sold.

(c) Find the equilibrium price and the quantities that are produced and sold at equilibrium.

Solution

CONCEPTS

✓ **2.7 Exercises – Concepts – Fundamentals** Complete the Fundamentals section of the exercises at the end of Section 2.7. Compare your answers to those at the back of the textbook, and make corrections as necessary.

📖 **Read the Concept Check for Section 2.7**, in the Chapter Review at the end of Chapter 2.

2.7 Solutions to ✐ Exercises

✐ Section 2.7 Exercise 5

(a) From the graph, we see that the x-coordinate of the intersection point is about $x = 2.3$ and the y-coordinate is about $y = 0.5$. Thus the point of intersection is approximately $(2.3, 0.5)$.

(b) From the graph, we can accurately read the coordinates of two points: the x-intercept at $(3, 0)$, and the y-intercept at $(0, 2)$. We use these to calculate the slope:

$$m = \frac{y_2 - y_1}{x_2 - x_1} = \frac{2 - 0}{0 - 3} = -\frac{2}{3}$$

So the slope is $m = -\frac{2}{3}$. We use the point-slope form of the line with $(x_1, y_2) = (3, 0)$.

$$y - y_1 = m(x - x_1) \qquad \text{Point-slope form}$$

$$y - 0 = -\frac{2}{3}(x - 3) \qquad \text{Replace } m \text{ by } -\frac{2}{3}, x_1 \text{ by 3 and } y_1 \text{ by 0}$$

$$y = -\frac{2}{3}x - \left(-\frac{2}{3}\right)3 \qquad \text{Distributive Property}$$

$$y = -\frac{2}{3}x + 2 \qquad \text{Simplify}$$

So an equation of the solid line is $y = -\frac{2}{3}x + 2$.

Similarly we write an equation for the dashed line. From the graph, we can accurately read the coordinates of two points: the x-intercept at $(2, 0)$, and the y-intercept at $(0, -2)$. We use these to calculate the slope:

$$m = \frac{y_2 - y_1}{x_2 - x_1} = \frac{-2 - 0}{0 - 2} = 1$$

So the slope is $m = 1$. We use the point-slope form of the line with $(x_1, y_2) = (2, 0)$.

$$y - y_1 = m(x - x_1) \qquad \text{Point-slope form}$$
$$y - 0 = 1(x - 2) \qquad \text{Replace } m \text{ by 1, } x_1 \text{ by 2 and } y_1 \text{ by 0}$$
$$y = x - 2 \qquad \text{Simplify}$$

So an equation of the solid line is $y = x - 2$.

(c) We solve the equation

$$-\frac{2}{3}x+2 = x-2 \qquad \text{Set functions equal}$$

$$-\frac{2}{3}x = x-4 \qquad \text{Subtract 2 from each side}$$

$$-\frac{2}{3}x-x = -4 \qquad \text{Subtract } x \text{ from each side}$$

$$-\frac{5}{3}x = -4 \qquad \text{Simplify}$$

$$x = 2.4 \qquad \text{Divide each side by } -\frac{5}{3}$$

So the graphs intersect when $x = 2.4$. To find the y-coordinate of the intersection point, we replace x by 2.4 in the equation of either line. Using the dashed line, we have: $y = 2.4-2 = 0.4$. So the coordinates of the intersection of the lines are (2.4, 0.4). Our original guess from the graph is an estimate; to get an accurate answer, we must use the algebraic expressions.

✎ Section 2.7 Exercise 27 Cell Phone Plan Comparison

(a) Since the first plan charges 20¢ per minute plus a monthly fee of $10, the cost of calling x minutes in a month is $f(x) = 10 + 0.2x$.

(b) Since the second plan offers unlimited minutes for a monthly fee of $100, the cost of calling x minutes in a month is $g(x) = 100$. Note that the amount you pay in this plan does not depend on how many minutes you call, so there is no x in the formula.

(c) To find the number of minutes for which the two plans have the same monthly cost, we find the value of x where $f(x) = g(x)$.

$$10 + 0.2x = 100 \qquad \text{Set } f(x) = g(x)$$

$$0.2x = 90 \qquad \text{Subtract 10 from each side}$$

$$x = 450 \qquad \text{Divide each side by 0.2}$$

So the two plans have the same monthly cost when Dietmar spends 450 minutes on the phone during the month.

✎ Section 2.7 Exercise 33 Commute to Work

Note that in the model in both parts (a) and (b), x is the length of time in hours since Jade leaves home. Since Jade leaves at 6:50 and Jari leaves 10 minutes later, x is the time in hours since 7:00 A.M..

(a) Since Jari drives at an average speed of 72 mi/h, the rate of change in her distance from home is $m = 72$. Jari has traveled 0 mi at 7:00 A.M., so when $x = 0$, we know $y = 0$. Thus the initial value is $b = 0$. An equation that models the distance Jari travels after x hours is $y = 72x$.

(b) Since Jade's average speed is 65 mi/h, the rate of change in her distance from home is $m = 65$. Jade has traveled 0 mi. at 6:50, which is

$$10 \text{ minutes} = 10 \text{ minutes} \frac{1 \text{ hour}}{60 \text{ minutes}} = \frac{1}{6} \text{ hours}$$

before 7:00 A.M.. So $y = 0$ when $x = -\frac{1}{6}$. We use the point-slope form, with $m = 65$, and $(x_1, y_1) = \left(-\frac{1}{6}, 0\right)$ to write the equation for the distance Jade travels as a function of time since 7:00 A.M..

$y - y_1 = m(x - x_1)$	Point-slope form
$y - 0 = 65\left(x - \left(-\frac{1}{6}\right)\right)$	Replace m by 65, x_1 by $-1/6$ and y_1 by 0
$y = 65x + \dfrac{65}{6}$	Distributive Property

So an equation that models the distance Jade travels x hours after 7:00 A.M. is $y = 65x + \dfrac{65}{6}$.

(c) We need to find the time x when the y-value in Jade's equation equals the y-value in Jari's equation. We set the formulas equal to each other and solve for x.

$72x = 65x + \dfrac{65}{6}$	Set y-values equal
$7x = \dfrac{65}{6}$	Subtract $65x$ from each side
$x = \dfrac{65}{42}$	Divide each side by 7

So Jari catches up to Jade after driving for $\frac{65}{42} \approx 1.5$ hours. They had traveled $y = 72\left(\frac{65}{42}\right) \approx 111.43$ miles from home at that time.

✎ Section 2.7 Exercise 35 Supply and Demand for Corn

(a) If no corn is produced, then $y = 0$ in the supply equation.

$0 = 4.18p - 11.5$	Set $y = 0$ in the supply equation
$11.5 = 4.18p$	Add 11.5 to each side
$p \approx 2.75$	Divide each side by 4.18

So when the price is $2.75 or lower, the suppliers will not produce any corn.

(b) If no corn is sold, then $y = 0$ in the demand equation.

$$0 = -1.06p + 19.3 \qquad \text{Set } y = 0 \text{ in the demand equation}$$
$$1.06p = 19.3 \qquad \text{Add } 1.06p \text{ to each side}$$
$$p \approx 18.21 \qquad \text{Divide each side by } 1.06$$

So when the price is $18.21 per bushel, consumers will not buy any corn.

(c) To find the equilibrium point, we set the supply and demand equations equal to each other and solve.

$$4.18p - 11.5 = -1.06p + 19.3 \qquad \text{Set functions equal}$$
$$5.24p = 30.8 \qquad \text{Add } 11.5 \text{ and } 1.06p \text{ to each side}$$
$$p \approx 5.88 \qquad \text{Divide each side by } 5.24$$

So the equilibrium price is $5.88. Evaluating the supply equation for $p = 5.88$, we have $y = 4.18(5.88) - 11.5 \approx 13.08$ million bushels. So for the equilibrium price of $5.88 per bushel, suppliers will produce and sell about 13 million bushels of corn.

3.1 Exponential Growth and Decay

OBJECTIVES

☑ *Check the box when you can do the exercises addressing the objective, such as those included in this study guide, which are listed here.*

Exercises

An Example of Exponential Growth

☐ Create a table of values from a verbal description of exponential growth or decay. Sketch the graph and use it to describe how the population grows or decays. **45**

Modeling Exponential Growth: The Growth Factor

☐ Recognize the shape of the graph of a function that models exponential growth. Identify the *y*-intercept of the graph from the model's formula. **17**

☐ Identify the growth factor, given in a verbal description of a population that grows exponentially. **25, 33**

☐ Find the growth factor for a certain time period (for example, the three-year growth factor), given the population at a particular time and at one time period later. **21(a), 39**

Modeling Exponential Growth: The Growth Rate

☐ Find the growth factor and an exponential growth model, given the percentage that the population increases each time period. **31**

☐ Find the growth rate, the percentage that the population increases each time period, or the growth factor, given any one of these. **25, 31, 33 (extended), 39**

☐ Use a model of exponential growth to make predictions. **39**

Modeling Exponential Decay

☐ Recognize the shape of the graph of a function that models exponential decay. Identify the *y*-intercept of the graph from the model's formula. **19**

☐ Identify a decay factor given, in a verbal description of a population that decays exponentially. **27**

☐ Calculate the decay factor for a certain time period (for example, the three-year decay factor), given the population at a particular time and at one time period later. **21(c)**

☐ Find the decay factor and an exponential decay model, given the percentage that the population decreases each time period. **29, 41**

☐ Find the decay rate, the percentage that the population decreases each time period, or the decay factor, given any one of these. **27, 29 (extended), 41**

☐ Identify the decay factor when the time period is the half-life. **45**

☐ Use a model of exponential decay to make predictions. **41, 45**

GET READY...

This section involves functions with exponents, and requires you to evaluate expressions with integer exponents. In Section 3.2, you continue this discussion, manipulating these expressions using the laws of exponents and coping with fractional exponents. Refresh your understanding of these in Algebra Toolkits A.3 and A.4.

✓ **Algebra Checkpoint** Test yourself by completing the Algebra Checkpoint at the end of this section of the textbook, and comparing your answers to those in the back of the textbook. Refer to Algebra Toolkits A.3 and A.4 as necessary.

The relationship between two concepts in this section, rate and factor, hinges upon the distributive law. Review the distributive law in Algebra Toolkit A.1.

SKILLS

Modeling Exponential Growth: The Growth Factor

In these exercises, you model exponential growth in terms of a growth factor and an initial value. Find the exponential growth model in a blue box in the textbook in Section 3.1.

[1] *In the exponential growth model $f(x) = Ca^x$, C is the _____, a is the _____ and x is the _____.*

📖 **Read Section 3.1 Example 1**, in which the authors model the growth of two populations, one of bacteria and one of fish.

[2] *In part (a), to find the number of bacteria after one time period (which in this case is _____), you multiply the initial value of _____ bacteria by the factor _____.*

In part (b), to find the number of fish after one time period (which in this case is _____), you multiply the initial value _____ fish by the factor _____.

✏ **Section 3.1 Exercises 25 and 33** A population P is initially 350. Find an exponential growth model in terms of the number of time periods x if in each time period, the population P

25. quadruples.

Solution

33. is multiplied by 1.7.

Solution

---

[1] The C is the initial value of f (i.e. C equals $f(0)$), a is the growth factor (i.e. a is the factor by which $f(x)$ is multiplied when x increases by one time period), and x is the number of time periods.

[2] In part (a), one time period is one hour, the initial value is $C = 100$ bacteria, and to find the population after one hour, you multiply the initial value by the one-hour growth factor $a = 3$. In part (b), one time period is one year, the initial value is $C = 5800$ fish, and to find the population after one year, you multiply the initial value by the one-year growth factor $a = 1.2$.

📖 **Read Section 3.1 Example 2**, in which the authors find the 3-year growth factor for a population of chinchillas.

[3]*A chinchilla is _____ (an animal / a vegetable / a mineral). To find the number of chinchillas after one time period (which in this case is _____), you multiply the initial value _____ chinchillas by the factor _____.*

✏ **Section 3.1 Exercises 21(a)** A population P is initially 3750 and six hours later reaches 7250. Find the six-hour growth factor.

Solution

Modeling Exponential Growth: The Growth Rate

In these exercises, you describe exponential growth using the growth rate instead of the growth factor. You convert the growth rate to a growth factor to write the model. Find the definition of the growth rate in bold typeface and the formula for converting the rate to a factor and vice versa in a blue box in Section 3.1 of the textbook.

📖 **Read Section 3.1 Example 3**, in which the authors model the growth of a rabbit population.

[4]*The distributive law tells us that $50 + 0.6 \times 50 = 50$ (____ + ____) $= 50$ (____). In words, this says that adding 60% of 50 to 50 is the same as multiplying 50 by _____. That is to say, increasing the population by 60% is the same as multiplying the population by a factor of _____.*

In summary, the population increases by _____% each year, which means the growth rate is $r =$_____, and the growth factor is $a =$ _____.

✏ **Section 3.1 Exercise 31** A population P is initially 350. Find an exponential growth model in terms of the number of time periods x if in each time period the population P increases by 300%.

Solution

---

[3] A chinchilla is a small furry rodent. To find the number of chinchillas after one time period (which is three years in this case), multiply the initial value $C = 20$ chinchillas by $a = 6.4$. The answer is of course 128 chinchillas.

[4] By the distributive law, we factor out 50 from $50 + 0.6 \times 50$. We get
$$50 + 0.6 \times 50 = 50 (1 + 0.6) = 50 (1.6)$$
In words, adding 60% of 50 to 50 is the same as multiplying 50 by 1.6. So increasing the population by 60% is the same as multiplying the population by a factor of 1.6. In summary, the population increases by 60% each year, so the growth rate is $r = 0.6$, and the growth factor is $a = 1.6$.

Modeling Exponential Decay

In these exercises, you model exponential decay in terms of a decay factor and an initial value. As for exponential growth, you sometimes use a decay rate to describe the model's behavior, instead of a decay factor. Find the exponential decay model in a blue box in Section 3.1 of the textbook. Compare the exponential decay and exponential growth models, and the general shapes of their graphs.

[5]*In the exponential decay model $f(x) = Ca^x$, C is the _____, a is the _____ and x is the _____, The graph of an exponential decay function is _____ (increasing / decreasing). If $f(x) = Ca^x$ models exponential decay, then a is _____ (greater than 1 / less than 1), and r is _____ (positive / negative).*

📖 **Read Section 3.1 Example 4**, in which the authors model the decay of a drug in a patient's bloodstream.

[6]*The distributive law tells us that $75 - 0.3 \times 75 = 75$ (____ – ____) = 75 (____). In words, this says that subtracting 30% of 75 from 75 is the same as multiplying 75 by _____. That is to say, decreasing the amount by 30% is the same as multiplying the amount by a factor of _____.*

In summary, the amount of drug decreases by _____% each hour, which means the decay rate is r = _____, and the decay factor is a = _____.

✏ **Section 3.1 Exercises 27 and 29** A population P is initially 350. Find an exponential decay model in terms of the number of time periods x if in each time period the population P

27. decreases by $\frac{1}{2}$.   29. decreases by 35%

       Solution   **Solution**

📖 **Read Section 3.1 Example 2 again**, paying particular attention to how the authors find the three-year growth factor.

[7]*Fill in each blank with "the population after 0 time periods" or "the population after 1 time period." The growth factor is _____ divided by_____.*

---

[5] The C is the initial value of f (i.e. C equals $f(0)$), a is the decay factor (i.e. a is the factor by which $f(x)$ is multiplied when x increases by one time period), and x is the number of time periods. The graph of an exponential decay function is decreasing. In a model of exponential decay, a is less than 1, and r is negative.

[6] By the distributive law, we factor out 75 from $75 - 0.3 \times 75$. We get
$$75 - 0.3 \times 75 = 75 (1 - 0.3) = 75 (0.7)$$
In words, this says that subtracting 30% of 75 from 75 is the same as multiplying 75 by 0.7. That is to say, decreasing the amount by 30% is the same as multiplying the amount by a factor of 0.7. In summary, the amount of drug decreases by 30% each hour, so the decay rate is $r = -0.3$, and the decay factor is $a = 0.7$.

[7] The growth factor is $a = \dfrac{\text{the population after 1 time period}}{\text{the population after 0 time periods}}$.

✎ **Section 3.1 Exercise 21(c)** A population P is initially 3750 and six hours later reaches 2500. Find the six-hour decay factor.

Solution

The following exercises ask you to identify and compare exponential growth and exponential decay, as well as the corresponding rates and factors.

✎ **Section 3.1 Exercises 25-33 odd (extended)** Refer to these exercises in this section of the study guide, above. Complete the following table.

	Decay or Growth Factor a	Decay or Growth Rate r	In each time period the population (increases / decreases) by ____%.
25.	*4*	*3*	*increases by 300%*
27.			
29.			
31.			
33.			

✎ **Section 3.1 Exercises 17 and 19 (extended)** Fill in the blanks, and match the exponential function with its graph.

17. The function $f(x) = 10 \cdot 2^x$ exhibits exponential _____ (growth / decay). The graph of this function crosses the y-axis at $y =$ _____. Graph number _____ (I / II)

19. The function $f(x) = 5 \cdot \left(\frac{1}{4}\right)^x$ exhibits exponential _____ (growth / decay). The graph of this function crosses the y-axis at $y =$ _____. Graph number _____ (I / II)

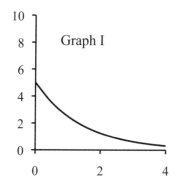

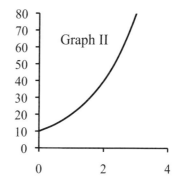

CONTEXTS

An Example of Exponential Growth

Motivate the study of exponential functions and gain insight into their behavior in this example.

📖 **Read the subsection, 'An Example of Exponential Growth,'** in which the authors describe how a bacterial infection develops.

[8]*Why can contracting even a few streptococcus bacteria be a threat to your health? Answer in your own words:*

Modeling Exponential Growth: The Growth Factor

In these exercises, you construct models for exponential growth and use them to make predictions.

📖 **Read Section 3.1 Example 3 again**, paying particular attention to how the authors use the model to predict future rabbit populations.

[9]*The input for the model P(x) is _____ and the output is _____. The number 8 given in part (b) is an _____ (input / output).*

✎ **Section 3.1 Exercise 39 Health-Care Spending** The Centers for Medicare and Medicaid Services report that health-care expenditures per capita were $2813 in 1990 and $3329 in 1993. Assume that this rate of growth continues.

(a) Find the three-year growth factor a, and find an exponential growth model $E(x) = Ca^x$ for the annual health care expenditures per capita, where x is the number of three-year time periods since 1990.

(b) Use the model found in part (a) to predict health care expenditures per capita in 1996 and 2005.

(c) Search the Internet to find the actual health-care expenditures for 1996 and 2005. Were your predictions reasonable?

Solution

---

[8] The authors write, "Although bacteria are invisible to the naked eye, their huge impact in our world is due to their ability to multiply rapidly." Indeed, in this example, although it began with only 10 bacteria, the infection grew to over 167 million bacteria in 24 hours. Rapid growth characterizes the exponential behavior you study in this chapter.

[9] The input for $P(x)$ is the number of time periods (years), and the output is the population of rabbits. The number 8 is the number of years (which is the number of time periods in this model), so 8 is an input.

Modeling Exponential Decay

In these exercises, you construct models for exponential decay and use them to make predictions.

📖 **Read Section 3.1 Example 4 again**, paying particular attention to how the authors use the model to predict the amount of drug that remains in a patient's body after a given amount of time.

[10] *The input for the model m(x) is _____ and the output is _____. The number 4 given in part (b) is an _____ (input / output).*

🖊 **Section 3.1 Exercise 41 Drug Absorption** A patient is administered 100 mg of a therapeutic drug. It is known that 25% of the drug is expelled from the body each hour.

(a) Find an exponential decay model for the amount of drug remaining in the patient's body after *t* hours.
(b) Use the model to predict the amount of the drug that remains in the patient's body after 6 hours.

Solution

📖 **Read Section 3.1 Example 5**, in which the authors model the amount of radium-226 that remains after it has radioactively decayed.

[11] *The time period in this model is _____. After one half-life, how much of the 50 g sample remains? _____ How much remains after two half-lives? _____ What does "half-life of a radioactive substance" mean?*
Answer in your own words:

Refer back to page 12
Hints and Tips 1.2c: Describing Trends in Relations

Hints and Tips 1.2c explains the authors' solution to Section 3.1 Example 5(d).

---

[10] The input for *m(x)* is the number of time periods (hours), and the output is the amount of drug (in milligrams) in the patient's body. The number 4 is the number of hours (which is the number of time periods in this model), so 4 is an input.

[11] The time period is one half-life, or 1600 years. After one half-life, the amount of radium decrease to half its original mass, or 25 g. After 2 half-lives, the amount of radium remaining is half of 25 g, which is 12.5 g. The half-life is the time required for half of the mass of a radioactive substance to decay, or change into a nonradioactive substance.

⇨ **Section 3.1 Example 5(d)** What does the graph tell us about how radium-226 decays?

Solution

Using Hints and Tips 1.2c, we describe the trend as the inputs increase. The input for this model is the number of half-lives, which is a measure of time, so we consider the graph as time increases, from left to right along the x-axis. The authors follow this tip, describing first the left hand side of the graph, and then the right, saying, "From the graph, we see that radium-226 decays rapidly at first, but that the rate of decay slows down as time goes on."

✎ **Section 3.1 Exercise 45 Radioactive Cesium** The half-life of cesium-137 is 30 years. Suppose we have a 10-gram sample.

(a) Find a function that models the mass $m(x)$ of cesium-137 remaining after x half-lives.
(b) Use the model to predict the amount of cesium-137 remaining after 270 years.
(c) Make a table of values for $m(x)$ with x varying between 0 and 5.
(d) Graph the function $m(x)$ and the entries in the table. What does the graph tell us about how $m(x)$ changes?

Solution

(c)

Number of half-lives x	Mass of cesium remaining (g) $m(x)$
0	
1	
2	
3	
4	
5	

CONCEPTS

✓ **3.1 Exercises – Concepts – Fundamentals** Complete the Fundamentals section of the exercises at the end of Section 3.1. Compare your answers to those at the back of the textbook, and make corrections as necessary.

📖 **Read the Concept Check for Section 3.1**, in the Chapter Review at the end of Chapter 3.

3.1 Solutions to ✎ Exercises

✎ Section 3.1 Exercise 17 (extended)

The function $f(x) = 10 \cdot 2^x$ corresponds to Graph II. It exhibits exponential growth. The graph of this function crosses the y-axis at $y = 10$.

✎ Section 3.1 Exercise 19 (extended)

The function $f(x) = 5\left(\frac{1}{2}\right)^x$ corresponds to Graph I. It exhibits exponential decay. The graph of this function crosses the y-axis at $y = 5$.

✎ Section 3.1 Exercise 21(a, c)

(a) The six-hour growth factor is the population after 6 hours divided by the initial population.

$$\text{six-hour growth factor} = \frac{\text{population in hour 6}}{\text{population in hour 0}} = \frac{7250}{3750} = \frac{29}{15} \approx 1.93$$

(c) The six-hour growth factor is the population after 6 hours divided by the initial population.

$$\text{six-hour growth factor} = \frac{\text{population in hour 6}}{\text{population in hour 0}} = \frac{2500}{3750} = \frac{2}{3} \approx 0.67$$

✎ Section 3.1 Exercise 25, 27, 29, 31 and 33

In each of these exercises, since the population grows exponentially, we are looking for a model of the form $m(x) = Ca^x$. The initial population is 350, so $C = 350$.

25. The population quadruples each time period, so the growth factor a is 4. Thus the model we seek is $f(x) = 350 \cdot 4^x$, where x is the number of time periods.

27. The population is divided by 2 each time period, so the population is multiplied by $\frac{1}{2}$ each time period. Thus the decay factor a is $\frac{1}{2}$. Thus the model we seek is $f(x) = 350\left(\frac{1}{2}\right)^x$, where x is the number of time periods.

29. The population decreases by 35% each time period, so the decay rate r is -0.35. This means that the decay factor is $a = 1 + r = 1 + (-0.35) = 0.65$. Thus the exponential decay model that we seek is $m(x) = 350(0.65)^x$, where x is the number of time periods.

31. The population grows by 300% each time period, so the growth rate r is 3. This means the growth factor is $a = 1 + r = 1 + 3 = 4$. Thus the exponential decay model that we seek is $m(x) = 350 \cdot 4^x$, where x is the number of time periods.

33. The population is multiplied by 1.7 each time period, so the growth factor a is 1.7. Thus the exponential growth model that we seek is $m(x) = 350(1.7)^x$, where x is the number of time periods.

✐ Section 3.1 Exercises 25-33 odd (extended)

	Decay or Growth Factor a	Decay or Growth Rate r	In each time period the population (increases / decreases) by ___%.
25.	4	3	increases by 300%
27.	0.5	-0.5	decreases by 50%
29.	0.65	-0.35	decreases by 35%
31.	4	3	increases by 300%
33.	1.7	0.7	increases by 70%

✐ Section 3.1 Exercise 39 Health-Care Spending

(a) The three-year growth factor is the expenditures after 3 years (in 1993) divided by the initial expenditures (in 1990).

$$\text{three-year growth factor} = \frac{\text{expenditures in year 3}}{\text{expenditures in year 0}} = \frac{3329}{2813} \approx 1.18$$

Thus $a = 1.18$. The initial expenditures were $C = 2813$, so the exponential model $P(x) = Ca^x$ that we seek is $P(x) = 2813(1.18)^x$, where x is measured in three-year periods.

(b) In 1996, two three-year time periods have passed since 1990, so $x = 2$.

$$P(2) = 2813(1.18)^2 \approx 3916.82$$

So the model predicts that in 1996, the health care expenditures per capita will be $3916.82.

In 2005, fifteen years have passed since 1990. So since 1990, 5 three-year time periods have passed, and $x = 5$.

$$P(5) = 2813 \cdot (1.18)^5 \approx 6435.46$$

So the model predicts that in 2005, the health care expenditures per capita to be $6435.46.

✐ Section 3.1 Exercise 41 Drug Absorption

(a) The amount of drug remaining in the body decreases by 25% each hour, so the decay rate r is -0.25. So the decay factor is $a = 1 + r = 1 + (-0.25) = 0.75$. Since the initial amount C is 100, the exponential decay model $m(x) = Ca^x$ that we seek is $m(x) = 100(0.75)^x$, where x is the number of hours since the drug was administered.

(b) Replacing x by 6 in our model, we get $m(6) = 100(0.75)^6 \approx 17.80$. So after six hours, approximately 17.8 mg of the drug remains.

✐ Section 3.1 Exercise 45 Radioactive Cesium

The initial mass C is 10 g, and the decay factor is $a = \frac{1}{2}$. So the model $m(x) = Ca^x$ that we seek is $m(x) = 10\left(\frac{1}{2}\right)^x$, where x is the number of 30-year time periods.

(b) To find the number of time periods that 270 years represents, we divide: $\frac{270}{30} = 9$. So 270 years represents 9 time periods. Replacing x by 9 in the model, we get $m(9) = 10\left(\frac{1}{2}\right)^9 \approx 0.0195$. So the mass remaining after 270 years is approximately 0.0195 g.

(c)

Number of half-lives x	Mass of cesium remaining (g) $m(x)$
0	10
1	5
2	2.5
3	1.25
4	0.625
5	0.3125

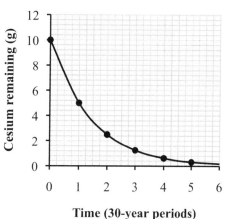

(d) Following **Hints and Tips 1.2c**, we describe the trend as the inputs increase. The inputs for this model is the number of half-lives, which is time, so we describe the trend as time increases, from left to right along the x-axis. The mass of cesium remaining decays rapidly at first, but the average rate of change of the mass decreases as time goes on.

3.2 Exponential Models: Comparing Rates

OBJECTIVES

☑ *Check the box when you can do the exercises addressing the objective, such as those included in this study guide, which are listed here.*

Exercises

Changing the Time Period

☐ Find the growth or decay factor and an exponential model corresponding to a particular time period, given the percentage that the output changes over a different time period.   **9, 43**

☐ Find the growth or decay factor and an exponential model corresponding to a particular time period, given the factor by which the output changes over a different time period.   **13, 47**

☐ Find the growth or decay factor and an exponential model corresponding to a particular time period, given the populations at two times.   **5(a), 45**

☐ Compare the rates of growth or decay of two exponential models by changing them to have factors relative to the same time period.   **43**

☐ Use a model of exponential growth to make predictions.   **45, 47**

Growth of an investment: Compound Interest

☐ Compare investments with different interest rates and compounding.   **41**

☐ Calculate the APY for a given interest rate and compounding.   **37**

GET READY...

This Section involves solving power equations. Refresh your understanding of these in Algebra Toolkit C.1.

✓ **Algebra Checkpoint** Test yourself by completing the Algebra Checkpoint at the end of this section of the textbook, and comparing your answers to those in the back of the textbook. Refer to Algebra Toolkit C.1 as necessary.

SKILLS

Changing the Time Period

These exercises ask you to construct exponential models corresponding to a given time period. In contrast, in Section 3.1, you constructed exponential models relative to the time period that was the most convenient, using the information at hand.

📖 **Read the introduction to the subsection, 'Changing the Time Period,'** in which the authors explain how to use exponents to find one growth factor from another.

[12] *Suppose a is the daily growth factor, b is the weekly growth factor and d is the yearly growth factor for a population. There are _____ days in a week, so to find the population after a week, you can multiply the initial population by a _____ times, which is the same as multiplying by _____. This means the weekly growth factor is b = _____. Solving for a, you find that the daily growth factor is a = _____.*

There are _____ weeks in a year, so to find the population after a year, you can multiply the initial population by b _____ times, which is the same as multiplying by _____. This means the yearly growth factor is d = _____. Solving for b, you find that the weekly growth factor is b = _____.

📖 **Read Section 3.2 Example 1**, in which the authors convert a 30-minute growth rate to a 1-hour growth rate.

[13] *What is the first thing the authors do to solve this problem?*
Answer in your own words:

✎ **Section 3.2 Exercise 9** The bacteria population in a certain culture grows exponentially. Find the one-hour growth rate if the three-hour growth rate is 57%.

Solution

---

[12] There are 7 days in a week, so to find the population after a week, you can multiply the initial population by a 7 times, which is the same thing as multiplying by a^7. So the weekly growth factor $b = a^7$, and the daily growth factor is $a = b^{1/7}$.
There are 52 weeks in a year, so to find the population after a year, you can multiply the initial population by b 52 times, which is the same thing as multiplying by b^{52}. So the yearly growth factor $d = b^{52}$, and the weekly growth factor is $b = d^{1/52}$.

[13] The authors begin by finding the growth factor corresponding to the given rate.

📖 **Read Section 3.1 Example 2 again**, paying particular attention to the role played by the three-year time period.

[14]*Since the given populations are _____ years apart, the most convenient growth factor to calculate is the _____ growth factor.*

📖 **Read Section 3.2 Example 2**, in which the authors find the one-year growth factor for the chinchillas studied in Section 3.1 Example 2.

[15]*Suppose a is the one-year growth factor. There are _____ one-year time periods in a three-year time period, so to find the population after a three-year period, you can multiply the initial population by a _____ times, which is the same as multiplying by _____. This means the three-year growth factor is 6.4 = _____. Solving for a, you find that the one-year growth factor is a = _____.*

✏️ **Section 3.2 Exercise 5(a)** A population P starts at 2000, and four years later the population reaches 8000.

(i) Find the four-year growth factor.

(ii) Find the annual growth factor.

Solution

📖 **Read Section 3.2 Example 3**, in which the author constructs two models for the growth of a bacterial infection, one with a five-hour time period, and the other with a one-hour time period.

[16]*In part (a), the input of the model is the number of five-hour time periods, so the growth factor is the _____ growth factor. In part (b), the input of the model is the number of one-hour time periods (i.e. the number of hours), so the growth factor is the _____ growth factor.*

Referring to the second solution to part (b), there are _____ five-hour time periods in t hours. For example, there are _____ five-hour time periods in 15 hours, and _____ five-hour time periods in 50 hours .

---

[14] Since the populations 20 and 128 are three years apart, the three-year growth rate is the most convenient to calculate.

[15] There are 3 one-year time periods in a three-year time period, so to find the population after a three-year period, you can multiply the initial population by a 3 times, which is the same thing as multiplying by a^3. So the three-year growth factor is $6.4 = a^3$, and the one-year growth factor is $a = 6.4^{1/3} \approx 1.86$.

[16] In part (a), the input is the number of five-hour periods, so the growth factor is the five-hour growth factor. In part (b), the input is the number of hours, so the growth factor is the one-hour growth factor. In the second solution to part (b), there are $t/5$ five-hour time periods in t hours. For example, there are 3 five-hour time periods in 15 hours, and 10 five-hour time periods in 50 hours.

✎ **Section 3.2 Exercise 13** A bacterial infection starts with 2000 bacteria, and the bacteria count doubles in 20 minutes. Find an exponential growth model for the number of bacteria

(a) x 20-minute time periods after infection.
(b) t hours after infection.

Solution

CONTEXTS

Changing the Time Period

These exercises ask you to construct exponential models corresponding to a given time period for the purpose of comparing growth or decay rates, or predicting the future.

📖 **Read the introduction to Section 3.2 in the textbook,** in which the authors describe an experiment through which a biologist estimates the 40-minute growth factor of a population of bacteria.

[17]*If a is the 40-minute growth factor for a population modeled by $f(x) = Ca^x$, then the input x of the model represents _____. If b is the one-hour growth factor for a population modeled by $g(z) = Cb^z$, then the input z of the model represents _____.*

----

[17] If the growth factor in an exponential model is a 40-minute growth factor, then the input of the model is the number of 40-minute time periods. If the growth factor in an exponential model is a one-hour growth factor, then the input of the model is the number of one-hour time periods, or more simply, the number of hours.

✎ **Section 3.2 Exercise 45 Population of India** Although India occupies only a small portion of the world's land area, it is the second most populous country in the world, and its population is growing rapidly. The population was 846 million in 1990 and 1148 million in 2000. Assume that India's population grows exponentially.

(a) Find the 10-year growth factor and the annual growth factor for India's population.
(b) Find an exponential growth model P for the population t years after 1990.
(c) Use the model found in part (b) to predict the population of India in 2010.
(d) Graph the function P for t between 0 and 25, using a graphing calculator.

Solution

📖 **Read Section 3.2 Example 4**, in which the authors compare the growth rates of two types of bacteria.

[18]*The growth of these two types of bacteria are challenging to compare as given because both the factors and the time periods are different. If only one of these differ, you can make the comparison without calculation.*
If A doubles every 5 hours, and C triples every 5 hours, then _____ (A / C) has the largest growth rate.
If A doubles every 5 hours, and D doubles every 7 hours, then _____ (A / D) has the largest growth rate.
If B triples every 7 hours, and C triples every 5 hours, then _____ (B / C) has the largest growth rate.
If B triples every 7 hours, and D doubles every 7 hours, then _____ (B / D) has the largest growth rate.

---

[18] Bacteria C has a larger growth rate than A. Bacteria A has a larger growth rate than D. Bacteria C has a larger growth rate than B. Bacteria B has a larger growth rate than D.

✐ **Section 3.2 Exercise 43 Bacteria** Bacterioplankton that occur in large bodies of water are a powerful indicator of the status of aquatic life in the water. One team of biologists tests a certain location, and their data indicate that the bacterioplankton are increasing by 19% every 20 minutes. Another team of biologists test a different location, and this data indicate that bacterioplankton are increasing by 40% every 3 hours.

(a) Find the one-hour growth factor for each location.
(b) Which location has the larger growth rate?

Solution

📖 **Reread Section 3.1 Example 5**, paying particular attention to the input of the model the authors construct in part (a).

[19]*The input for the model the authors construct in part (a) is _____.*

📖 **Read Section 3.2 Example 5**, in which the authors again use the half-life to construct a model for the decay of radium-226.

[20]*The input for the model the authors construct in part (a) is _____, so the decay factor in that model is the _____ decay factor. The 1600-year decay factor is _____. To find the yearly decay factor, the authors use that t years is the same as _____ 1600-year time periods. For example, 3200 years is the same as _____ 1600-year time periods.*

📖 **Read the paragraph after Section 3.2 Example 5**, in which the authors write a formula for finding an exponential decay model using the half-life.

[21]*The authors suppose the half-life is h years; t years is the same as _____ h-year time periods.*

---

[19] The input is the number of half-lives during which the sample has been stored, or in this case the number of 1600-year periods.

[20] The input is the number of years during which the sample has been stored. The decay factor in the model is the yearly (annual) decay factor. The 1600-year decay factor is 1/2. To find the yearly decay factor, the authors used that t years is the same as $\frac{t}{1600}$ 1600-year time periods. For example, 3200 years is the same as $\frac{3200}{1600} = 2$ 1600-year time periods.

[21] A period of t years is the same as t/h h-year time periods.

✏ **Section 3.1 Exercise 45 Radioactive Cesium** *Find this exercise on page 164 of this study guide.*

✏ **Section 3.2 Exercise 47 Radioactive Plutonium (excerpt)** Nuclear power plants produce radioactive plutonium-239, which has a half-life of 24,360 years. A 700-gram sample of plutonium-239 is placed in an underground waste disposal facility.

(a) Find a function that models the mass $m(t)$ of plutonium-239 remaining in a sample after t years. What is the decay factor?

(b) Use the model to predict the amount of plutonium-239 remaining in a sample after 500 years.

Solution

Growth of an Investment: Compound Interest

These exercises ask you to compute how much money is in an investment account after a given period of time, when the interest is compounding periodically. Find the formula for compound interest in a blue box in Section 3.2 of the textbook.

📖 **Read the introduction to the subsection of Section 3.2 entitled, 'Growth of an Investment: Compound Interest,'** in which the authors discuss an investment with a 6% annual interest rate compounding monthly.

22 *The CD pays 6% interest annually, compounding monthly. The monthly growth rate is _____, which the authors obtain by dividing _____ by _____.*

📖 **Read Section 3.2 Example 6**, in which the authors compare an investment compounded daily to an investment compounded semiannually.

23 *For the investment that is compounding daily, the variable n = _____, and for the investment compounding semiannually, the variable n = _____.*

---

22 The monthly growth rate is 0.5%, which is obtained by dividing 6% by 12.

23 When compounding daily, $n = 365$. When compounding semiannually, $n = 2$.

✐ **Section 3.2 Exercise 41** Kai wants to invest $5000, and he is comparing two different investment options:

(i) 3¼% interest each year, compounded semiannually
(ii) 3% interest each year, compounded daily
Which of he two options would prove the better investment?

Solution

📖 **Read Section 3.2 Example 7**, in which the authors find the annual percentage yield for an investment with interest compounding daily.

[24]*The acronym APY stands for* _____. *To find the APY, evaluate the model at t =* ____.

✐ **Section 3.2 Exercise 37** Find the annual percentage yield for an investment that earns 2.5% each year, compounded daily.

Solution

CONCEPTS

✓ **3.2 Exercises – Concepts – Fundamentals** Complete the Fundamentals section of the exercises at the end of Section 3.2. Compare your answers to those at the back of the textbook, and make corrections as necessary.

📖 **Read the Concept Check for Section 3.2**, in the Chapter Review at the end of Chapter 3.

---

[24] APY stands for annual percentage yield. To find the APY, substitute *t = 1 into the formula.*

3.2 Solutions to ✏ Exercises

✏ Section 3.2 Exercise 5(a)

(i) The four-year growth factor is the population after 4 years divided by the initial population.

$$\text{four-year growth factor} = \frac{\text{population in year 4}}{\text{population in year 0}} = \frac{8000}{2000} = 4$$

(ii) Let a be the one-year growth factor. Since there are four one-year time periods in four years, we have

$$a^4 = 4 \qquad \text{(The 4 on the right hand side is the four-year growth factor from part (a).)}$$

$$a = (1.57)^{1/4}$$

$$a \approx 1.12$$

So the annual growth factor is approximately 1.12.

✏ Section 3.2 Exercise 9

The three-hour growth factor is $1 + 0.57 = 1.57$. Let a be the one-hour growth factor. Since there are three one-hour time periods in three hours, we have

$$a^3 = 1.57$$

$$a = (1.57)^{1/3}$$

$$a \approx 1.16$$

So the one-hour growth factor is approximately 1.16. It follows that the one-hour growth rate is $r = a - 1 = 1.16 - 1 = 0.16$.

✏ Section 3.2 Exercise 13

(a) The initial number of bacteria is 2000, and the growth factor for a 20-minute time period is 2. So a model for the number of bacteria is $P(x) = 2000 \cdot 2^x$, where x is the number of 20-minute time periods since infection.

(b) Let a be the one-hour growth factor. Since there are three 20-minute time periods in one hour, we have

$$a = 2^3 \qquad \text{20-minute growth factor is 2}$$

$$a = 8 \qquad \text{Calculator}$$

So the one-hour growth factor is 8. An exponential model is obtained as follows:

$$P(t) = Ca^t \qquad \text{Model}$$

$$P(t) = 2000 \cdot 8^t \qquad \text{Replace } C \text{ by 2000, } a \text{ by 8l}$$

The model is $P(t) = 2000 \cdot 8^t$, where t is measured in hours.

✒ Section 3.2 Exercise 37

The rate r is 0.025, and the number of compounding periods n is 365. Using the formula for compound interest, replacing r by 0.025 and n by 365, we get

$$A(t) = P\left(1 + \frac{0.025}{365}\right)^{365t} \qquad \text{Replace } r \text{ by } 0.03, \text{ and } n \text{ by } 365$$

$$\approx P\left[\left(1 + \frac{0.025}{365}\right)^{365}\right]^t \qquad \text{Property of exponents}$$

$$\approx P(1.0253)^t \qquad \text{Calculator}$$

So after 1 year, any principal P will grow to the amount $A(1) = P(1.0253)$. So the annual growth rate is $1.0253 - 1 = 0.0253$, and the annual percentage yield is 2.53%.

✒ Section 3.2 Exercise 41

The principal P is \$5000.
(i) The rate r is 0.0325, and the number n of compounding periods is 2. Using the formula for compound interest, replacing r by 0.0325, and n by 2, we get

$$A(t) = 5000\left(1 + \frac{0.0325}{2}\right)^{2t} \qquad \text{Replace } r \text{ by } 0.0325, \text{ and } n \text{ by } 2$$

$$= 5000\left[\left(1 + \frac{0.0325}{2}\right)^{2}\right]^t \qquad \text{Property of exponents}$$

$$\approx 5000(1.03276)^t \qquad \text{Calculator}$$

(ii) The rate r is 0.03, and the number n of compounding periods is 365. Using the formula for compound interest, replacing r by 0.03, and n by 365, we get

$$A(t) = 5000\left(1 + \frac{0.03}{365}\right)^{365t} \qquad \text{Replace } r \text{ by } 0.03, \text{ and } n \text{ by } 365$$

$$\approx 5000\left[\left(1 + \frac{0.03}{365}\right)^{365}\right]^t \qquad \text{Property of exponents}$$

$$\approx 5000(1.03045)^t \qquad \text{Calculator}$$

Option (i) has a growth factor of 1.03276, while option (ii) has a growth factor of 1.03045. Thus option (i) is a better investment.

✐ Section 3.2 Exercise 43 Bacteria

(a) Location I: The 20-minute growth rate is 19%, so $r = 0.19$. Thus the 20-minute growth factor is $1 + 0.19 = 1.19$.

Let a be the one-hour growth factor. Since there are three 20-minute growth factors in one hour, we have $a = (1.19)^3 \approx 1.69$. So the one-hour growth factor is 1.69, and the one-hour growth rate is $1.69 - 1 = 0.69$, or 69% per hour.

Location II: The three-hour growth rate is 40%, so $r = 0.4$. Thus the three-hour growth factor is $1 + 0.4 = 1.4$. Let a be the one-hour growth factor.

$$a^3 = 1.4 \qquad \text{The three-hour growth factor is 1.4}$$

$$a = \sqrt[3]{1.4} \qquad \text{Solve for } a$$

$$\approx 1.12 \qquad \text{Calculator}$$

The one-hour growth factor is 1.12, so the one-hour growth rate is $1.12 - 1 = 0.12$, or 12% per hour.

(b) The first location has the larger growth rate.

✐ Section 3.2 Exercise 45 Population of India

(a) The 10-year growth factor is the population after 10 years (in 2000) divided by the initial population (in 1990).

$$10\text{-year growth factor} = \frac{\text{population in year 2000}}{\text{population in year 1990}} = \frac{1148 \text{ million}}{846 \text{ million}} \approx 1.36$$

(b) Let a be the annual growth factor.

$$a^{10} = 1.36 \qquad \text{The 10-year growth factor is 1.36}$$

$$a = (1.36)^{1/10} \qquad \text{Solve for } a$$

$$\approx 1.031 \qquad \text{Calculator}$$

So the annual growth factor is 1.031. So an exponential model is obtained s follows:

$$P(t) = Ca^t \qquad \text{Model}$$

$$P(t) = 846(1.031)^t \qquad \text{Replace } C \text{ by 846, } a \text{ by 1.031}$$

The model is $P(t) = 846(1.031)^t$, where t is measured in years and the output is measured in millions of people.

(c) In 2010, t is 20 years. Replacing t with 20 in the model, we have $P(20) = 846(1.031)^{20} = 1558$. So the model predicts that in 2010, the population of India to be approximately 1558 million.

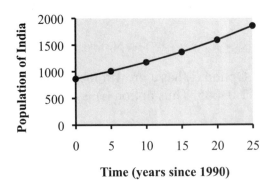

✐ **Section 3.2 Exercise 47 Radioactive Plutonium (excerpt)**

(a) The initial mass is 700 g, and the 24,360-year decay factor a is $\frac{1}{2}$.

Let t be the number of years. Since each time period consists of 24,360 years, after x time periods we have $t = 24{,}360x$. Solving for x, we get $x = t/24{,}360$. So an exponential growth model is

$$m(t) = 700\left(\frac{1}{2}\right)^{t/24360} \qquad \text{Replace } x \text{ by } t/24360$$

$$= 700(0.5^{1/24360})^{t} \qquad \text{Property of exponents}$$

$$\approx 700(0.999972)^{t} \qquad \text{Calculator}$$

where t is measured in years. Thus the annual decay factor is approximately $a = 0.999972$.

(b) Replacing t by 500 in the model, we get $m(t) = 700(0.999972)^{500} \approx 690.27$. So the mass remaining after 500 years is about 690.27 g.

3.3 Comparing Linear and Exponential Growth

OBJECTIVES

☑ *Check the box when you can do the exercises addressing the objective, such as those included in this study guide, which are listed here.*

Exercises

Average Rate of Change and Percentage Rate of Change

☐ Calculate average rates of change and percentage rates of change. — 9, 13, 17

☐ Use percentage rate of change to determine if an exponential model is appropriate for given data, and if so, construct a model. — 13, 17

Comparing Linear and Exponential Growth

☐ Create data that changes linearly, using addition, and data that changes exponentially using multiplication. — 19, 29

☐ Construct a linear and an exponential model given an initial value and a value after one time period. — 19, 29

☐ Compare the behavior of linear and exponential functions from their graphs. — 19

☐ Compare models of the real world that grow exponentially and linearly. — 29

Logistic Growth: Growth with Limited Resources

☐ Use logistic models to describe growth under limited resources. — 31

☐ Identify the carrying capacity and initial value from a logistic growth model or from its graph. — 31

GET READY...

This section involves average rates of change. Refresh your understanding in Section 2.1 of the textbook. Also review Section 1.3 in which you used first differences to determine if data are linear. This section also uses vocabulary pertaining to a graphing calculator, like viewing rectangle, and skills, like graphing two functions on the same screen. Refresh your understanding of these in Algebra Toolkit D.3 Example 2.

✓ **Check Yourself** Use your calculator to verify the following computations:

(a) $\dfrac{110}{1+6\cdot(3)^{-4}} \approx 102.413793$

(b) $\dfrac{25.7}{1+7.8\cdot(2.4)^{-12}} \approx 25.694512$

Troubleshooting: Make sure to put parentheses around the denominator. Also, note your calculator has two "minus" keys, one for subtraction $\boxed{-}$ and one for typing negative numbers $\boxed{(-)}$; for the negative numbers in the exponents, use the $\boxed{(-)}$ key. For example, in part (a), type: $\boxed{1}\boxed{1}\boxed{0}\boxed{\div}\boxed{(}\boxed{(}\boxed{1}\boxed{+}\boxed{6}\boxed{\times}\boxed{3}\boxed{\wedge}\boxed{(-)}\boxed{4}\boxed{)}$.

SKILLS

Average Rate of Change and Percent Rate of Change

These exercises ask you to compute percentage rates of change and average rates of change in order to compare exponential and linear growth. Find the definition of the average rate of change of a function in a blue box in Section 2.1 of the textbook, and find the definition of percentage rates of change in a blue box in Section 3.3 of the textbook.

📖 **Read Section 3.3 Example 1, and the paragraph that follows it**, in which the authors calculate average rates of change and percent rates of change for an exponential function.

[25]*How do the authors find each of the values in the fourth row of the table, corresponding to x = 3? Answer in your own words:*

If the _____ (average rate of change / percentage rate of change) is constant, then there is a linear model that fits the data exactly.

If the _____ (average rate of change / percentage rate of change) is constant for consecutive outputs, then there is an exponential model that fits the data exactly.

---

[25] The value in the $f(x)$ column is $f(3) = 10(3^3) = 270$. The value in the average rate of change column is the average rate of change from $x = 2$ to $x = 3$, or $\dfrac{f(3)-f(2)}{3-2}=\dfrac{270-90}{1}=180$. The value in the percent rate of

change column is $\dfrac{f(3)-f(2)}{f(2)}=\dfrac{270-90}{90}=2$, which is 200% in decimal form.

If the average rate of change is constant, then there is a linear model that fits the data exactly. If the percentage rate of change is constant then there is an exponential model that fits the data exactly.

✎ **Section 3.3 Exercise 9** A population is modeled by the function $f(t) = 5000(1.3)^t$, where t is measured in hours. Make a table of values for f, the average rate of change of f, and the percentage rate of change of f in each time period.

Solution

Hours t	Population $f(t)$	Average rate of change	Percentage rate of change
0	**5000**	–	–
1	**6500**	**1500**	**30%**
2			
3			
4			

📖 **Read Section 3.3 Example 2**, in which the authors determine if a linear or exponential function is an appropriate model for a set of data, by calculating the average rates of change and percentage rates of change.

[26]*In part (b), how do the authors find the a and the C in the exponential model for this data?*
Answer in your own words:

---

[26] The initial value C is $f(0)$, which is given on the table to be 10,000. The decay rate r is the same as the percent rate of change over one time period, which the authors find in part (a) to be –0.3. So the decay factor $a = 1 + (-0.3) = 0.7$.

✎ **Section 3.3 Exercises 13 and 17** A data set with equally spaced inputs is given.

(a) Find the average rate of change and the percentage rate of change for consecutive outputs.

(b) Is an exponential model appropriate? If so, find the model and sketch a graph of the model together with a scatter plot of the data.

Solution

An exponential function is appropriate to model one of these two data sets; construct your graph for that data set on the coordinate axis below the tables.

13.

x	y	Average rate of change	Percentage rate of change
0	30,000	–	–
1	57,000		
2	108,300		
3	205,770		
4	390,963		

17.

x	y	Average rate of change	Percentage rate of change
0	3000	–	–
1	2440		
2	1880		
3	1320		
4	760		

An exponential model is appropriate for the data in Exercise _____ (13 / 17). Sketch its graph here.

Comparing Linear and Exponential Growth

These exercises ask you to create data with linear growth using addition, and exponential growth using multiplication. You construct models in each case and compare their graphs.

📖 **Read the introduction to the subsection, 'Comparing Linear and Exponential Growth,'** in which the authors compare linear and exponential models.

27
For a linear model, $f(x) = b + mx$, increasing x by one unit has the effect of
_____. *For example, if $f(x) = 4 + 10x$, then f(6) =_____, and f(7) =_____. So increasing x from 6 to 7 has the effect of adding _____ to f(6).*
For an exponential model, $f(x) = Ca^x$, increasing x by one unit has the effect of _____.

For example if $f(x) = 4{\cdot}10^x$, then f(6) =_____, and f(7) =_____. So increasing x from 6 to 7 has the effect of multiplying f(6) by _____.

📖 **Read Section 3.3 Example 3**, in which the authors compare estimates for the growth of the population of the city of Newburgh.

28
In completing the table of values for the linear model proposed by Planner A, how can you obtain the population in year 3 from the population in year 2?
Answer in your own words:

In completing the table of values for the exponential model proposed by Planner B, how can you obtain the population in year 3 from the population in year 2?
Answer in your own words:

27
For $f(x) = b + mx$, increasing x by one unit has the effect of adding m to $f(x)$. If $f(x) = 4 + 10x$, then $f(6) = 64$, and $f(7) = 74$. So increasing x from 6 to 7 has the effect of adding 10 to $f(6)$.

For an exponential model, $f(x) = Ca^x$, increasing x by one unit has the effect of multiplying $f(x)$ by a.

For example if $f(x) = 4{\cdot}10^x$, then $f(6) = 4,000,000$ and $f(7) = 40,000,000$. So increasing x from 6 to 7 has the effect of multiplying $f(6)$ by 10.

28
In the table of values for the linear model proposed by Planner A, find the population in year 3 by adding 500 to the population in year 2. In the table of values for the exponential model proposed by Planner B, find the population in year 3 by multiplying the population in year 2 by a factor of 1.05, which corresponds to a 5% growth rate.

✎ **Section 3.3 Exercise 19** Suppose that the function *f* is a model for linear growth and that *g* is a model for exponential growth, with both *f* and *g* functions of time *t*.

(a) Use the assumptions that *f* is linear and *g* is exponential to complete the tables.

(b) Construct equations for the functions *f* and *g* in terms of *t*.

(c) Use a graphing calculator to graph the functions *f* and *g* on the same screen for $0 \le t \le 10$. What does the graph tell us about the differences between the two functions' behaviors?

Solution

(a)

t	f(t)
0	200
1	350
2	
3	
4	

t	g(t)
0	200
1	350
2	
3	
4	

Copy your graph from part (c) here.

CONTEXTS

Comparing Linear and Exponential Growth

These exercises ask you to construct linear and exponential models, make predictions using the models, and compare the results.

📖 **Read Section 3.3 Example 3 again**, paying particular attention to the verbal descriptions of the population's growth.

[29] *How did the authors know to use a linear model for Planner A's estimate, and an exponential model for Planner B's estimate?*

Answer in your own words:

---

[29] For Planner A's estimate, increasing the year by one has the effect of adding 500 to the population, so the model is linear. For Planner B's estimate, increasing the year by one has the effect of multiplying the population by the factor 1.05 (corresponding to a growth rate of 5%), so the model is exponential.

🖉 **Section 3.3 Exercise 29 Salary Comparison** Suppose you are offered a job that lasts three years and you are to be very well paid. Which of the following methods of payment is more profitable for you?

> **Offer A:** You are paid $10,000 in the starting month (month 0) and get a $1000 raise each month.

> **Offer B:** You are paid 2 cents in month 0, 4 cents in month 1, 8 cents in month 2, and so on, doubling your pay each month.

(a) Complete the salary tables for Offer A and Offer B.

(b) Find a function f_A that models your salary in month t if you accept Offer A. What kind of function is f_A?

(c) Find a function f_B that models your salary in month t if you accept Offer B. What kind of function is f_B?

(d) Find the salary for the last month (month 35) for each offer. Which offer gives the highest salary in the last month?

Solution

Offer A			Offer B		
Month	Monthly salary ($)	Average rate of change ($/mo)	Month	Monthly salary ($)	Percentage rate of change
0	10,000	–	0	0.02	–
1	11,000	1000	1	0.04	100%
2			2		
3			3		
4			4		
5			5		

Logistic Growth: Growth with Limited Resources

These exercises use logistic models to represent growth that is bounded by a limited carrying capacity. Find the definitions of a logistic model and its carrying capacity in a blue box in Section 3.3 of the textbook.

📖 **Read the introduction to the subsection, 'Logistic Growth: Growth with Limited Resources,'** in which the authors discuss the growth of a cell phone company.

[30]*Why is an exponential growth model inappropriate to model the sale of cell phones?*
Answer in your own words:

In context, constant C represents the _____ . *Graphically, the horizontal line y = C is an* _____ *for the logistic growth model.*

📖 **Read Section 3.3 Example 4**, in which the authors model the number of fish in a small pond.

[31]*The maximum number of fish that the pond can support is* _____ . *To find the initial fish population, the authors calculate f(*_____*). For practice, use your calculator to find f(11), and compare your answer to that on the table.*

---

[30] An exponential growth model is unbounded, which means it gets very large. In this example, after a year, the exponential growth model predicts that the number of cell phones exceeds the number of people on the planet. In context, C represents the carrying capacity, which is the maximum population the resources can support. Graphically, the horizontal line $y = C$ is an asymptote for the logistic growth model.

[31] The maximum number of fish that the pond can support, or the carrying capacity of the pond, is 100. To find the initial population, the authors calculate $f(0)$.

✎ **Section 3.3 Exercise 31 Limited Fox Population** The fox population on a small island behaves according to the logistic growth model

$$n(t) = \frac{1200}{1 + 11 \cdot (1.7)^{-x}}$$

where t is the number of years since the fox population was first observed.
(a) Find the initial fox population.
(b) Make a table of values of f for t between 0 and 20. From the table, what can you conclude happens to the fox population as t increases?
(c) Use a graphing calculator to draw a graph of the function $n(t)$. From the graph, what can you conclude happens to the fox population as t increases?
(d) What is the carrying capacity? Does this answer agree with the table in part (b) and the graph in part (c)?

Solution

(b)

t (years)	$n(t)$ (foxes)	t (years)	$n(t)$ (foxes)	t (years)	$n(t)$ (foxes)
0		7		14	
1		8		15	
2		9		16	
3		10		17	
4		11		18	
5		12		19	
6		13		20	

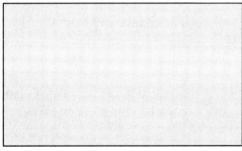

Copy your graph from part (c) here.

CONCEPTS

✓ **3.3 Exercises – Concepts – Fundamentals** Complete the Fundamentals section of the exercises at the end of Section 3.3. Compare your answers to those at the back of the textbook, and make corrections as necessary.

📖 **Read the Concept Check for Section 3.3**, in the Chapter Review at the end of Chapter 3.

3.3 Solutions to ✎ Exercises

✎ Section 3.3 Exercise 9

The following shows how we calculate the entries in the third row of the table.

After 2 hours, $t = 2$, and the population is

$$f(2) = 5000(1.3)^2 = 8450$$

The average rate of change of f from $t = 1$ to $t = 2$ is

$$\frac{f(2) - f(1)}{2 - 1} = \frac{8450 - 6500}{1} = 1950$$

The change in f from $t = 1$ to $t = 2$ as a proportion of the value of f at 1 is

Hours t	Population $f(t)$	Average rate of change	Percentage rate of change
0	5000	–	–
1	6500	1500	30%
2	8450	1950	30%
3	10985	2535	30%
4	14280.5	3295.5	30%

$$r = \frac{f(2) - f(1)}{f(1)} = \frac{1950}{6500} = 0.3$$

So the percentage rate of change is 30%. The remaining rows of the table are calculated in the same way. The average rate of change of f appears to increase with increasing values of t. The percentage rate change is constant for all intervals of length 1.

✎ Section 3.3 Exercises 13 and 17

The following shows how we calculate the entries in the second row of the table for Section 3.3 Exercise 13. The average rate of change in f from $x = 0$ to $x = 1$ is

$$\frac{f(1) - f(0)}{1 - 0} = \frac{57,000 - 30,000}{1} = 27,000$$

The change in f from 0 to 1 as a proportion of the value of f at 0 is

$$r = \frac{f(1) - f(0)}{f(0)} = \frac{27,000}{30,000} = 0.9$$

So the percentage rate of change is 90%. The remaining rows in the tables are calculated in the same way.

13.

x	y	Average rate of change	Percentage rate of change
0	30,000	–	–
1	57,000	27,000	90%
2	108,300	51,300	90%
3	205,770	97,470	90%
4	390,963	185,193	90%

17.

x	y	Average rate of change	Percentage rate of change
0	3000	–	–
1	2440	−560	−18.67%
2	1880	−560	−22.95%
3	1320	−560	−29.79%
4	760	−560	−42.42%

(b) An exponential model is appropriate for the data in Exercise 13, since the percentage rates of change are constant. The initial value C is the y-value 30,000 corresponding to $x = 0$. The percentage rate of change is a positive 90%, so the growth rate r is 0.90, and the growth factor a is 1.9. The model is $y = 30{,}000(1.9)^x$. The graph shown here shows the model for Exercise 13.

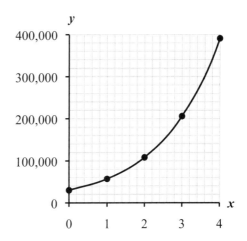

In Exercise 17, the percentage rates of change are not constant, so an exponential model is not appropriate.

✎ Section 3.3 Exercise 19

(a) Since f is linear, increasing t by one unit has the effect of adding m to $f(t)$. In the table, we see that increasing t from 0 to 1 had the effect of adding 150 to $f(0)$, thus $m = 150$. To calculate $f(2)$, we add 150 to $f(1)$. We compute the rest of the values for f in that column similarly.

Since g is exponential, increasing t by one unit has the effect of multiplying $f(t)$ by the growth factor a. From the table, we see that increasing t from 0 to 1 has the effect of multiplying $f(0)$ by $\frac{350}{200} = 1.75$, thus $a = 1.75$. To calculate $g(2)$, we multiply $g(1)$ by 1.75. We compute the rest of the values for g in that column similarly.

(a)

t	$f(t)$
0	200
1	350
2	350 + 150 = 500
3	500 + 150 = 650
4	650 + 150 = 800

t	$g(t)$
0	200
1	350
2	350 × 1.75 = 612.5
3	612.5 × 1.75 = 1071.875
4	1071.875 × 1.75 = 1875.78125

(b) We find in part (a) that the slope m of the linear function f is 150. Since the initial value b is 200, we have the linear model $f(t) = 200 + 150t$.

We find in part (b) that the growth factor a of the exponential function g is 1.75. Since the initial value C is 200, we have the exponential growth model $g(t) = 200(1.75)^t$.

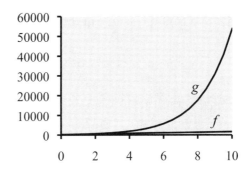

(c) From the graph, we see that the exponential growth function g eventually grows much faster than the linear function f.

✐ Section 3.3 Exercise 29 Salary Comparison

(a) With Offer A, your salary grows by $1000 per month, so after t months, $1000t$ dollars is added to your initial salary. So in the second row of the table under Offer A, the monthly salary is $10,000 + 1000(1) = \$11,000$. We find the average rate of change in the second row of the table by calculating

$$\frac{f_A(1) - f_A(0)}{1 - 0} = \frac{11,000 - 10,000}{1} = 1000$$

With Offer B, your salary doubles every month, so after t months, your initial salary will be multiplied by 2^t. So in the second row of the table under Offer B, the monthly salary is $0.02(2^1) = 0.04$. We find the percentage rate of change in the second row of the table by calculating

$$r = \frac{f_B(1) - f_B(0)}{f_B(0)} = \frac{0.04 - 0.02}{0.02} = 1$$

So the percentage rate of change is 100%. The remaining rows in the table are calculated in the same way.

Offer A			Offer B		
Month	Monthly salary ($)	Average rate of change ($/mo)	Month	Monthly salary ($)	Percentage rate of change
0	10,000	–	0	0.02	–
1	11,000	1000	1	0.04	100%
2	12,000	1000	2	0.08	100%
3	13,000	1000	3	0.16	100%
4	14,000	1000	4	0.32	100%
5	15,000	1000	5	0.64	100%

(b) The function $f_A(t) = 10,000 + 1000t$ is a linear model with initial value 10,000 and rate of change 1000.

(c) The function $f_B(t) = 0.02(2)^t$ is an exponential model with initial value 0.02 and growth factor $a = 2$.

(d) We find the salary in the last month by replacing t by 35 in each model. We have

$$f_A(35) = 10,000 + 1000 \cdot 35 = 45,000$$
$$f_B(35) = 0.02(2)^{35} = 687,194,767.40$$

Thus, with Offer A, you make $45,000 in the last month, and with Offer B, you make $687,194,767.40 in the last month. So Offer B gives the far greatest salary in the last month.

🖉 Section 3.3 Exercise 31 Limited Fox Population

(a) To find the initial fox population, we replace t by 0 in the model:

$$n(0) = \frac{1200}{1+11(1.7)^{-(0)}} = \frac{1200}{1+11\cdot 1} = 100$$

So initially there are 100 foxes.

(b) We use a calculator to find the values of the function $n(t)$ shown in the table below. It appears that the population stabilizes at 1200 foxes.

(b)	t (years)	$n(t)$ (foxes)	t (years)	$n(t)$ (foxes)	t (years)	$n(t)$ (foxes)
	0	100	7	946	14	1192
	1	161	8	1037	15	1195
	2	250	9	1098	16	1197
	3	370	10	1138	17	1198
	4	518	11	1163	18	1199
	5	676	12	1178	19	1199
	6	824	13	1187	20	1200

(c) From the graph, it appears that the population gets closer and closer to 1200 foxes but will never exceed that number. The line $y = 1200$ is a horizontal asymptote of $n(t)$.

(d) Comparing the general form of the logistic equation with the model, we see that the carrying capacity C is 1200. This means that, according to the model, the maximum number of foxes the island can support is 1200, so the population never exceeds this capacity. The table of values and the graph of n confirm this property of the model.

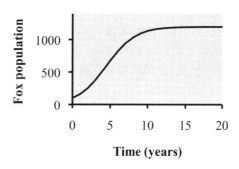

3.4 Graphs of Exponential Functions

OBJECTIVES

☑ *Check the box when you can do the exercises addressing the objective, such as those included in this study guide, which are listed here.*

Exercises

Graphs of Exponential Functions

☐ Create a table of values and sketch the graphs of exponential functions.

5, 6, 35

The Effect of Varying *a* or *C*

☐ Describe changes in the graph of an exponential function due to changes in *a* or C.

15, 17, 19, 33, 35

☐ Connect differences in *a* or *C* in two exponential models to differences in the real world quantities being modeled.

33, 35

Finding an Exponential Function from a Graph

☐ Construct a formula for an exponential function given its graph.

29, 31

GET READY...

This Section involves manipulating exponents. Review this in Algebra Toolkits A.3 and A.4. Also, as in Section 3.2, you solve power equations to find the parameters in an exponential model. Review this in Algebra Toolkit C.1.

SKILLS

Graphs of Exponential Functions

These exercises ask you to graph exponential functions. Find the definition of an exponential function and a description of its graph in blue boxes in Section 3.4 of the textbook. Horizontal asymptote, is a term used to describe a feature of the graph of an exponential function when the input is large. Find horizontal asymptote in italics in Section 3.4 of the textbook.

📖 **Read Section 3.4 Example 1**, in which the authors evaluate and graph the exponential function $f(x) = 2^x$.

[32] *Use properties of exponents (see Algebra Toolkit A.3) to rewrite the following expressions as fractions or whole numbers:* $2^0 = $ _____, $2^{-1} = $ _____, $2^{-2} = $ _____, *and* $2^{-3} = $ _____.

---

[32] We have $2^0 = 1$, $2^{-1} = 1/2$, $2^{-2} = (1/2)^2 = 1/4$, and $2^{-3} = (1/2)^3 = 1/8$.

📖 **Read Section 3.4 Example 2**, in which the authors compare the graphs of $f(x) = 3^x$ and $g(x) = 3^{-x}$.

[33] *The coordinates of the point of intersection of these two graphs are (_____, _____). The function f(x) is in the form f(x) = a^x, where a = _____. The function g(x) can also be written in the form g(x) = a^x, this time with a = ___. The function f is _____ (increasing / decreasing), since the base a is _____ (less than 1 / greater than 1). The function g is _____ (increasing / decreasing), since the base a is _____ (less than 1 / greater than 1).*

✏ **Section 3.4 Exercises 5 and 6 (modified)** Fill in the table and sketch the graphs of the given exponential functions on the same coordinate axis.

Solution

5. $f(x) = 6^x$ 　　6. $g(x) = 6^{-x}$

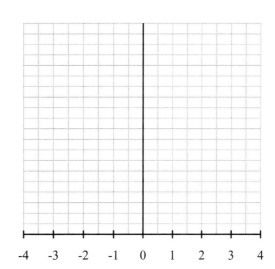

x	$f(x)$
−3	
−2	
−1	
0	
1	
2	
3	

x	$g(x)$
−3	
−2	
−1	
0	
1	
2	
3	

The effect of varying *a* or C

These exercises ask you to graph exponential functions with a variety of values for *a* and *C* and compare the results.

📖 **Read Section 3.4 Example 3**, in which the authors graph exponential functions with a variety of bases.

[34] *In part (a), in which the base a is greater than 1, the function with the largest base is increasing more _____ (slowly / rapidly). In part (b), in which the base a is less than 1, the function with the largest base is decreasing more _____ (slowly / rapidly).*

---

[33] The graphs of *f* and *g* intersect at (0, 1). We have $f(x) = a^x$ where $a = 3$. Also, since $g(x) = 3^{-x} = (1/3)^x$ by the Rules of Exponents (see Algebra Toolkit A.3), we have $g(x) = a^x$, where $a = 1/3$. The function *f* is increasing since the base $a = 3$ is greater than 1. The function *g* is decreasing since the base $a = 1/3$ is less than 1.

[34] In part (a), the function with the largest base, $f(x) = 4^x$, is increasing more rapidly (i.e. has a steeper positive slope). In part (b), the function with the largest base, $g(x) = 0.4^x$, is decreasing more slowly (i.e. has a shallower negative slope).

✎ **Section 3.4 Exercises 15 and 17** Use a graphing calculator to graph the family of exponential functions for the given value of *a*. Explain how changing the value of *a* affects the graph.

Solution

Copy the graphs from your calculator into the spaces provided.

15. $f(x) = a^x$; $a = 4, 6, 8$

17. $f(x) = a^x$; $a = 0.5, 0.7, 0.9$

📖 **Read Section 3.4 Example 5**, in which the authors graph exponential functions with a variety of initial values.

[35]*The initial value of y = 20(3^x) is _____ times bigger than the initial value of y = 10(3^x). This means that the graph of y = 20(3^x) is _____ times higher than the graph of y = 10(3^x) throughout. The graph of the function with the largest value of C lies _____ (above / below) the graphs of the other functions.*

✎ **Section 3.4 Exercise 19** Use a graphing calculator to graph the family of exponential functions $f(x) = C \cdot 5^x$, for $C = 10$, 20, and 100. Explain how changing the value of C affects the graph.

Solution
Copy the graph from your calculator into the space provided.

Finding an Exponential Function from a Graph

These exercises ask you to find formulas for exponential functions given in graphs.

📖 **Read Section 3.4 Example 7**, in which the authors find the formula for an exponential function from its graph.

[36]*Seeing (–1, 1/5) on the graph in part (a) means that f(_____) = _____. How can you find the value of C directly from the graph?*
Answer in your own words:

―――――――――――――――

[35] The initial value $C = 20$ of $y = 20(3^x)$ is 2 times bigger than the initial value $C = 10$ of $y = 10(3^x)$. This means that the graph of $y = 20(3^x)$ is 2 times higher than the graph of $y = 10(3^x)$ throughout. The graph of the function with the largest value of C lies above the graphs of the other functions.
[36] Seeing (–1, 1/5) on the graph means that $f(-1) = 1/5$. The value C is the y-intercept of the graph, since the y-intercept is $f(0)$, and $f(0) = Ca^0 = C$.

✐ **Section 3.4 Exercises 29 and 31** Find the function $f(x) = Ca^x$, whose graph is given.

29.

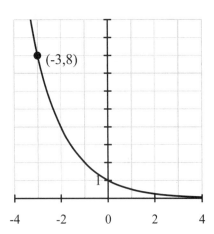

31.

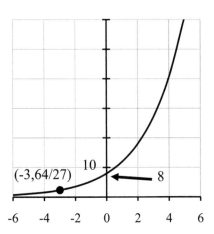

Solution

CONTEXTS

The Effect of Varying *a* of *C*

These exercises demonstrate how changes in the rate (and hence *a*) and the initial value (*C*) affect models of the real world.

📖 **Read Section 3.4 Example 4,** in which the authors analyze the growth rate of the world's population.

[37] *When t > 0, $(1.014)^t$ is _____ (greater than / less than) $(1.010)^t$. So the graph of y = $6.1(1.014)^t$ lies _____ (above / below) the graph of y = $6.1(1.010)^t$. Thus the model with annual growth rate 0.014 predicts _____ (higher / lower) populations than the model with annual growth rate 0.010.*

---

[37] When $t > 0$, $(1.014)^t$ is greater than $(1.010)^t$. So the graph of y = $6.1(1.014)^t$ lies above the graph of $y = 6.1(1.010)^t$. Thus the model with annual growth rate 0.014 predicts higher populations than the model with annual growth rate 0.010.

✎ **Section 3.4 Exercise 35 Investment Value** On January 1, 2005, Marina invested $2000 in each of three mutual funds: a growth fund with an estimated yearly return of 10%, a bond fund with an estimated yearly return of 6%, and a real estate fund with an estimated yearly return of 17%.

(a) Complete the table for the yearly value of each investment at the beginning of the given year.
(b) Find exponential growth models for each of Marina's investments t years since 2005.
(c) Graph the functions found in part (b) for t between 0 and 10. Explain how the expected yearly returns affect the graphs.
(d) Predict the value of each investment in 2011.

Solution
Copy the graphs in part (c) into the space provided.

Year	Growth fund value ($)	Bond fund value ($)	Real estate fund value ($)
2005	*2000*	*2000*	*2000*
2006			
2007			
2008			

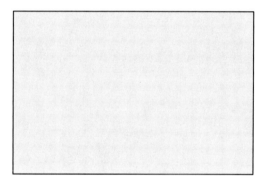

📖 **Read Section 3.4 Example 6**, in which the authors graph the results of radioactive decay.

[38]*Will there ever be the same amount of radioactive material remaining in the two samples? How can you tell from the graph?*
Answer in your own words:

When the rate is negative, the factor a is _____ *(less than 1 / greater than 1), and the graph is* _____ *(increasing / decreasing).*

_____

[38] There will never be the same amount of radioactive material in the two samples. In fact the sample starting with 200 g will always have twice the mass of the sample starting with 100 g. On the graph, we can guess that this is the case because in this viewing rectangle, the graphs do not cross.

✎ **Section 3.4 Exercise 33 Bacteria** To prevent bacterial infections, it is recommended that you wash your hands and cooking utensils as often as possible. At a family barbeque, Jim prepares his meat without using proper sanitary precautions. There are about 100 colony-forming units (CFU/mL) of a certain type of bacteria on the meat. Harold prepares his meat using proper sanitary precautions, so there are only 2 CFU/mL of the bacteria on his meat. It is known that the doubling time for this type of bacteria is 20 minutes.

(a) For each situation, find an exponential model for the number of bacteria on the meat t hours since it was prepared.

(b) Graph the models for t between 0 and 6. What do the graphs tell us about the number of bacteria present on the meat?

(c) For each model, find the predicted number of bacteria after 6 hours.

Solution
In part (b), copy the graph from your calculator into the space provided.

CONCEPTS

✓ **3.4 Exercises – Concepts – Fundamentals** Complete the Fundamentals section of the exercises at the end of Section 3.4. Compare your answers to those at the back of the textbook, and make corrections as necessary.

📖 **Read the Concept Check for Section 3.4**, in the Chapter Review at the end of Chapter 3.

3.4 Solutions to ✐ Exercises

✐ Section 3.4 Exercises 5 and 6 (modified)

We calculate the values of $f(x)$ and plot points to get the graph below. For instance, to calculate $f(-3)$, we use the Rules of Exponents to get

$$f(-3) = 6^{-3} = \left(\frac{1}{6}\right)^3 = \frac{1}{216}$$

The other entries in the tables are calculated similarly.

5.

x	$f(x) = 6^x$
−3	1/216
−2	1/36
−1	1/6
0	1
1	6
2	36
3	216

6.

x	$g(x) = 6^{-x}$
−3	216
−2	36
−1	6
0	1
1	1/6
2	1/36
3	1/216

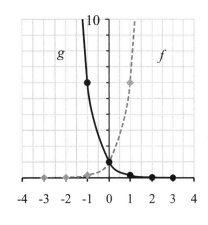

✐ Section 3.4 Exercise 15

We see from the graph that if the base a is greater than 1, then the larger the value of a, the more rapidly the function increases.

The growth is slower when the value of a is closer to 1.

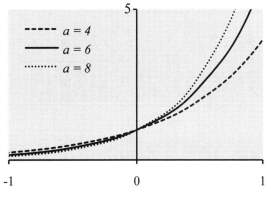

✐ Section 3.4 Exercise 17

We see from the graph that if the base a is less than 1, then the smaller the value of a, the more rapidly the function decreases.

The growth is slower when the value of a is closer to 1.

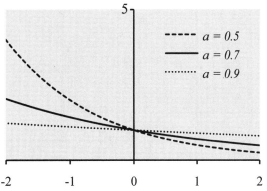

✏ **Section 3.4 Exercise 19**

From the graphs we see that the value C is the y-intercept of the exponential function $f(x) = C \cdot 5^x$. Increasing the value of C has the effect of "stretching" the graph vertically.

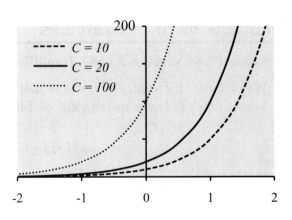

✏ **Section 3.4 Exercise 29**

From the graph, $f(0) = Ca^0 = 1$, so C is 1. Also, since $(-3, 8)$ lies on the graph, $f(-3) = Ca^{-3} = 8$. We solve for a:

$$Ca^{-3} = 8 \qquad\qquad \text{Given}$$

$$a^{-3} = 8 \qquad\qquad \text{Replace } C \text{ by 1 and simplify}$$

$$\left(a^{-3}\right)^{-1/3} = (8)^{-1/3} \qquad\qquad \text{Raise both sides to the } -1/3 \text{ power}$$

$$a = \frac{1}{2} \qquad\qquad \text{Rules of Exponents}$$

So $f(x) = \left(\frac{1}{2}\right)^x$.

✏ **Section 3.4 Exercise 31**

From the graph, $f(0) = Ca^0 = 8$, so C is 8. Also, since $\left(-3, \frac{64}{27}\right)$ lies on the graph,

$f(-3) = Ca^{-3} = \frac{64}{27}$. We solve for a:

$$Ca^{-3} = \frac{64}{27} \qquad\qquad \text{Given}$$

$$8a^{-3} = \frac{64}{27} \qquad\qquad \text{Replace } C \text{ by 8}$$

$$a^{-3} = \frac{8}{27} \qquad\qquad \text{Divide by 8}$$

$$\left(a^{-3}\right)^{-1/3} = \left(\frac{8}{27}\right)^{-1/3} \qquad\qquad \text{Raise both sides to the } -1/3 \text{ power}$$

$$a = \left(\frac{8}{27}\right)^{-1/3} \qquad\qquad \text{Rules of Exponents}$$

Now $\left(\frac{8}{27}\right)^{-1/3} = \left(\frac{27}{8}\right)^{1/3} = \frac{27^{1/3}}{8^{1/3}} = \frac{3}{2}$, so we have $a = \frac{3}{2}$. Thus the model is $f(x) = 8\left(\frac{3}{2}\right)^x$.

✐ Section 3.4 Exercise 33 Bacteria

(a) Since the doubling time is 20 minutes, the 20-minute growth factor is 2. Let a be the one-hour growth factor. Since there are three 20-minute periods in one hour, $a = 2^3 = 8$. So the one-hour growth factor is 8. The models we seek are $n(t) = 100 \cdot 8^t$ and $m(t) = 2 \cdot 8^t$, where t is measured in hours and $n(t)$ and $m(t)$ are measured in CFU/mL.

(b) The graph shows that Harold's 100 CFU/mL of bacteria will quickly grow into a much larger quantity than Jim's 2 CFU/mL of bacteria.

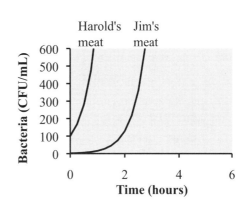

(c) Replace t by 6 in each model. We have $n(6) = 100 \cdot 8^6 = 26,214,400$ and $m(6) = 2 \cdot 8^6 = 524,288$. So the models predicts that after 6 hours, Jim's meat will have 26,214,400 CFU/mL of bacteria and Harold's meat will have 524,288 CFU/mL of bacteria.

✐ Section 3.4 Exercise 35 Investment Value

(a) The growth factor for the growth fund is $1 + 0.1 = 1.1$; the growth factor for the bond fund is $1 + 0.06 = 1.06$; and, the growth factor for the real estate fund is $1 + 0.17 = 1.17$. To find the value of the funds in 2006 for the second row of the table, we multiply 2000 by each of these factors. The other entries in the table are found in the same way.

Year	Growth fund value ($)	Bond fund value ($)	Real estate fund value ($)
2005	*2000*	*2000*	*2000*
2006	*2000(1.1) = 2200*	*2000(1.06) = 2120*	*2000(1.17) =2340*
2007	*2200(1.1) = 2420*	*2120(1.06) =2247.20*	*2340(1.17) =2737.80*
2008	*2420(1.1) = 2662*	*2247.20(1.06) = 2382.03*	*2737.80(1.17) = 3203.23*

(b) The models we seek are $f(t) = 2000(1.1)^t$, $g(t) = 2000(1.06)^t$ and $h(t) = 2000(1.17)^t$, where t is measured in years since 2005 and $f(t)$, $g(t)$ and $h(t)$ are measured in dollars.

(c) The graph shows that a small change in the relative rate of growth will, over time, make a big difference in the value of the investment.

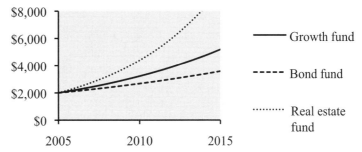

(d) The year 2011 is 6 years after 2005. So we replace t by 6 in each model, we get $f(6) = 2000(1.1)^6 \approx 3543.12$, $g(6) = 2000(1.06)^6 \approx 2837.04$ and $h(6) = 2000(1.17)^6 \approx 5130.33$. So the models predict that in 2011, Marina will have $3543.12 in the growth fund, $2837.04 in the bond fund and $5130.33 in the real estate fund.

3.5 Fitting Exponential Curves to Data

OBJECTIVES

☑ *Check the box when you can do the exercises addressing the objective, such as those included in this study guide, which are listed here.*

Exercises

Finding Exponential Models for Data

	Exercises
☐ Determine if a linear model is appropriate to model data by looking at a scatter plot.	13
☐ Use the exponential regression function on a calculator to find an exponential model for data.	9, 13
☐ Determine qualitatively if a model fits data by looking at a graph.	9, 13
☐ Use real-world data to create a model, and then use the model to make predictions about the real world.	13

Is an Exponential Model Appropriate?

☐ Use average rate of change and percentage rate of change to determine whether a linear or an exponential model is more appropriate.	9

Modeling Logistic Growth

☐ Use the logistic regression function on a calculator to find a logistic model for data.	19
☐ Predict the carrying capacity of a population from a logistic model.	19

GET READY...

This section involves using a graphing calculator to find an exponential model for data. Review Section 2.5 in which you used a graphing calculator to find linear models for data.

SKILLS

Finding Exponential Models for Data

📖 **Read Section 3.5 Example 1**, in which the authors model the world population in the 20th century.

[39]*How do the authors determine whether or not a linear model would be appropriate? Answer in your own words:*

The ExpReg command yields the yearly growth factor _____, *and the initial population* _____.

---

[39] Looking at the graph, the authors note that the data do not lie in a line, so a linear model would not fit well. The ExpReg command yields a yearly growth factor of $b = 1.013718645$ and an initial value of $a = 1444.793012$.

✐ **Section 3.5 Exercise 13 (a, b, c) U.S. Population** The U.S. Constitution requires a census every 10 years. The census data for 1790-2000 are given in the table.

(a) Make a scatter plot of the data. Is a linear model appropriate?

(b) Use a calculator to find an exponential curve $f(x) = b \cdot a^x$ that models the population x years since 1790.

(c) Draw a graph of the function that you found together with the scatter plot. How well does the model fit the data?

Solution
The model you find in part (b) is:

Complete this table to help you add the graph of your model to the scatter plot.

x	model output
0	
50	
100	
150	
200	
250	

Year	x	Population (millions)
1790	0	3.9
1800	10	5.3
1810	20	7.2
1820	30	9.6
1830	40	12.9
1840	50	17.1
1850	60	23.2
1860	70	31.4
1870	80	38.6
1880	90	50.2
1890	100	63.0
1900	110	76.2
1910	120	92.2
1920	130	106.0
1930	140	123.2
1940	150	132.2
1950	160	151.3
1960	170	179.3
1970	180	203.3
1980	190	226.5
1990	200	248.7
2000	210	281.4

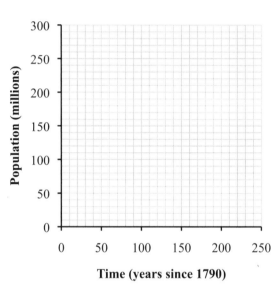

Population (millions) vs. Time (years since 1790)

Is an Exponential Model Appropriate?

📖 **Read Section 3.5 Example 2**, in which the authors determine if a linear model or an exponential model best fits a collection of data.

[40] How do the authors determine whether or not a linear model would be appropriate?
Answer in your own words:

How do the authors determine that an exponential model would not be appropriate?
Answer in your own words:

---

[40] The authors calculate the average rates of change, and since the results are close to constant, a linear model would be appropriate. The authors calculate the percentage rate of change, and find that the percentage rate of change is decreasing, rather than constant, so an exponential model would not be appropriate.

✎ **Section 3.5 Exercise 9** A data set is given, with equally spaced inputs.

(a) Fill in the table to find the average rate of change and the percentage rate of change between successive data points.

(b) Use your results to determine whether a linear model or an exponential model is appropriate, and use a calculator to find the appropriate model.

(c) Make a scatter plot of the data and graph the model you found in part (b). Does your model appear to be appropriate?

Solution

x	$f(x)$	Average rate of change	Percentage rate of change
0	210	–	–
1	281	*71*	*34%*
2	379		
3	512		
4	689		
5	932		

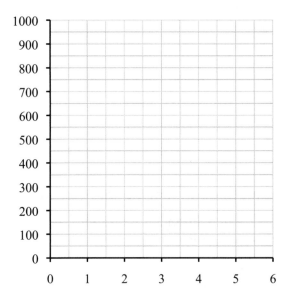

The model you find in part (b) is:

Complete this table to help you add the graph of your model to the scatter plot.

x	model output		x	model output
0			3	
1			4	
2			5	

Modeling Logistic Growth

These exercises ask you to use your calculator to produce logistic models for data. Review the logistic function in Section 3.3 of the textbook.

📖 **Read Section 3.5 Example 3**, in which the authors model the number of fish in a pond.

[41] *Why do the authors decide to use a logistic model?*
Answer in your own words:

From the model, the carrying capacity of the pond is _____ catfish. The model indicates that the initial population is _____ catfish.

✏ **Section 3.5 Exercise 19(a, b) Logistic Population Growth** The table gives the population of black flies in a closed laboratory container over an 18-day period.

(a) Use the Logistic command on your calculator to find a logistic model for these data.

(b) Graph the model you found in part (a) together with a scatter plot of the data.

Time (days)	0	2	4	6	8	10	12	16	18
Number of flies	10	25	66	144	262	374	446	494	498

Solution
The model you find in part (a) is:

Complete this table to help you add the graph of your model to the scatter plot

x	model output		x	model output
0			10	
2			12	
4			14	
6			16	
8			18	

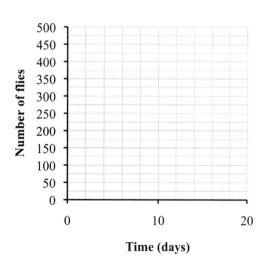

---

[41] The authors note that the catfish population is restricted by its habitat, so a logistic model is appropriate, from the context. The model indicates that the carrying capacity of the pond is approximately 7,925 catfish, and the initial population is approximately $P(0) = 911$ catfish.

CONTEXTS

Finding Exponential Models for Data

These exercises remind you that we find models in order to make predictions.

🖊 **Section 3.5 Exercise 13(d) U.S. Population** *Find this exercise on page 203.* Use your model to predict the population at the 2010 census.

Solution

Modeling Logistic Growth

These exercises remind you that we find models in order to make predictions, and in the case of logistic functions, estimate the carrying capacity of a population.

📖 **Read Section 3.5 Example 3(c) again**, paying particular attention to features of the graph that the authors choose to discus.

Refer back to page 12 (see also pages 163-164)
Hints and Tips 1.2c: Describing Trends in Data

The authors follow Hints and Tips 1.2c in their answer to Section 3.5 Example 3(c), describing how the logistic model predicts the fish population will change with time.

⇨ **Section 3.5 Example 3(c)** What does the model predict about how the fish population will change with time?

Solution

The authors describe the growth of the fish population as time increases, from left to right along the *x*-axis. "From the graph of *P* in Figure 4(b), we see that the catfish population increases rapidly until about 80 weeks. Then growth slows down, and at about 120 weeks the population levels off and remains more or less constant at slightly over 7900."

🖊 **Section 3.5 Exercise 19(c) Logistic Population Growth** *Find this exercise on page 205.* According to the model how does the fly population change with time? At what value will the population level off?

Solution

CONCEPTS

✓ **3.5 Exercises – Concepts – Fundamentals** Complete the Fundamentals section of the exercises at the end of Section 3.5. Compare your answers to those at the back of the textbook, and make corrections as necessary.

📖 **Read the Concept Check for Section 3.5**, in the Chapter Review at the end of Chapter 3.

3.5 Solutions to ✎ Exercises

✎ Section 3.5 Exercise 9

(a) We first note that the inputs x are equally spaced, so we calculate the net change and the percent rate of change between consecutive data points. Here is how the entries in the second row are calculated:

The average rate of change in $f(x)$ from 0 to 1 is

$$\frac{f(1)-f(0)}{1-0} = \frac{281-210}{1-0} = 71$$

The percentage rate of change in $f(x)$ from 0 to 1 is

$$\frac{f(1)-f(0)}{f(0)} = \frac{281-210}{210} = \frac{71}{210} \approx 0.34$$

x	$f(x)$	Average rate of change	Percentage rate of change
0	210	–	–
1	281	**71**	**34%**
2	379	**98**	**35%**
3	512	**133**	**35%**
4	689	**177**	**35%**
5	932	**243**	**35%**

Expressed in percentage form, the percent rate of change from 0 to 1 is 34%. The remaining rows of the table are calculated in the same way.

(b) Since the percentage rate of change is approximately constant (but the average rate of change is increasing), an exponential model is more appropriate. Using a graphing calculator and the ExpReg command, we get the exponential model $P(t) = 209.2(1.3476)^t$.

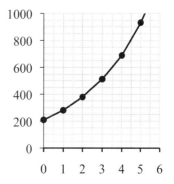

(c) From the graph we see that the model appears to fit the data very well.

✎ Section 3.5 Exercise 13 (a, b, c) U.S. Population

(a) Note from the scatter plot that a linear model is not appropriate.

(b) Using a graphing calculator and the ExpReg command, we get the model $P(t) = 6.051(1.0204)^t$.

(c) From the graph we see that the model appears to fit the data fairly well until $x = 160$ (in year 1950). After that point the model grows much faster than the actual data.

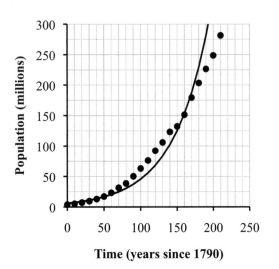

(d) The model predicts that the U.S. population in 2010 (when t is 220) will be

$$P(220) = 6.051(1.0204)^{220} \approx 514.4$$

So the model predicts a population of about 514.4 million people in 2010.

Time (years since 1790)

✎ Section 3.5 Exercise 19 Logistic Population Growth

(a) Since the black fly population is restricted by its habitat (the laboratory container), a logistic model is appropriate. Using the Logistic command on a calculator, we find the following model for the black fly population $P(t)$:

$$P(t) = \frac{501}{1 + 50.5e^{-0.5t}}, \text{ where } e \text{ is a number}$$

approximately equal to 2.718.

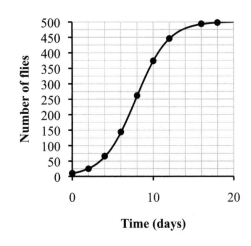

Time (days)

(c) Following **Hints and Tips 1.2c**, we describe the graph as time increases, from left to right along the x-axis. The fly population increases rapidly until about 10 hours. Then the growth slows down, and at about 16 hours the population levels off and remains more or less constant around 500 flies.

4.1 Logarithmic Functions

OBJECTIVES

☑ *Check the box when you can do the exercises addressing the objective, such as those included in this study guide, which are listed here.*

Exercises

Logarithms Base 10

☐ Calculate logarithms base 10 of some numbers without a calculator. **7**

☐ Calculate logarithms base 10 with a calculator. **11**

Logarithms Base *a*

☐ Calculate logarithms base *a* of some numbers without using a calculator. **17**

☐ Convert expressions from exponential form to logarithmic form, and vice versa. **27**

Basic Properties of Logarithms

☐ Recognize the properties of log, and apply them to simplify expressions involving logarithms. **21**

Logarithmic Functions and Their Graphs

☐ Create a table of values and graph a logarithmic function. **41**

☐ Compare the graphs of logarithmic functions for different bases. **51**

GET READY...

This section involves exponential equations the properties of exponents. Refresh your understanding of these in Algebra Toolkits A.3 and A.4.

✓ **Algebra Checkpoint** Test yourself by completing the Algebra Checkpoint at the end of this section of the textbook, and comparing your answers to those in the back of the textbook. Refer to Algebra Toolkits A.3 and A.4 as necessary.

SKILLS

Logarithms Base 10

These exercises ask you to use the relationship between powers of 10 and the logarithms base 10. Find the definition of a logarithm base 10 in a blue box in Section 4.1 of the textbook.

📖 **Read Section 4.1 Example 1,** in which the authors calculate some logarithms without using a calculator.

[1]*Since 10 to the power _____ equals 1000, $\log_{10} 1000 = $ _____. Since 10 to the power _____ equals $\sqrt{10}$, $\log_{10} \sqrt{10} = $ _____.*

---

[1] Since $10^3 = 1000$, $\log_{10} 1000 = 3$. Since $10^{1/2} = \sqrt{10}$, $\log_{10} \sqrt{10} = \frac{1}{2}$.

✎ **Section 4.1 Exercise 7** Find the given common logarithm.

(a) $\log_{10} 10{,}000$

(b) $\log_{10} \dfrac{1}{10}$

(c) $\log_{10} 0.01$

(d) $\log_{10} \sqrt[3]{10}$

📖 **Read the text between Section 4.1 Examples 1 and 2**, in which the authors discuss finding common logarithms of numbers that are not easily written as powers of 10.

[2] *Since 100 < 735 <1000, we have _____ < log_{10} 735 < _____. Similarly, since _____ < 50 < _____, we have _____ < log_{10} 50 < _____. Since 0.01 < 0.056 < 0.1, we have _____ < log_{10} 0.056 < _____.*

📖 **Read Section 4.1 Examples 2 and 3**, in which the authors use a calculator to calculate common logs.

[3] *What is the difference between log x and log_{10} x?*
Answer in your own words:

✎ **Section 4.1 Exercise 11** Find the logarithm of the given numbers.

(a) The diameter of a bacterium: 0.00012 centimeters
(b) The height of a human: 173 centimeters

Solution

Logarithms Base *a*

These exercises ask you to use the relationship between powers of *a* and the logarithms base *a*. Find the definition of a logarithm base *a* in a blue box in Section 4.1 of the textbook. To solve logarithms without a calculator you take advantage of the relationship between logarithms and exponents. Find the definitions of logarithmic form and exponential form in Section 4.1 of the textbook.

---

[2] Since 100 < 735 < 1000, we have 2 < $\log_{10}$735 < 3. Since 10 < 50 < 100, we have 1 < $\log_{10}$ 50 < 2. Since 0.01 < 0.056 <0.1, we have −2 < $\log_{10}$ 0.056 < −1.

[3] The common log can be expressed as $\log_{10} x$ or log x; $\log_{10}x$ and log x mean the same thing.

[4]After the blue box entitled *Logarithms Base 10*, there is a table showing powers of 10 and their logs. After the blue box entitled *Logarithms Base a*, there is a table showing powers of 2 and their logs. Make a similar table for powers of 5.

x	5^{-4}	5^{-3}							
$\log_5 x$	-4								

📖 **Read Section 4.1 Example 4**, in which the authors compute base a logarithms of numbers that can be written as powers of a.

[5]*Since 2 to the power* ___ *equals 32,* $\log_2 32 =$ ___ . *Since 16 to the power* ___ *equals 4,* $\log_{16} 4 =$ ___ .

✎ **Section 4.1 Exercise 17** Find the given logarithm.

(a) $\log_3 3$ (b) $\log_3 1$ (c) $\log_3 3^2$

📖 **Section 4.1 Example 5**, in which the authors give examples of exponential and logarithmic forms.

[6]*In the expression log 100,000 = 5, the base is* ___ *; if it were written in exponential form, the exponent would be* ___ . *In the expression* $\log_2 1/8 = -3$, *the base is* ___ *; if it were written in exponential form, the exponent would be* ___ .

✎ **Section 4.1 Exercise 27** Fill in the table by finding the appropriate logarithmic or exponential form of the equation.

Logarithmic form	Exponential form	Logarithmic form	Exponential form
$\log_8 8 = 1$			$8^3 = 512$
$\log_8 64 = 2$		$\log_8\left(\dfrac{1}{8}\right) = -1$	
	$8^{2/3} = 4$		$8^{-2} = \dfrac{1}{64}$

---

[4]The logarithm base 5 of 5^n is n:

x	5^{-4}	5^{-3}	5^{-2}	5^{-1}	5^0	5^1	5^2	5^3	5^4
$\log_5 x$	-4	-3	-2	-1	0	1	2	3	4

[5] Since 2 to the power 5 equals 32, $\log_2 32 = 5$. Since 16 to the power 1/2 equals 4, $\log_{16} 4 = 1/2$.

[6] In the expression log 100,000 = 5, the base is 10; if it were written in exponential form, the exponent would be 5. In the expression $\log_2 1/8 = -3$, the base is 2; if it were written in exponential form, the exponent would be -3.

Basic Properties of Logarithms

These exercises ask you to use the properties of logarithms to calculate logarithms without a calculator. Find the basic properties of logarithms in a blue box in Section 4.1 of the textbook.

📖 **Read Example 6**, in which the authors identify which properties of logarithms you use to make various calculations.

[7]*In part (d), the authors use property 4, with a =_____ and x =_____. Write the expression in part (d) with a = 33 and x = 4: _____.*

🖉 **Section 4.1 Exercise 21** Find the given logarithm.

(a) $2^{\log_2 3}$

(b) $3^{\log_3 8}$

(c) $4^{\log_4 5}$

Logarithmic Functions and Their Graphs

In these exercises you graph functions that include logarithms. Find the definition of a logarithmic function and a description of its graph in blue boxes in Section 4.1 of the textbook.

📖 **Read Section 4.1 Example 7**, in which the authors graph a logarithmic function.

[8]*How do the authors choose the values of x in the table? Why is that a good choice?*
Answer in your own words:

Why does the table of values only include positive numbers?
Answer in your own words:

---

[7] In part (d), take $a = 5$ and $x = 12$. Property 4, with $a = 33$ and $x = 4$, becomes $33^{\log_{33} 4} = 4$.

[8] The authors choose the x-values to be powers of 2 so that they can easily find the base 2-logarithms. The table only includes positive numbers because the domain of the logarithmic function is $(0, \infty)$, which does not include any negative numbers or zero.

✎ **Section 4.1 Exercise 41** Fill in the table and sketch the graph of the function by plotting points.

Solution

x	$f(x)$
$\dfrac{1}{27}$	
$\dfrac{1}{9}$	
$\dfrac{1}{3}$	
1	
3	
9	

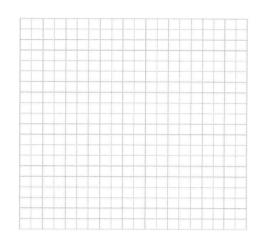

📖 **Read Section 4.1 Example 8**, in which the authors compare the graphs logarithmic functions with a variety of bases.

[9]*Since 4 is larger than 2, the graph of* $\log_4 x$ *increases more* _____ *(slowly / rapidly) than the graph of* $y = \log_2 x$.

_____

[9] The logarithmic function increases more slowly for larger values of the base a; so $\log_4 x$ increases more slowly than $\log_2 x$.

✎ **Section 4.1 Exercise 51** Graph the family of logarithmic functions $f(x) = \log_a x$ for $a = 3, 5, 7$ on the coordinate axis provided. How are the graphs related?

Solution

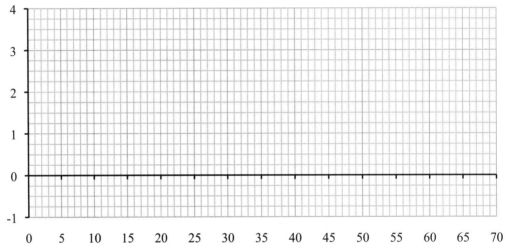

x	$\log_3 x$		x	$\log_5 x$		x	$\log_7 x$
$\dfrac{1}{3}$			$\dfrac{1}{5}$			$\dfrac{1}{7}$	
1			1			1	
3			5			7	
9			25			49	
27			125			343	

CONCEPTS

✓ **4.1 Exercises – Concepts – Fundamentals** Complete the Fundamentals section of the exercises at the end of Section 4.1. Compare your answers to those at the back of the textbook, and make corrections as necessary.

📖 **Read the Concept Check for Section 4.1**, in the Chapter Review at the end of Chapter 4.

4.1 Solutions to ✎ Exercises

✎ **Section 4.1 Exercise 7**

To find the logarithm of a number, it is helpful to express the number as a power of 10.

(a) $\log_{10} 10{,}000 = \log_{10} 10^4 = 4$

(b) $\log_{10}\left(\dfrac{1}{10}\right) = \log_{10} 10^{-1} = -1$

(c) $\log_{10} 0.01 = \log_{10} 10^{-2} = -2$

(d) $\log_{10} \sqrt[3]{10} = \log_{10} 10^{1/3} = \dfrac{1}{3}$

✎ Section 4.1 Exercise 11

We use a calculator to find the logarithms.

(a) Bacterium: $\log 0.00012 \approx -3.9$ (b) Human: $\log 173 \approx 2.2$

✎ Section 4.1 Exercise 17

To find the logarithm of a number, it is helpful to express the number as a power of a the base.

(a) $\log_3 3 = \log_3 3^1 = 1$ (b) $\log_3 1 = \log_3 3^0 = 0$ (c) $\log_3 3^2 = 2$

✎ Section 4.1 Exercise 21

We use the fourth Property of Logarithms (see page 328 of the textbook) to evaluate the expression.

(a) $2^{\log_2 3} = 3$ (b) $3^{\log_3 8} = 8$ (c) $4^{\log_4 5} = 5$

✎ Section 4.1 Exercise 27

Logarithmic form	Exponential form
$\log_8 8 = 1$	$8^1 = 8$
$\log_8 64 = 2$	$8^2 = 64$
$\log_8 4 = \dfrac{2}{3}$	$8^{2/3} = 4$

Logarithmic form	Exponential form
$\log_8 512 = 3$	$8^3 = 512$
$\log_8\left(\dfrac{1}{8}\right) = -1$	$8^{-1} = \dfrac{1}{8}$
$\log_8\left(\dfrac{1}{64}\right) = -2$	$8^{-2} = \dfrac{1}{64}$

✎ Section 4.1 Exercise 41

x	$f(x) = \log_3 x$
$\dfrac{1}{27}$	-3
$\dfrac{1}{9}$	-2
$\dfrac{1}{3}$	-1
1	0
3	1
9	2

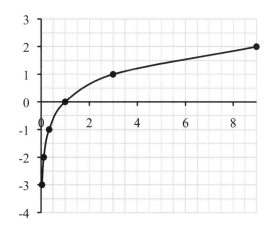

✎ Section 4.1 Exercise 51

We notice from these graphs that the logarithmic function increases more slowly for larger values of the base a.

x	$\log_3 x$
$\frac{1}{3}$	*−1*
1	*0*
3	*1*
9	*2*
27	*3*

x	$\log_5 x$
$\frac{1}{5}$	*−1*
1	*0*
5	*1*
25	*2*
125	*3*

x	$\log_7 x$
$\frac{1}{7}$	*−1*
1	*0*
7	*1*
49	*2*
343	*3*

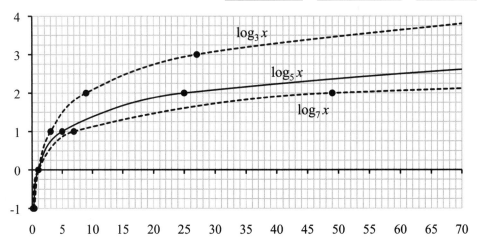

4.2 Laws of Logarithms

OBJECTIVES

☑ *Check the box when you can do the exercises addressing the objective, such as those included in this study guide, which are listed here.*

Exercises

Laws of Logarithms

☐ Evaluate expressions with logarithms by first using the Laws of Logarithms to simplify.

9

Expanding and Combining Logarithmic Expressions

☐ Expand expressions with logarithms of products, quotients and powers, using the Laws of Logarithms.

11, 15

☐ Combine expressions with sums, differences and constant multiplies of logarithms, using the Laws of Logarithms.

19

☐ Compare the logarithms of unknown quantities, given that one is a multiple of the other (for example that one is 10 times bigger than the other).

25, 29

Change of Base Formula

☐ Use the Change of Base Formula to convert logarithms to base 10, so that you can evaluate them on a standard scientific calculator.

33

☐ Use the Change of Base Formula to convert logarithms to base 10, so that you can graph them on a graphing calculator.

39

GET READY...

This Section involves exponential equations and the properties of exponents. Refresh your understanding of these in Algebra Toolkits A.3 and A.4. This section also uses vocabulary pertaining to a graphing calculator, like viewing rectangle, and skills, like graphing two functions on the same screen. Refresh your understanding of these in Algebra Toolkit D.3 Example 2.

SKILLS

Laws of Logarithms

These exercises ask you to apply the Laws of Logarithms to evaluate complicated expressions. Find the Laws of Logarithms in a blue box in Section 4.2 of the textbook.

📖 **Read the introduction to the subsection, 'Laws of Logarithms,'** in which the authors derive relationships between logarithms from relationships between exponents.

[1] *The authors use the property of exponents that says if $A = 10^x$ and $B = 10^y$, then $AB =$ _____ , to demonstrate that $\log_{10}AB =$ _____ .*

📖 **Read Section 4.2 Example 1**, in which the authors use the Laws of Logarithms to simplify expressions with logarithms.

[2] *What do you look for in an expression with logarithms that would tell you that you might be able to use Law 1 to simplify? What about Laws 2 and 3?*
Answer in your own words:

✎ **Section 4.2 Exercise 9** Evaluate the given expressions.

(a) $\log_{10} 4 + \log_{10} 25$ (b) $\log_2 160 - \log_2 5$ (c) $-\frac{1}{2}\log_2 64$

---

[1] The authors use the property of exponents that says if $A = 10^x$ and $B = 10^y$, then $AB = 10^{x+y}$ to demonstrate that $\log_{10}AB = \log_{10} A + \log_{10} B$

[2] To identify that you might use Law 1, look for a product inside a logarithm (like $\log_a AB$), or a sum of logarithms (like $\log_a A + \log_a B$). For Law 2, look for a quotient inside a logarithm (like $\log_{20}(A/B)$, or a difference of logarithms (like $\log_a A - \log_a B$). To identify Law 3, look for an exponent inside a logarithm (like $\log_a A^C$), or a constant multiple of a logarithm (like $C\log_{10}A$).

Expanding and Combining Logarithmic Expressions

These exercises ask you to expand and combine expressions with logarithms of products, quotients and powers.

📖 **Read Section 4.2 Example 2**, in which the authors use the Laws of Logarithms to expand logarithms of quotients and products.

[3]*The logarithm of a product is the _____ (sum / difference / product / quotient) of logarithms. The logarithm of a quotient is the _____ (sum / difference / product / quotient) of logarithms.*

✎ **Section 4.2 Exercises 11 and 15** Use the Laws of Logarithms to expand the given expression.

11. (a) $\log_2 2x$
    (b) $\log_3 \dfrac{x}{y^3}$

15. (a) $\log_2 s^2 \sqrt{t}$
    (b) $\log \dfrac{r^3 s^4}{\sqrt[4]{t}}$

📖 **Read Section 4.2 Example 3**, in which the authors use the Laws of Logarithms to combine sums and differences of logarithms into a single logarithm.

[4]*In part (a), what did the authors have to do before they could combine the logarithms using Law 1? Answer in your own words:*

✎ **Section 4.2 Exercise 19** Use the Laws of Logarithms to combine the given expression.

(a) $\log_2 A + \log_2 B - \log_2 C$
(b) $4\log_6 y - \dfrac{1}{4}\log_6 z$

---

[3] By Law 1, the logarithm of a product is the sum of logarithms. By Law 2, the logarithm of a quotient is a difference of logarithms.

[4] In part (a), the authors could not use Law 1 right away because of the coefficients (3 and 2) multiplying each term. They used Law 3 to change the coefficients into exponents, before combining the logarithms with Law 1.

📖 **Read Section 4.2 Example 4**, in which the authors compare the logarithms of the weights of various animals.

[5]*Since the weight of the beetle is 10 (which is 10 to the power ___) times the weight of the ant, the logarithm of the weight of the beetle is ___ plus the logarithm of the weight of the ant. Since the weight of the mouse is 1/1000 (which is 10 to the power ___) times the weight of the wolf, the logarithm of the weight of the mouse is the logarithm of the weight of the wolf minus ___.*

✐ **Section 4.2 Exercises 25 and 29** Find the difference in the logarithms of the given weights, heights or lengths.

25. The weight of an elephant is 100,000 times as much as the weight of a mouse.

Solution

29. The length of a ladybug is one thousandth of the height of a giraffe.

Solution

Change of Base Formula

In these exercises, you use the change of base formula to convert logarithms to base 10, so that you can calculate them on a standard scientific calculator. Find the change of base formula in a blue box in Section 4.2 of the textbook.

📖 **Read Section 4.2 Example 5**, in which the authors change a logarithm to a logarithm base 10, in order to calculate its value on a scientific calculator.

[6]*To change $\log_b x$ to base a, divide _____ by _____.*

✐ **Section 4.2 Exercise 33** Use the Change of Base Formula and a calculator to evaluate the logarithm, correct to six decimal places.

(a) $\log_2 5$ (b) $\log_5 2$

📖 **Read Section 4.2 Example 6**, in which the authors compare the graphs of logarithms with a variety of bases.

The graphs of $\log_a x$, for a = 2, 3, 5, 10 all intersect at (_____, _____). The graph that is increasing the most steeply has the _____ (highest / lowest) base a.

---

[5] The weight of the beetle is 10^1 times the weight of the ant, so the logarithm of the weight of the beetle equals 1 plus the logarithm of the weight of the ant. The weight of the mouse is 10^{-3} times the weight of the wolf, so the logarithm of the weight of the mouse is the logarithm of the weight of the wolf minus 3.

[6] To change $\log_b x$ to base a, divide $\log_a x$ by $\log_a b$.

✎ **Section 4.2 Exercise 39** Graph the family of logarithmic functions $f(x) = \log_a x$ for $a = 2, 4, 6$, all in the same viewing rectangle on a graphing calculator. How are these graphs related?

Solution
Copy the graphs from your calculator into the space provided.

CONCEPTS

✓ **4.2 Exercises – Concepts – Fundamentals** Complete the Fundamentals section of the exercises at the end of Section 4.2. Compare your answers to those at the back of the textbook, and make corrections as necessary.

📖 **Read the Concept Check for Section 4.2**, in the Chapter Review at the end of Chapter 4.

4.2 Solutions to ✎ Exercises

✎ **Section 4.2 Exercise 9**

To find the logarithm of a number, it is helpful to express the number as a power of the base.

(a) $\log_{10} 4 + \log_{10} 25 = \log_{10}(4 \cdot 25)$ Law 1

$$= \log_{10} 100 = 2 \quad \text{Because } 100 = 10^2$$

(b) $\log_2 160 - \log_2 5 = \log_2\left(\dfrac{160}{5}\right)$ Law 2

$$= \log_2(32) = 5 \quad \text{Because } 32 = 2^5$$

(c) $-\dfrac{1}{2}\log_2 64 = \log_2 64^{-1/2}$ Law 3

$$= \log_2\left(\frac{1}{8}\right) \quad \text{Rules of Exponents}$$

$$= -3 \quad \text{Because } \frac{1}{8} = 2^{-3}$$

✎ Section 4.2 Exercise 11

(a) $$\log_2 2x = \log_2 2 + \log_2 x \qquad \text{Law 1}$$
$$= 1 + \log_2 x \qquad \text{Because } 2^1 = 2$$

(b) $$\log_3 \frac{x}{y^3} = \log_3 x - \log_3 y^3 \qquad \text{Law 2}$$
$$= \log_3 x - 3\log_3 y \qquad \text{Law 3}$$

✎ Section 4.2 Exercise 15

(a) $$\log_2 s^2 \sqrt{t} = \log_2 s^2 + \log_2 \sqrt{t} \qquad \text{Law 1}$$
$$= 2\log_2 s + \log_2 t^{1/2} \qquad \text{Law 3}$$
$$= 2\log_2 s + \frac{1}{2}\log_2 t \qquad \text{Law 3}$$

(b) $$\log \frac{r^3 s^4}{\sqrt[4]{t}} = \log r^3 s^4 - \log \sqrt[4]{t} \qquad \text{Law 2}$$
$$= \log r^3 + \log s^4 - \log t^{1/4} \qquad \text{Law 1}$$
$$= 3\log r + 4\log s - \frac{1}{4}\log t \qquad \text{Law 3}$$

✎ Section 4.2 Exercise 19

(a) $$\log_2 A + \log_2 B - \log_2 C = \log_2 AB - \log_2 C \qquad \text{Law 1}$$
$$= \log_2 \frac{AB}{C} \qquad \text{Law 2}$$

(b) $$4\log_6 y - \frac{1}{4}\log_6 z = \log_6 y^4 - \log_6 z^{1/4} \qquad \text{Law 3}$$
$$= \log_6 \frac{y^4}{z^{1/4}} \qquad \text{Law 2}$$
$$= \log_6 \frac{y^4}{\sqrt[4]{z}} \qquad \text{Replace the radical}$$

✎ Section 4.2 Exercise 25

Let A be the weight of the mouse. Then the weight of the elephant is $100{,}000A$. The difference in the logarithms is

$$\log(100{,}000A) - \log(A) = \log\left(\frac{100{,}000A}{A}\right) = \log(100{,}000) = 5$$

The difference in the logarithms is 5.

✒ Section 4.2 Exercise 29

Let G be the height of the giraffe. Then the height of the ladybug is $G/1000$. The difference in the logarithms is

$$\log(G) - \log(G/1000) = \log(G) - (\log G - \log 1000) \qquad \text{Law 2}$$

$$= \log(100,000) \qquad \text{Simplify}$$

$$= 3 \qquad \text{Because } 1000 = 10^3$$

The difference in the logarithms is 3.

✒ Section 4.2 Exercise 33

(a) We use the Change of Base Formula with $a = 10$, $b = 2$, and $x = 5$.

$$\log_2 5 = \frac{\log_{10} 5}{\log_{10} 2} \qquad \text{Change of Base Formula}$$

$$\approx 2.32193 \qquad \text{Calculator}$$

(b) We use the Change of Base Formula with $a = 10$, $b = 5$, and $x = 2$.

$$\log_5 2 = \frac{\log_{10} 2}{\log_{10} 5} \qquad \text{Change of Base Formula}$$

$$\approx 0.43068 \qquad \text{Calculator}$$

✒ Section 4.2 Exercise 39

To draw a graph of $y = \log_2 x$ using a graphing calculator, we first use the Change of Base Formula:

$$\log_2 x = \frac{\log x}{\log 2} \qquad \text{Change of Base Formula}$$

$$= \left(\frac{1}{\log 2}\right) \log x \qquad \text{Rewrite the fraction}$$

$$\approx (3.32) \log x \qquad \text{Calculator}$$

Similarly, you can check that

$$f(x) = \log_4 x \approx (1.66) \log x \text{, and}$$

$$f(x) = \log_6 x \approx (1.29) \log x$$

Notice on the graph that log functions with increasing values of the base a remain closer to the x-axis. Also notice that regardless of the base a, each graph passes through the point $(1,0)$.

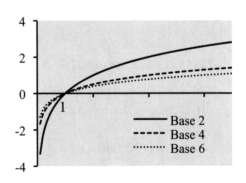

4.3 Logarithmic Scales

OBJECTIVES

☑ *Check the box when you can do the exercises addressing the objective, such as those included in this study guide, which are listed here.*   **Exercises**

Logarithmic Scales

☐ Graph real world quantities on logarithmic scales in order to better compare them.	7

The pH Scale

☐ Calculate the pH given the concentration of hydrogen ions, and calculate the concentration of hydrogen ions given the pH.	**11, 13, 17**
☐ Determine if a substance is considered acidic or basic from its pH.	**11, 13**

The Decibel Scale

☐ Calculate the decibel level given the intensity of a sound.	**15, 21**

The Richter Scale

☐ Calculate the magnitude of an earthquake on the Richter scale given its intensity, and calculate the intensity of an earthquake given its magnitude on the Richter scale.	**23, 25**
☐ Compare earthquakes of different magnitudes on the Richter scale by discussing how many times more intense one earthquake is than another.	**23, 25**

GET READY...

This section involves formulas for measures of intensity. Review how we write and use formulas in Section 1.9 of the textbook. In particular, reread the introduction to Section 1.9 and the subsection entitled, 'Reading and Using Formulas.'

SKILLS

Logarithmic Scales

📖 **Read the introduction to the subsection, 'Logarithmic Scales,'** in which the authors define a logarithmic scale.

[1]*In a linear scale, each successive mark on the scale corresponds to _____ (adding / multiplying by) a fixed constant. In a logarithmic scale, each successive mark on the scale corresponds to _____ (adding / multiplying by) a fixed factor. You would plot the number 1000 on the mark at ___ on the logarithmic scale, since 1000 equals 10 to the power ___.*

---

[1] In a linear scale, each successive mark on the scale corresponds to adding a fixed constant. In a logarithmic scale, each successive mark on the scale corresponds to multiplying by a fixed factor. The number 1000 corresponds to the number 3 on the logarithmic scale, since 1000 equals 10 to the power 3.

📖 **Read Section 4.3 Example 1**, in which the authors plot the weights of animals on a logarithmic scale to better compare their relative sizes.

[2]*What do you calculate to compare the weights of the animals on a logarithmic scale?*
Answer in your own words:

✏ **Section 4.3 Exercise 7** Three quantities are given. Graph the quantities on a logarithmic scale.

Weight of the smallest primate: 1 ounce
Weight of a monkey: 12 pounds
Weight of a gorilla: 350 pounds

Solution

CONTEXTS

The pH Scale

These exercises ask you to discuss the acidity of various solutions using the pH scale. Find the definition of the pH scale highlighted in yellow in Section 4.3 of the textbook.

📖 **Read the introduction to the subsection, 'The pH Scale,'** in which the authors introduce the pH scale used to describe the acidity of a solution.

[3]*Why is it more convenient to express the hydrogen ion concentration of a solution on a logarithmic scale than on a linear scale?*
Answer in your own words:

Acidic substances have a pH _____ (greater than / less than) 7. More acidic substances have a_____ (lower / higher) hydrogen ion concentration and a _____ (lower / higher) pH.

---

[2] To compare values on a logarithmic scale, first find the common logarithms of the values.

[3] The hydrogen ion concentration varies greatly from substance to substance and involves very small numbers; using a log scale avoids very small numbers and negative exponents. Acidic substances have pH less than 7. More acidic substances have a higher hydrogen ion concentration and a lower pH.

📖 **Read Section 4.3 Example 2**, in which the authors find the pH level of blood and acid rain.

[4]*Human blood is _____ (acidic / basic), since the pH is _____ (greater than / less than) 7.*

✏ **Section 4.3 Exercise 11 (extended)** The hydrogen ion concentration of a substance is given. Calculate the pH of the substance, and classify it as an acid or a base.

(a) Lemon juice: $[H^+] = 5.0 \times 10^{-3}$ M   (b) Tomato juice: $[H^+] = 3.2 \times 10^{-4}$ M

Solution

✏ **Section 4.3 Exercise 13 (extended)** The pH reading of a sample of each substance is given. Classify the substance as an acid or a base, and calculate its hydrogen ion concentration.

(a) Soda: pH = 2.6   (b) Milk: pH = 6.5

Solution

---

[4] Human blood is basic, since the pH is greater than 7.

✎ **Section 4.3 Exercise 17 pH of Wine** The pH of wines varies from 2.8 to 3.8. If the pH of a wine is too high, say 4.0 or above, the wine becomes unstable and has a flat taste.

(a) A certain California red wine has a pH of 3.8, and a certain Italian white wine has a pH of 2.8. Find the corresponding hydrogen ion concentration of the two wines.

(b) Which wine has the lower hydrogen ion concentration?

Solution

The Decibel Scale

These questions ask you to calculate the decibel level of various sounds. Find the formula for the decibel level highlighted in yellow in Section 4.3 of the textbook.

📖 **Read the introduction to the subsection, 'The Decibel Scale,'** in which the authors define the decibel scale.

[5]*What does threshold intensity refer to, what notation do we use for it, and what is the value of the threshold intensity for sound? What is the decibel level of the threshold intensity for sound? Answer in your own words:*

The louder the sound the _____ (lower / higher) the decibel level.

📖 **Read Section 4.3 Example 3**, in which the authors calculate the decibel level of a jet taking off.

[6]*Since the sound intensity of the jet taking off is 100 W/m², the authors substitute _____ for I in the formula for the decibel level. If the intensity you are interested in is not an easy power of 10, what would you do to find the decibel level? Answer in your own words:*

---

[5] The threshold intensity I_0 is the intensity at which the stimulus is barely felt. For sound, the threshold intensity is 10^{-12} W/m², which is 0 dB on the decibel scale. The louder the sound the higher the decibel level.

[6] The authors substitute $I = 10^2$. If the intensity were not an easy power of 10, use a calculator to find the logarithm.

✐ **Section 4.3 Exercise 15** The intensity of a sound is given. Calculate the decibel level of the sound.

(a) Power mower: 1.3×10^{-3} W/m^2 (b) Circular saw: 7.9×10^{-3} W/m^2

Solution

✐ **Section 4.3 Exercise 21 Hearing Loss at Work (extended)** The National Campaign for Hearing Health states that hearing loss results from prolonged exposure to noise of 85 dB or higher. People who work with power tools must take precautionary measures, like wearing earmuffs, to protect their hearing.

(a) Mika uses a snow blower to clean yards. The intensity of the sound from the blower is measured at 3.2×10^{-2} W/m^2. Find the intensity level in decibels. According to the National Campaign for Hearing Health, should Mika take precautionary measures?

(b) Mika's friend Jerri uses a hair dryer in her salon. The intensity of the sound from her dryer is measured at 3.1×10^{-4} W/m^2. Find the intensity level in decibels. According to the National Campaign for Hearing Health, should Jerri take precautionary measures?

Solution

The Richter Scale

These exercises ask you to use and manipulate a formula for the magnitude of an earthquake. Find the formula for the magnitude of an earthquake highlighted in yellow in Section 4.3 of the textbook. Also, review the ideas in the subsection, 'Reading and Using Formulas,' in Section 1.9 of the textbook.

📖 **Read the introduction to the subsection, 'The Richter Scale,'** in which the authors define the Richter scale for measuring the magnitude of earthquakes.

[7] *An earthquake of magnitude 6 on the Richter scale is _____ times bigger than an earthquake of magnitude 5, and _____ times bigger than an earthquake of magnitude 4.*

📖 **Read Section 4.3 Example 4**, in which the authors find the magnitude of the Colombian-Ecuador earthquake of 1906.

[8] *In the solution, the authors use the variable I to represent _____. Then 4I in the solution represents _____. In the calculation, the authors use this strategy: they write the magnitude of the Colombia-Ecuador earthquake, which is given by the formula _____, in terms of the magnitude of the San Francisco earthquake, which is given by the formula _____ (and is equal to 8.3).*

✐ **Section 4.3 Exercise 23 Earthquake Magnitude** The 1995 Mexico City earthquake had a magnitude of 8.1 on the Richter scale. The 1976 earthquake in Tangshan, China, was 1.26 times as intense. What was the magnitude of the Tangshan earthquake?

Solution

---

[7] An earthquake of magnitude 6 on the Richter scale is 10 times bigger than an earthquake of magnitude 5, and 100 times bigger than an earthquake of magnitude 4.

[8] The variable I represents the intensity of the San Francisco earthquake; and 4I represents the intensity of the Colombia-Ecuador earthquake. They write the magnitude of the Colombia-Ecuador earthquake, which is given by $M = \log(4I/S)$, in terms of the magnitude of the San Francisco earthquake, which is given by the formula $\log(I/S)$.

📖 **Read Section 4.3 Example 5**, in which the authors compare the intensities of the 2004 Sumatra earthquake to the 2008 earthquake in Sichuan China.

[9] *To find how many times more intense the Sumatra earthquake, with intensity I_1, is than the China earthquake, with intensity I_2, we calculate _____.*

✎ **Section 4.3 Exercise 25 Earthquake Magnitude** The Gansu, China, earthquake of 1920 had a magnitude of 8.6 on the Richter scale. How many times more intense was this than the 1906 San Francisco earthquake, which had magnitude 8.3.

Solution

CONCEPTS

✓ **4.3 Exercises – Concepts – Fundamentals** Complete the Fundamentals section of the exercises at the end of Section 4.3. Compare your answers to those at the back of the textbook, and make corrections as necessary.

📖 **Read the Concept Check for Section 4.3**, in the Chapter Review at the end of Chapter 4

---

[9] To find how many times more intense the Sumatra earthquake, with intensity I_1, is than the China earthquake, with intensity I_2, we calculate I_1/I_2.

4.3 Solutions to ✐ Exercises

✐ Section 4.3 Exercise 7

First, the weight of the smallest primate is given as 1 ounce, which is equivalent to 0.0625 pounds. On a logarithmic scale, these weights are represented by their logarithms. So on a logarithmic scale, the weights are represented by −1.2, 1.1, and 2.5, respectively.

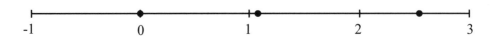

✐ Section 4.3 Exercise 11 (extended)

(a) The definition of pH gives

$$pH = -\log[H^+] \qquad \text{Definition of pH}$$

$$= -\log(5.0 \times 10^{-3}) \qquad \text{Replace } [H^+] \text{ by } 5.0 \times 10^{-3}$$

$$\approx 2.3 \qquad \text{Calculator}$$

So the pH is 2.3. Since this is less than 7, the lemon juice is acidic.

(b) The definition of pH gives

$$pH = -\log[H^+] \qquad \text{Definition of pH}$$

$$= -\log(3.2 \times 10^{-4}) \qquad \text{Replace } [H^+] \text{ by } 3.2 \times 10^{-4}$$

$$\approx 3.5 \qquad \text{Calculator}$$

So the pH is 3.5. Since this is less than 7, the tomato juice is acidic.

✐ Section 4.3 Exercise 13 (extended)

(a) Since the pH of soda is 2.6, which is less than 7, the soda is acidic. We use the definition of pH and express it in exponential form.

$$\log[H^+] = -pH \qquad \text{From the definition of pH}$$

$$[H^+] = 10^{-pH} \qquad \text{Exponential form}$$

$$[H^+] = 10^{-2.6} \qquad \text{Replace pH by 2.6}$$

$$\approx 0.0025 \qquad \text{Calculator}$$

The hydrogen ion concentration of soda is approximately 2.5×10^{-3} moles per liter.

(b) Since the pH of milk is 6.5, which is less than 7, the milk is acidic. We use the definition of pH and express it in exponential form.

$$\log[H^+] = -pH \qquad \text{From the definition of pH}$$

$$[H^+] = 10^{-pH} \qquad \text{Exponential form}$$

$$[H^+] = 10^{-6.5} \qquad \text{Replace pH by 6.5}$$

$$\approx 0.00000032 \qquad \text{Calculator}$$

The hydrogen ion concentration of milk is approximately 3.2×10^{-7} moles per liter.

✐ **Section 4.3 Exercise 15**

From the definition of decibel level we see that

$$B = 10 \log \frac{I}{I_0} \qquad \text{Definition of decibel}$$

(a) $$B = 10 \log \frac{1.3 \times 10^{-3}}{10^{-12}} \qquad \text{Replace } I \text{ by } 1.3 \times 10^{-3} \text{ and } I_0 \text{ by } 10^{-12}$$

$$= 10 \log(1.3 \times 10^{9}) \qquad \text{Rules of Exponents}$$

$$\approx 91 \qquad \text{Calculator}$$

Thus, the decibel level of a power mower is 91 dB.

(b) $$B = 10 \log \frac{7.9 \times 10^{-3}}{10^{-12}} \qquad \text{Replace } I \text{ by } 7.9 \times 10^{-3} \text{ and } I_0 \text{ by } 10^{-12}$$

$$= 10 \log(7.9 \times 10^{9}) \qquad \text{Rules of Exponents}$$

$$\approx 99 \qquad \text{Calculator}$$

Thus, the decibel level of a circular saw is 99 dB.

✐ **Section 4.3 Exercise 17 pH of Wine**

(a) We will use the formula derived in the solution to Exercise 13.
California red wine:

$$[H^+] = 10^{-pH} \qquad \text{Exponential form}$$

$$[H^+] = 10^{-3.8} \qquad \text{Replace pH by 3.8}$$

$$\approx 0.00016 \qquad \text{Calculator}$$

The hydrogen ion concentration of the California red wine is approximately 1.6×10^{-4} moles per liter.

Italian white wine:

$$[H^+] = 10^{-pH} \qquad \text{Exponential form}$$

$$[H^+] = 10^{-2.8} \qquad \text{Replace pH by 2.8}$$

$$\approx 0.0016 \qquad \text{Calculator}$$

The hydrogen ion concentration of the California red wine is approximately 1.6×10^{-3} moles per liter.

(b) The California red wine has the lower hydrogen ion concentration.

Section 4.3 Exercise 21 Hearing Loss at Work (extended)

From the definition of decibel level we see that

$$B = 10 \log \frac{I}{I_0} \qquad \text{Definition of decibel}$$

(a)
$$B = 10 \log \frac{3.2 \times 10^{-2}}{10^{-12}} \qquad \text{Replace } I \text{ by } 3.2 \times 10^{-2} \text{ and } I_0 \text{ by } 10^{-12}$$

$$= 10 \log(3.2 \times 10^{10}) \qquad \text{Rules of Exponents}$$

$$\approx 105 \qquad \text{Calculator}$$

Thus, the decibel level of a snow blower is 105 dB, so Mika should take precautionary measures while cleaning yards.

(b)
$$B = 10 \log \frac{3.1 \times 10^{-4}}{10^{-12}} \qquad \text{Replace } I \text{ by } 3.1 \times 10^{-4} \text{ and } I_0 \text{ by } 10^{-12}$$

$$= 10 \log(3.1 \times 10^{8}) \qquad \text{Rules of Exponents}$$

$$\approx 84.9 \qquad \text{Calculator}$$

Thus, the decibel level of a hair dryer is 84.9 dB. Since this is at the top of the sound levels acceptable to the National Campaign for hearing Health's safe range, Jerri should consider taking precautionary measures, just to be sure to avoid hearing loss.

✐ Section 4.3 Exercise 23 Earthquake Magnitude

If I is the intensity of the Mexico City earthquake, then from the definition of magnitude we have $M = \log \dfrac{I}{S} = 8.1$.

The intensity of the Tangshan earthquake was $1.26I$, so its magnitude is

$$M = \log \frac{1.26I}{S} \qquad \text{Definition of magnitude}$$

$$= \log 1.26 + \log \frac{I}{S} \qquad \text{Laws of logarithms}$$

$$= \log 1.26 + 8.1 \qquad \text{Replace } \log \frac{I}{S} \text{ by 8.1}$$

$$\approx 8.2 \qquad \text{Calculator}$$

An earthquake 1.26 times more intense measures about 0.1 higher on the Richter scale.

✐ Section 4.3 Exercise 25 Earthquake Magnitude

From Section 4.3 Example 4(a), if I is the intensity of an earthquake and M is the magnitude of the earthquake on the Richter scale, then $I = S \cdot 10^M$. If I_1 and I_2 are the intensities of the Gansu and San Francisco earthquakes respectively, then $I_1 = S \cdot 10^{8.6}$, and $I_2 = S \cdot 10^{8.3}$. We find I_1/I_2.

$$\frac{I_1}{I_2} = \frac{S \cdot 10^{8.6}}{S \cdot 10^{8.3}} \qquad \text{Replace } I_1 \text{ by } S \cdot 10^{8.6} \text{ and } I_2 \text{ by } S \cdot 10^{8.3}$$

$$= 10^{8.6 - 8.3} \qquad \text{Rules of Exponents}$$

$$\approx 2 \qquad \text{Calculator}$$

The Gansu earthquake was about 2 times as intense as the San Francisco earthquake.

4.4 The Natural Exponential and Logarithmic Functions

OBJECTIVES

☑ *Check the box when you can do the exercises addressing the objective, such as those included in this study guide, which are listed here.*

	Exercises
What is the Number *e*?	
☐ Evaluate expressions that include *e*, using a calculator.	**9, 11**
The Natural Exponential and Logarithmic Functions	
☐ Graph both decreasing and increasing natural exponential functions.	**11**
☐ Evaluate expressions that include the natural logarithm, using the properties of logarithms and a calculator.	**13**
Continuously Compounded Interest	
☐ Calculate the amount of investments with interest compounded periodically and continuously.	**43**
Instantaneous Rates of Growth or Decay	
☐ Use the instantaneous growth or decay rates to construct an exponential growth or decay model.	**27, 29, 49**
Expressing Exponential Models in Terms of *e*	
☐ Express exponential growth and decay models in terms of the base *e*.	**31, 53**
☐ Find the growth or decay factor from the instantaneous growth or decay rates and vice versa.	**53**

GET READY...

This section involves *e*, a special base for exponential functions. Review the effect of changing the base of an exponential function in Section 3.4. This section also involves continuously compounding interest. Review the subsection of Section 3.2 entitled, 'The Growth of an investment: Compound Interest.'

SKILLS

What is the Number *e*?

These exercises ask you to become familiar with the $\boxed{e}$ button on your calculator.

📖 **Read the subsection, 'What is the Number *e*?' including Example 1**, in which the authors define the value of the symbol *e*.

[1] *The symbol e represents _____ (a number / a variable / a function / an equation).*

_____

[1] The symbol *e* represents a number that is approximately equal to 2.71828182845904523536.

✎ **Section 4.4 Exercise 9** Evaluate the expression using a calculator.

(a) $3e$ (b) $e^{1.4}$ (c) $3e^{-3.1}$ (d) $5 + 2e^2$

Solution

The Natural Exponential and Logarithmic Functions

These exercises ask you to compare the natural exponential and logarithmic functions to exponential and logarithmic functions with other bases. Find the definition of the natural exponential function and natural logarithmic functions in blue boxes in Section 4.4 of the textbook. Also find the basic properties of logarithms written specifically in the case of the natural logarithm in a blue box Section 4.4 of the textbook. Compare these to the blue box in Section 4.1 entitled, 'Basic Properties of Logarithms.'

📖 **Read the introduction to the subsection, 'The Natural Exponential and Logarithmic Functions,' and the text between Section 4.4 Examples 2 and 3**, in which the authors compare the natural exponential and logarithmic functions to exponential and logarithmic functions with other bases.

^{2} *Since* _____ $< e <$ _____ *, we have that* _____ $< e^x <$ _____ *, and* _____ $< \ln x <$ _____ *.*

📖 **Read Section 4.4 Example 2**, in which the authors compare the graphs of two exponential functions with base e.

^{3} *The function e^x is* _____ *(increasing / decreasing). The function e^{-x} is* _____ *(increasing / decreasing).*

✎ **Section 4.4 Exercise 11** Complete the table and sketch the graphs of the functions $f(x) = e^{3x}$ and $g(x) = e^{-3x}$ on the same coordinate axis.

Solution

x	$f(x)$
-2	
-1	
0	
1	
2	

x	$g(x)$
-2	
-1	
0	
1	
2	

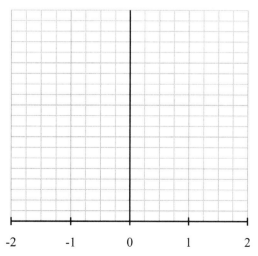

^{2} Since $2 < e < 3$, we have $2^x < e^x < 3^x$ and $\log_3 x < \ln x < \log_2 x$.

^{3} The function e^x is increasing, while the function e^{-x} is decreasing.

📖 **Read Section 4.4 Example 3**, in which the authors use the basic properties of the natural logarithm to evaluate expressions.

[4]*Review Section 4.1 Example 5. Write the expression ln a = b in exponential form:* _____
Write the expression $e^c = d$ in logarithmic form: _____

✎ **Section 4.4 Exercise 13** Evaluate the expression. Use a calculator if necessary.

(a) $\ln e^4$ (b) $\ln(1/e)$ (c) $\ln 2$ (d) $5 + \ln(e-2)$

Expressing Exponential Models in Terms of *e*

These exercises ask you to convert between models written in terms of a growth or decay factor for a given time period and models written in terms of a continuous growth or decay rate. Find the correlation between these two representations in a blue box in Section 4.4 of the textbook.

📖 **Read the introduction to the section, 'Expressing Exponential Models in Terms of *e*,'** in which the authors derive the relationship between growth factors and instantaneous growth rates.

[5]*In the derivation, the authors use two properties of natural logarithms. What are they?*
Answer in your own words:

📖 **Read Section 4.4 Example 7**, in which the authors express an exponential growth model in terms of *e*.

[6]*In this exercise, the growth factor is* _____; *the instantaneous growth rate is* _____. *In general when the growth factor is a, the instantaneous growth rate is* _____.

_____

[4] The expression $\ln a = b$ is $e^b = a$ in exponential form. The expression $e^c = d$ is $\ln d = c$ in logarithmic form.

[5] To write $\ln 2^x = x \ln 2$, the authors used the property, $\log_a A^C = C \log_a A$ (the third Law of Logarithms; see Section 4.2). To write $\ln e^{rx} = rx$, the authors use the basic property of logarithms, $\ln e^x = x$ (the third on the list of Basic Properties of the Natural Logarithm). In exponential form, the expression $r = \ln a$ becomes $e^r = a$.

[6] The growth factor is 1.4, and the instantaneous growth rate is $r = \ln 1.4 \approx 0.34$. In general, when the growth factor is a, the instantaneous growth rate is $r = \ln a$.

🖉 **Section 4.4 Exercise 31** Express the given model in term of the base e. What is the instantaneous growth rate?

(a) $f(t) = 2500 \cdot 2^t$

(b) $g(x) = 2500 \cdot 2^{-x}$

Solution

CONTEXTS

Continuously Compounded Interest

These exercises ask you to model continuously compounded interest. Find its definition in a blue box in Section 4.4 of the textbook. Also find the formula for interest compounded periodically in a blue box in Section 3.2 of the textbook.

📖 Read the introduction to the subsection, 'Continuously Compounded Interest,' in which the authors model investments with interest compounding over smaller and smaller time periods.

[7]*If interest is compounded n times per year, for a very large value of n, the account balance after t years for an investment of P dollars, with interest rate r is approximately equal to _____.*

📖 Read Section 4.4 Example 4, in which the authors compute the balance in an account compounded continuously and compares the results to periodic compounding.

[8]*With an interest rate r, compounding interest more often leads to _____ (more / less) interest per year. Again at a fixed interest rate r, compounding continuously yields _____ (more / less) interest per year than compounding periodically.*

🖉 **Section 4.4 Exercise 43 Compound Interest** If $25,000 is invested at an interest rate of 7% per year, find the amount of the investment at the end of 5 years for the following compounding methods.

(a) Semiannual   (b) Quarterly   (c) Monthly   (d) Continuously

Solution

---

[7] If the interest is compounded n times per year and n is very large, the account balance is approximately equal to $A(t) = Pe^{rt}$, which is the balance if interest is compounded continuously.

[8] With an interest rate r, compounding interest more often leads to more interest per year. With in interest rate r compounding continuously yields more interest per year than compounding periodically.

Instantaneous Rates of Growth or Decay

These exercises ask you to model the real world using exponential functions with base e, governed by an instantaneous growth rate. Find the formula for exponential growth and decay with base e, and the definition of instantaneous growth and decay rates in a blue box in Section 4.4 of the textbook.

📖 **Read Section 4.4 Example 5**, in which the authors model the growth of bacteria.

[9] *CFU/mL stands for* _____

📖 **Read Section 4.4 Example 6**, in which the authors model exponential decay of bacteria in the presence of an antibiotic.

[10] *The decay rate is given as –0.12 per hour. What would be different in this model if the rate were per day instead of per hour?*
Answer in your own words:

✎ **Section 4.4 Exercise 27 and 29** The size of a population is 1000, and it has the given instantaneous growth or decay rate per year.

(a) Find an exponential model $f(x) = Ce^{rx}$ for the population, where t is measured in years.
(b) Sketch the graph of the function found in part (a).

27. 2.4%

Solution
Copy the graph from your calculator into the space provided.

---

[9] CFU/mL stands for colony-forming units per milliliter.

[10] If the decay rate were given per day, then the time t in the model would be measured in days.

29. –0.36

Solution
Copy the graph from your calculator into the space provided.

📖 **Read Section 4.4 Example 6 again,**
paying particular attention to the decay rate.

[11] *The instantaneous decay rate (in percent form) is* _____.

✏ **Section 4.4 Exercise 49 Population of Minneapolis** The population of Minneapolis was 383,000 in 2000 with an instantaneous decay rate r of –0.0029 per year.

(a) Express the instantaneous decay rate in percentage form. Is the population increasing or decreasing?

(b) Find an exponential model $f(t) = Ce^{rt}$ for the population t years after 2000.

(c) Sketch a graph of the function you found in part (b).

Solution
Copy the graph from your calculator into the space provided.

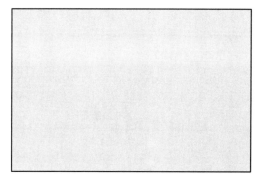

---

[11] The instantaneous decay rate is –0.12, which is 12%.

Expressing Exponential Models in Terms of *e*

These exercises ask you to write models of the real world in terms of the natural exponential function, and find growth or decay factors for models written in terms of the natural exponential function.

📖 **Read Section 4.4 Example 8**, in which the authors model the amount of radioactive substance that remains in a sample as a function of time.

[12]*Since the substance is decreasing at 10% per day, the daily decay rate r =_____ per day, and the daily decay factor is _____. The instantaneous decay rate is _____.*

Why is there only one graph in the answer to part (c)?
Answer in your own words:

✎ **Section 4.4 Exercise 53 Internet Usage in China** Internet World Stats reports that the number of Internet users in China increased by 1024% from 2000 to 2008 (so the growth rate is 10.24 and the growth factor is 11.24 for this time period). The number of Internet users in 2000 was about 22.5 million. Assume that the number of Internet users increases exponentially.

(a) Find the yearly growth factor a, and find an exponential model $f(t) = Ca^t$ for the number of Internet users t years since 2000.

(b) Find the instantaneous growth rate r, and find an exponential model $g(t) = Ce^{rt}$ for the number of users t years since 2000.
(c) Use each of the models you found to predict the number of Internet users in China in 2010. Do your models give the same result? Does this result seem reasonable?

Solution

---

[12] The daily decay rate is $r = -0.10$, and the daily decay factor is $1 + r = 0.9$. The instantaneous decay rate is $r = \ln a = \ln 0.9 \approx 0.105$. The graphs of f and g coincide, since these functions are equal.

CONCEPTS

✓ **4.4 Exercises – Concepts – Fundamentals** Complete the Fundamentals section of the exercises at the end of Section 4.4. Compare your answers to those at the back of the textbook, and make corrections as necessary.

📖 **Read the Concept Check for Section 4.4**, in the Chapter Review at the end of Chapter 4.

4.4 Solutions to ✐ Exercises

✐ **Section 4.4 Exercise 9**

(a) $3e = 8.154845$

(b) $e^{1.4} = 4.055200$

(c) $3e^{-3.1} = 0.135148$

(d) $5 + 2e^2 = 19.778112$

✐ **Section 4.4 Exercise 11**

x	$f(x)$
-2	0.0025
-1	0.0498
0	1
1	20.086
2	403.43

x	$g(x)$
-2	403.43
-1	20.086
0	1
1	0.0498
2	0.0025

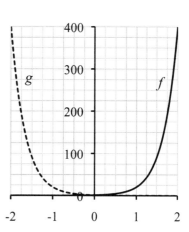

✐ **Section 4.4 Exercise 13**

(a) $\quad \ln e^4 = 4$ Property 3: $\ln e^x = x$

(b) $\quad \ln(1/e) = \ln e^{-1}$ Rules of Exponents

$\qquad\qquad = -1$ Property 3: $\ln e^x = x$

(c) $\quad \ln 2 \approx 0.693$ Calculator

(d) $5 + \ln(e - 2) \approx 4.669$ Calculator

✐ **Section 4.4 Exercise 27**

(a) We use the exponential growth model with $C = 1000$ and $r = 0.024$ to get $f(x) = 1000e^{0.024x}$, where x is measured in years.

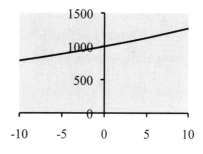

✎ Section 4.4 Exercise 29

(a) We use the exponential growth model with $C = 1000$ and $r = -0.36$ to get $f(x) = 1000e^{-0.36x}$, where x is measured in years.

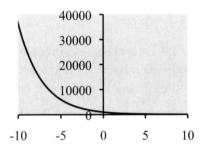

✎ Section 4.4 Exercise 31

(a) In the model $f(t) = 2500 \cdot 2^t$ the growth factor is 2, so the instantaneous growth rate is $r = \ln 2 \approx 0.69$. So the model that we seek is $f(t) = 2500e^{0.69t}$.

(b) Rewrite the model using the Rules for Exponents (see Algebra Toolkit A.3):

$g(x) = 2500 \cdot 2^{-x} = 2500 \cdot \left(\frac{1}{2}\right)^x$, we see the growth factor is $\frac{1}{2}$, so the instantaneous growth rate is $r = \ln\left(\frac{1}{2}\right) \approx -0.69$. The model that we seek is $g(x) = 2500e^{-0.69x}$.

✎ Section 4.4 Exercise 43 Compound Interest

For parts (a) – (c) we use the compound formula from Section 3.1 $A(t) = P\left(1 + \frac{r}{n}\right)^{nt}$ with $P = \$25,000$, $r = 0.07$, and $t = 5$.

(a) Semiannual compounding means that $n = 2$, so we get

$A(t) = 25,000\left(1 + \frac{0.07}{2}\right)^{2(5)} = 35,264.97$. Thus the amount at the end of 5 years for semiannual compounding is $\$35,264.97$.

(b) Quarterly compounding means that $n = 4$, so we get

$A(t) = 25,000\left(1 + \frac{0.07}{4}\right)^{4(5)} = 35,369.45$. Thus the amount at the end of 5 years for semiannual compounding is $\$35,369.45$.

(c) Monthly compounding means that $n = 12$, so we get

$A(t) = 25,000\left(1 + \frac{0.07}{12}\right)^{12(5)} = 35,440.63$. Thus the amount at the end of 5 years for semiannual compounding is $\$35,440.63$.

(d) We use the formula for continuously compounded interest with $P = \$25,000$, $r = 0.07$, and $t = 5$ to get $A(t) = 25,000e^{0.07(5)} = 35,476.69$. Thus the amount at the end of 5 years for continuous compounding is $\$35,476.69$.

✐ Section 4.4 Exercise 49 Population of Minneapolis

(a) The decay rate in percentage form is 0.29%. Since the rate of decay is given, the population is decreasing.

(b) We use the exponential growth model with $C = 383,000$ and $r = -0.0029$ to get $f(t) = 383,000e^{-0.0029t}$, where t is measured in years since 2000.

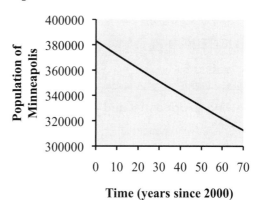

Time (years since 2000)

✐ Section 4.4 Exercise 53 Internet Usage in China

(a) Since the numbers of Internet users increases by 1024% over eight years, the eight-year growth rate is 10.24 and the eight-year growth factor is 11.24. To find the annual growth factor, we use the methods from Section 3.2 (see Section 3.2 Example 3). Since there are eight one-year time periods in eight years, we have $a^8 = 11.24$. Solving, we get $a = \sqrt[8]{11.24} \approx 1.35$. So the annual growth factor is 1.35. The exponential growth model that we seek is $f(t) = 22.5(1.35)^t$, where t is measured in years since 2000, and $f(t)$ is measured in millions of people.

(b) The instantaneous growth rate is $r = \ln a = \ln(1.35) \approx 0.3$. We use the exponential growth model, where C is 22.5 and r is 0.3, to get $g(t) = 22.5e^{0.3t}$, where t is measured in years since 2000, and $g(t)$ is measured in millions of people.

(c) In the year 2010 ten years have passed since 2000. By replacing t with 10 in $f(t)$ we get $f(10) = 22.5 \cdot (1.35)^{10} \approx 452.4$. Thus the first model predicts the number of Internet users in China in 2010 to be approximately 452.4 million people.

By replacing t with 10 in $g(t)$ we get $g(10) = 22.5e^{0.3(10)} \approx 451.9$. Thus the second model predicts the number of Internet users in China in 2010 to be approximately 451.9 million people.

Due to rounding errors the models give slightly different results. Since the population of China is over 1.3 billion, this large number of users does seem reasonable.

4.5 Exponential Equations: Getting Information from a Model

OBJECTIVES

☑ *Check the box when you can do the exercises addressing the objective, such as those included in this study guide, which are listed here.*

Exercises

Solving Exponential and Logarithmic Equations

☐ Solve exponential equations.

7, 15, 21, 49, 55, 57, 65

☐ Solve equations by graphing.

21, 41, 57

☐ Solve logarithmic equations.

29, 35, 37

Getting Information from Exponential Models: Population and Investment

☐ Use the formula for compound interest to construct a model for the amount of an investment as a function of time.

49

☐ Construct an exponential growth or decay model, given the growth or decay rate over a given time period.

55

☐ Use a model that is a function of time to predict the time it takes for the output to reach a given level (or to reach a given multiple of the current level, for example, to double). Here the output is the amount of an investment or a population.

49, 55

Getting Information from Exponential Models: Newton's Law of Cooling

☐ Use Newton's Law of Cooling to construct a model for the temperature of a cooling object as a function of time.

57

☐ Use a model that is a function of time to predict the time it takes for the output to reach a given level (or to reach a given multiple of the current level, for example, to halve). Here the output is the temperature of a cooling object.

57

Finding the Age of Ancient Objects: Radiocarbon Dating

☐ Construct an exponential decay model as a function of time given the half-life of the output.

65

☐ Use carbon-14 dating to determine the age of once living tissue.

65

☐ Use a model that is a function of time to predict the time it takes for the output to reach a given level (or to reach a given multiple of the current level, for example, to halve). Here the output is the amount of radioactive material present in a sample.

65

GET READY...

This section involves solving exponential equations, using the Rules for Exponents (Algebra Toolkit A.3) and Laws of Logarithms (Section 4.2). Prepare by reviewing these topics. This section also uses vocabulary pertaining to a graphing calculator, like *viewing rectangle,* and skills, like graphing two functions on the same screen. Refresh your understanding of these in Algebra Toolkit D.3 Example 2.

✓ **Check Yourself** Use your calculator to verify the following computations:

(a) $\dfrac{\log 7.2}{\log 2} \approx 2.8479969$

(b) $\dfrac{\log 512}{\log(\frac{8}{3})} \approx 6.3602555$

Troubleshooting: When you make calculations on your calculator, do not round until the last step of the calculation. For example, in part (a) type $\boxed{\log}$ $\boxed{7}$ $\boxed{.}$ $\boxed{2}$ $\boxed{\div}$ $\boxed{\log}$ $\boxed{2}$ $\boxed{\text{ENTER}}$ to get the most accurate answer. Note that if you round after each log calculation, in part (a), you get log 7.2 ≈ 0.857, and log 2 ≈ 0.301, and your final answer would be 0.857/0.301 ≈ 2.8471761, which is incorrect by nearly 0.001.

SKILLS

Solving Exponential and Logarithmic Equations

In these exercises, you solve exponential and logarithmic equations.

📖 **Read the introduction to the subsection, 'Solving Exponential and Logarithmic Equations,'** in which the authors solve an exponential equation.

[1] *What is an exponential equation?*
Answer in your own words:

The authors use the property of logarithms that says _____ in the process of solving the equation.

📖 **Read Section 4.5 Example 1**, in which the authors solve an exponential equation.

[2] *What do the authors do before taking the log of each sides of the equation?*
Answer in your own words:

✏ **Section 4.5 Exercise 7** Solve the exponential equation $25 \cdot 3^x = 7$ for x.

Solution

---

[1] An exponential equation is an equation in which the unknown occurs in the exponent. In solving the equation, the authors use the third Law of Logarithms (see Section 4.2) which says $\log_a A^C = C \log_a A$.

[2] Before taking log of each side of the equation, the authors put the exponential term alone on one side of the equation.

📖 **Read Section 4.5 Example 2**, in which the authors solve an exponential equation.

[3]*What are the authors trying to accomplish in the first steps of solving this exponential equation?*
Answer in your own words:

How do the authors check the answer is correct?
Answer in your own words:

✏ **Section 4.5 Exercise 15** Solve the exponential equation $6^x = 21 \cdot 2^x$ for x.

Solution

📖 **Read Section 1.7 Example 3 again**, paying particular to how the authors solve the equality $f(x) = g(x)$ graphically.

[4]*How do you find the solutions of f(x) = g(x) graphically? (I.e. how do you find the values of x at which*
f(x) = g(x) by looking at the graphs?
Answer in your own words:

---

[3] The first step is to manipulate the equation so that there is only one term with the unknown in the exponent and it is alone on one side of the equation. The authors check that the answer $x = 6.36$ is correct by evaluating the right hand side (RHS) and left and side (LHS) of the original equation at 6.36. Since the RHS and LHS both equal 554,000, the answer is correct.

[4] The x-value at which the graphs f and g intersect are solutions to the equation $f(x) = g(x)$.

📖 **Read Section 4.5 Example 3**, in which the authors solve an exponential equation both algebraically and graphically.

[5] *To solve this equation graphically, the authors graph two functions:* _____ *and* _____ .

✎ **Section 4.5 Exercise 21** Solve the exponential equation $2^{1-x} = 3$ (a) algebraically and (b) graphically.

Solution
Copy the graph from your calculator into the space provided.

📖 **Read Section 4.1 Example 5 again**, paying particular attention to the connection between equations with logarithms and equations with exponential terms.

[6] *Write the equation* $\log_a c = b$ *in exponential form.* _____

📖 **Read the paragraph before Section 4.5 Example 4**, in which the authors solve a logarithmic equation.

[7] *The authors convert a logarithmic form* $\log_a c = b$ *to an exponential form* $a^b = c$, *where in this case* $a =$ _____ , $b =$ _____ , *and* $c =$ _____ .

---

[5] The authors graph the functions $y = e^{3-2x}$ and $y = 4$.

[6] The logarithmic form $\log_a c = b$ corresponds to the exponential form $a^b = c$.

[7] The first equation corresponds to $\log_a c = b$, so $a = 2$, $b = 5$, and $c = x + 2$.

📖 **Read Section 4.5 Example 4**, in which the authors solve logarithmic equations.

[8]*In part (a), what do the authors do before rewriting the logarithmic form into exponential form?*
Answer in your own words:

In part (b), what are the authors trying to accomplish in the first steps of solving this logarithmic equation?
Answer in your own words:

In part (b), how do the authors check the answer is correct?
Answer in your own words:

In part (c), what do the authors do to solve $\dfrac{x+1}{x} = 100$ *? See Algebra Toolkit C.4 Examples 4 and 5 for*
examples of solving equations with fractions.
Answer in your own words:

✏ **Section 4.5 Exercise 29, 35 and 37** Solve the logarithmic equation for the unknown x.

29. $3\log x = 12$

Solution

---

[8] In part (a), before rewriting the logarithmic form into exponential form, the authors put the logarithmic term alone on one side of the equation.

In part (b), the first step is to manipulate the equation so that there is only one term with the logarithm of the unknown and it is alone on one side of the equation. The authors check that the answer $x = -0.46$ is correct by evaluating the right hand side (RHS) and left and side (LHS) of the original equation at -0.46. Since the RHS and LHS both equal 1, the answer is correct.

In part (c), the authors multiply both sides of the equation by x to eliminate the denominator. If you had more than one fraction in your equation, you would multiply by the l.c.d. of all of the denominators, as in Algebraic Toolkit C.4 Example 4.

35. $\log(2x+1) + \log 2 = 2$

Solution

37. $\log_2 x - \log_2(x-2) = 2$

Solution

📖 **Read Section 4.5 Example 5**, in which the authors solve a logarithmic equation graphically.

[9] *Instead of graphing the function on each side of the equation, the authors rearrange the equation, so that the function an the left hand side is _____, and the function on the right hand side is _____. How do the authors see the graph of the equation y = 0 in Figure 2? Answer in your own words:*

🖉 **Section 4.5 Exercise 41** Use a graphing calculator to find all solutions of the equation $\ln x = 3 - x$.

Solution

Copy the graph from your calculator into the space provided.

CONTEXTS

Getting Information from Exponential Models: Population and Investment

In these exercises you find the time it takes for investments and populations to reach particular levels. Find the formula for compound interest in Section 3.2 of the textbook.

📖 **Read Section 4.5 Example 6**, in which the authors model the balance of a high-yield uninsured certificate of deposit.

[10] *In part (a), how did the authors know to use n = 2? Answer in your own words:*

In part (b), the authors substituted 3 in for _____ (A / t). In part (c), the authors substituted 25,000 in for _____ (A / t).

---

[9] The function on the left hand side is $y = x^2 - 2\ln(x + 2)$ and the function on the right hand side is $y = 0$. The graph of $y = 0$ is the x-axis. So the solutions of this equation are the x-coordinates of the intersection of $y = x^2 - 2\ln(x + 2)$ with the x-axis.

[10] The variable n in the formula for the balance in an account with compound interest is the number of times the interest is compounded each year. In this example, the interest is compounded every 6 months, or twice per year, so $n = 2$. In part (b), 3 is a time in years, so the authors substitute 3 for t. In part (c), 25,000 is an account balance in dollars, so the authors substitute 25,000 for A.

✏ **Section 4.5 Exercise 49 Compound Interest** Aviel invests $6000 in a high-yield uninsured certificate of deposit that pays 4.5% interest per year, compounded quarterly.

(a) Find a formula for the amount A of the certificate after t years.
(b) What is the amount after 3 years?
(c) How long will it take for his investment to grow to $8000?

Solution

📖 **Read Section 4.5 Example 7**, in which the authors estimate the year in which the world's population will double.

[11]*In part (a), how do the authors find the growth factor a for the model?*
Answer in your own words:

In part (b), since 2015 – 2007 = 8, the authors substitute 8 in for ____ (P / t). In part (c), the authors substitute 13.2 (double the population in 2007) for ____ (P / t).

---

[11] In part (a), the growth rate 1.36% corresponds the $r = 0.0136$, so the growth factor is $a = 1 + r = 1.0136$. In part (b), the number 8 is a time since 2007 in years, so the authors substitute 8 for t. In part (c), the number 13.2 is a population in billions of people, so the authors substitute 13.2 for P.

✐ **Section 4.5 Exercise 55 Population of Ethiopia** In 2003 the United Nations estimated that the population of Ethiopia was about 70.7 million, with an annual growth rate of 2.9%. Assume that this rate of growth continues.

(a) Find the yearly growth factor a.

(b) Find an exponential growth model $f(t) = Ca^t$ for the population t years since 2003.

(c) How long will it take for the population to double?

(d) Use the model found in part (b) to predict the year in which the population will reach 90 million.

Solution

Getting Information from Exponential Models: Newton's Law of Cooling

These exercises ask you to apply Newton's Law of Cooling to describe how temperature changes as hot objects cool. Find its formula in Section 4.5 of the textbook.

📖 **Read the introduction to the subsection, 'Getting Information from Exponential Models: Newton's Law of Cooling,'** in which the authors discuss the rate at which hot objects cool.

[12]*What do each of the variables in Newton's law of cooling represent?*

T represents _____ A represents _____

I represents _____ k represents _____

t represents _____

---

[12] The variable T represents the temperature of the cooling object at time t, A represents the temperature of the surroundings, I represents the initial temperature of the object (at time 0), k represents the heat transfer coefficient, and t represents the time during which the object has been cooling.

📖 **Read Section 4.5 Example 8**, in which the authors use Newton's Law of Cooling to determine when a bathtub was filled, as part of a crime scene investigation.

[13] *The authors use Newton's Law of Cooling to construct a model with input variable _____, representing _____, and output variable _____, representing _____.*

✐ **Section 4.5 Exercise 57 Pot of Chili** Angela prepares a large pot of chili the night before a church potluck. The temperature of the chili is 212°F, and it must cool down to 70°F before it can be stored in the refrigerator. Assume that the ambient temperature is 65°F and the heat transfer coefficient is $k = 2.895$.

(a) Find a model for the temperature T of the pot of chili t hours after cooling.
(b) How long will it take for the pot of chili to cool down to the desired temperature of 70°F?
(c) Graph the function T to confirm your answers to part (b).

Solution
For part (c), copy the graph from your calculator into the space provided.

---

[13] The input of the model is t, which represents the time since the bathtub was filled. The output of the model is T, which is the temperature of the bath water at that time.

Finding the Age of Ancient Objects: Radiocarbon Dating

In these exercises you use carbon-14 measurements to determine the age of archeological finds.

📖 **Read the introduction of the subsection, 'Finding the Age of Ancient Objects: Radiocarbon Dating,'** in which the authors tell how hikers in 2009 discovered the frozen body of a man who lived in the Late Stone (Neolithic) Age.

[14]*How does radiocarbon dating work?*
Answer in your own words:

📖 **Read Section 3.1 Example 5**, paying particular attention to how the authors use the half-life of radium-226 to construct the model.

[15]*If a is the 10-year decay factor for the amount of radium-226 in a sample, and the sample has 50 grams of radium-226 today, then in 10 years, the sample will have _____ grams of radium-226, and in 20 years, the sample will have _____ grams of radium-226. In the model in Example 5, a =1/2 is the _____-year decay factor, since the half-life of radium-226 is _____ years.*

📖 **Read Section 4.5 Example 9**, in which the authors calculate the age of the iceman.

[16]*The input of the model is _____. The output of the model is _____.
The factor a = 1/2 is the _____-year decay factor, since the half-life of carbon-14 is _____ years. The scientists do not know the initial carbon-14 content of the tissue sample, so the authors called the initial value _____ grams. The scientists find that the tissue sample from the iceman currently contain _____% of the initial carbon-14, so the current mass of carbon-14 in the tissue sample is _____ grams.*

---

[14] A living thing contains the same ratio of carbon-12 to radioactive carbon-14 as that found in the atmosphere. After it dies, the carbon-14 decays exponentially, altering this ratio. By measuring the ratio of the carbon-12 to the carbon-14 present in a once living creature, scientists determine the length of time that the carbon-14 has been decaying, which is the length of time since the organism's death.

[15] Since a is the 10-year decay factor, in 10 years, the sample will have $50a$ grams of radium-226. In 20 years, the sample will have $50a^2$ grams of radium-226. In the model in Example 5, $a = 1/2$ is the 1600-year decay factor, since the half-life of radium-226 is 1600 years.

[16] The input is time in years and the output is the mass of carbon-14 in grams. The authors call the initial mass of carbon-14 C grams. The tissue sample currently contains 57.67% of the initial carbon-14. The current mass of carbon-14 in the tissue sample is $0.5767C$ grams.

✏ **Section 4.5 Exercise 65 Carbon-14 Dating** Archeologists find an ancient shard of pottery and use some burnt olive pits found in the same layer of the site to determine the age of the shard. The archeologists determine that the olive pits contain 69.32% of the carbon-14 that is present in a living olive. (The half-life of carbon-14 is 5730 years). How old is the shard of pottery?

Solution

CONCEPTS

✓ **4.5 Exercises – Concepts – Fundamentals** Complete the Fundamentals section of the exercises at the end of Section 4.5. Compare your answers to those at the back of the textbook, and make corrections as necessary.

📖 **Read the Concept Check for Section 4.5**, in the Chapter Review at the end of Chapter 4.

4.5 Solutions to ✏ Exercises

✏ **Section 4.5 Exercise 7**

We first put the exponential term alone on one side of the equation.

$25 \cdot 3^x = 7$	Given equation
$3^x = 0.28$	Divide by 25
$\log 3^x = \log 0.28$	Take the log of each side
$x \log 3 = \log 0.28$	Law 3: "Bring down the exponent"
$x = \dfrac{\log 0.28}{\log 3}$	Divide each side by $\log 3$
$x \approx -1.16$	Calculator

✐ Section 4.5 Exercise 15

To put the exponential term alone on one side of the equation, we divide each side by 2^x.

$$6^x = 21 \cdot 2^x \qquad\qquad \text{Given equation}$$

$$\frac{6^x}{2^x} = 21 \qquad\qquad \text{Divide each side by } 2^x$$

$$\left(\frac{6}{2}\right)^x = 21 \qquad\qquad \text{Rules of exponents}$$

$$\log 3^x = \log 21 \qquad\qquad \text{Simplify and take the log of each side}$$

$$x \log 3 = \log 21 \qquad\qquad \text{Law 3: "Bring down the exponent"}$$

$$x = \frac{\log 21}{\log 3} \qquad\qquad \text{Divide each side by } \log 3$$

$$x \approx 2.77 \qquad\qquad \text{Calculator}$$

✐ Section 4.5 Exercise 21

(a) Since the exponential term is alone on one side, we begin by taking the logarithm of each side.

$$2^{1-x} = 3 \qquad\qquad \text{Given equation}$$

$$\log 2^{1-x} = \log 3 \qquad\qquad \text{Take the log of each side}$$

$$(1-x)\log 2 = \log 3 \qquad\qquad \text{Law 3: "Bring down the exponent"}$$

$$1 - x = \frac{\log 3}{\log 2} \qquad\qquad \text{Divide each side by } \log 2$$

$$x = 1 - \frac{\log 3}{\log 2} \qquad\qquad \text{Solve for } x$$

$$x \approx -0.585 \qquad\qquad \text{Calculator}$$

(b) We graph the equations $y = 2^{1-x}$ and $y = 3$ in the same viewing rectangle. The solution occurs where the graphs intersect. Zooming in on the point of intersection of the two graphs with a calculator, we see that $x \approx -0.585$

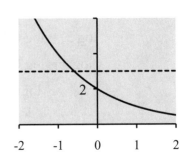

✎ Section 4.5 Exercise 29

We first isolate the logarithm term.

$3\log x = 12$ Given equation

$\log x = 4$ Divide each side by 3

$x = 10^4$ Exponential form

$x = 10,000$ Simplify

✎ Section 4.5 Exercise 35

We first combine the logarithm terms using the Laws of Logarithms.

$\log(2x+1)+\log 2 = 2$ Given equation

$\log 2(2x+1) = 2$ Laws of Logarithms

$2(2x+1) = 10^2$ Exponential form

$x = 24.5$ Solve for x

✎ Section 4.5 Exercise 37

We first combine the logarithm terms using the Laws of Logarithms.

$\log_2 x - \log_2(x-2) = 2$ Given equation

$\log_2 \dfrac{x}{x-2} = 2$ Laws of Logarithms

$\dfrac{x}{x-2} = 2^2$ Exponential form

$x = 4(x-2)$ Multiply each side by $(x-2)$

$x \approx 2.67$ Solve for x

✎ Section 4.5 Exercise 41

We first move all terms to one side of the equation: $\ln x - (3-x) = 0$.

Then we graph $y = \ln x - 3 + x$. The solutions are the x-intercepts of the graph. Zooming in on the x-intercepts, we see that there is one solution: $x \approx 2.20794$.

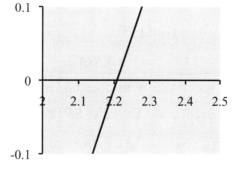

✎ **Section 4.5 Exercise 49 Compound Interest**

(a) We use the formula for compound interest from Section 3.1, where the principal P is 6000, the interest rate r is 0.045, and the number of compounding periods n is 4:

$$A(t) = P\left(1 + \frac{r}{n}\right)^{nt} \qquad \text{Compound interest formula}$$

$$A(t) = 6000\left(1 + \frac{0.045}{4}\right)^{4t} \qquad \text{Substitute the given values}$$

$$A(t) = 6000(1.01125)^{4t} \qquad \text{Simplify}$$

(b) Using the formula we found in part (a) and replacing t with 3, we get

$$A(3) = 6000(1.01125)^{4(3)} \qquad \text{Replace } t \text{ by 3}$$

$$\approx 6862.05 \qquad \text{Calculator}$$

So in 3 years the certificate is worth $6862.05.

(c) We need to find the time t it takes for the amount A to reach $8000. We use the formula from part (a):

$$A(t) = 6000(1.01125)^{4t} \qquad \text{Formula}$$

$$8000 = 6000(1.01125)^{4t} \qquad \text{Replace } A(t) \text{ by 8000}$$

$$\frac{4}{3} = (1.01125)^{4t} \qquad \text{Divide each side by 6000}$$

$$\log\frac{4}{3} = \log(1.01125)^{4t} \qquad \text{Take the logarithm of each side}$$

$$\log\frac{4}{3} = 4t\log(1.01125) \qquad \text{Law 3: "Bring down the exponent"}$$

$$\frac{\log(4/3)}{4\log(1.01125)} = t \qquad \text{Divide each side by } 4\log(1.01125)$$

$$t \approx 6.429 \qquad \text{Calculator}$$

It will take almost six and a half years for Aviel's investment to grow to $8000.

✎ **Section 4.5 Exercise 55 Population of Ethiopia**

(a) We use the exponential growth model $P(t) = Ca^t$ where C is the initial population and a is the growth factor. Since the population increases by 2.9% every year, the growth factor is $a = 1 + 0.029 = 1.029$.

(b) With the growth factor $a = 1.029$ and the initial population $C = 70.7$ the model we seek is $P(t) = 70.7(1.029)^t$, where t is the number of years after 2003, and $P(t)$ is measured in millions of people.

(c) We want to know when the population will double to $2(70.7) = 141.4$ million. In other words, we want to find the value of t for which $P(t)$ is 141.4 million. Substituting these values into the formula gives the following:

$P(t) = 70.7(1.029)^t$	Exponential growth model
$141.4 = 70.7(1.029)^t$	Replace $P(t)$ by 141.4
$2 = (1.029)^t$	Divide by 70.7
$\log 2 = \log(1.029)^t$	Take the logarithm of each side
$\log 2 = t \log(1.029)$	Law 3: "Bring down the exponent"
$\dfrac{\log 2}{\log(1.029)} = t$	Divide each side by $\log(1.029)$
$t \approx 24.2$	Calculator

So it will take about 24 years for the population of Ethiopia to double. This would happen in the year $2003 + 24 = 2027$.

(d) We want to know when the population will reach 90 million. In other words, we want to find the value of t for which $P(t)$ is 90 million. Substituting these values into the formula gives the following:

$P(t) = 70.7(1.029)^t$	Exponential growth model
$90 = 70.7(1.029)^t$	Replace $P(t)$ by 90
$\dfrac{90}{70.7} = (1.029)^t$	Divide each side by 70.7
$\log \dfrac{90}{70.7} = \log(1.029)^t$	Take the logarithm of each side
$\log \dfrac{90}{70.7} = t \log(1.029)$	Law 3: "Bring down the exponent"
$\dfrac{\log(90 / 70.7)}{\log(1.029)} = t$	Divide each side by $\log(1.029)$
$t \approx 8.4$	Calculator

So it will take about 8 years for the population of Ethiopia to reach 90 million. This would happen in the year $2003 + 8 = 2011$.

✎ Section 4.5 Exercise 57 Pot of Chili

(a) We use Newton's Law of Cooling, where the initial temperature I is 212°F, the ambient temperature A is 65°F, and the heat transfer constant k is 2.895.

$$T = A + (I - A)e^{-kt} \qquad\qquad \text{Newton's Law of Cooling}$$

$$T = 65 + (212 - 65)e^{-2.895t} \qquad \text{Replace } A \text{ by 65, } I \text{ by 212, and } k \text{ by 2.895}$$

$$T = 65 + 147e^{-2.895t} \qquad\qquad \text{Simplify}$$

(b) We want to know when the chili will reach 70°F, so we use the model from part (a), replace T by 70, and solve the resulting exponential equation for t.

$$T = 65 + 147e^{-2.895t} \qquad\qquad \text{Model}$$

$$70 = 65 + 147e^{-2.895t} \qquad\qquad \text{Replace } T \text{ by 70}$$

$$5 = 147e^{-2.895t} \qquad\qquad \text{Subtract 65 from each side}$$

$$\frac{5}{147} = e^{-2.895t} \qquad\qquad \text{Divide each side by 147}$$

$$\ln\frac{5}{147} = -2.895t \qquad\qquad \text{Take the natural logarithm of each side}$$

$$t = \frac{\ln(5/147)}{-2.895} \qquad\qquad \text{Divide each side by } -2.895$$

$$t \approx 1.17 \qquad\qquad \text{Calculator}$$

So the chili will cool to 70°F in approximately one hour and ten minutes.

(c) We see that the graph of the model (the solid line) crosses the horizontal line $y = 70$ (the dashed line) at a bit over one hour, as we found in part (b).

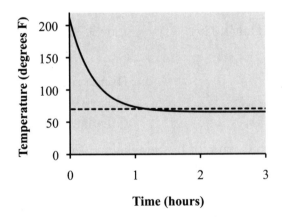

✐ Section 4.5 Exercise 65 Carbon-14 Dating

The formula for exponential decay is $m(t) = C\left(\dfrac{1}{2}\right)^{t/h}$ where C is the initial mass of carbon-14 in the olive pits, h is the half-life, and $m(t)$ is the mass of carbon-14 remaining in the artifact at time t. We are looking for the time t at which $m(t)$ is 69.32% of the initial mass of carbon-14 in the olive pits, that is, $m(t)$ is $0.6932C$. We now substitute the known values into the formula and solve for t.

$$m(t) = C\left(\frac{1}{2}\right)^{t/h} \qquad \text{Radioactive decay formula}$$

$$0.6932C = C\left(\frac{1}{2}\right)^{t/5730} \qquad \text{Replace } m(t) \text{ by } 0.6932C \text{ and } h \text{ by } 5730$$

$$0.6932 = \left(\frac{1}{2}\right)^{t/5730} \qquad \text{Divide by } C$$

$$\log 0.6932 = \log\left(\frac{1}{2}\right)^{t/5730} \qquad \text{Take the logarithm of each side}$$

$$\log 0.6932 = \frac{t}{5730}\log\left(\frac{1}{2}\right) \qquad \text{Law 3: ``Bring down the exponent''}$$

$$\frac{5730\log 0.6932}{\log\left(\frac{1}{2}\right)} = t \qquad \text{Multiply by 5730 and divide by } \log\left(\frac{1}{2}\right)$$

$$t \approx 3029 \qquad \text{Calculator}$$

So the olive pits and the shard of pottery are approximately 3029 years old.

4.6 Working with Functions: Composition and Inverse

OBJECTIVES

☑ *Check the box when you can do the exercises addressing the objective, such as those included in this study guide, which are listed here.*

Exercises

Functions of Functions

☐ Translate a formula for a composition of functions into a rule.	**15, 17**
☐ Write a formula for the composition of two functions.	**15, 17**
☐ Evaluate a composition of functions at a given value.	**15, 17**
☐ Use a composition of functions to model the real world.	**77**

Reversing the Rule of a Function

☐ Find the values of the inverse of a function from the function's graph.	**35**
☐ Find the formula for the inverse of a function.	**57, 61, 63, 83**
☐ Show that two given functions are inverses.	**37, 41**
☐ Use inverse functions to model the real world.	**83**

Which Functions Have Inverses?

☐ Determine if a given function is one-to-one (i.e. if it has an inverse), using its formula.	**53, 55**
☐ Determine if a given function is one-to-one (i.e. if it has an inverse), using its graph.	**71**

Exponential and Logarithmic Functions as Inverse Functions

☐ Connect exponential and logarithmic functions by calculating inverses.	**41**

GET READY...

This section involves composition of functions and inverses of functions. Review the definition of a function in Section 1.4 and function notation, evaluating a function, and finding the domain and range of a function in Section 1.5 of the textbook.

SKILLS

Functions of Functions

These exercises ask you to compose functions, which means use the output of one function as the input of another function. Find the definition of the composition of functions in a blue box in Section 4.6 of the textbook.

📖 **Read the introduction to the subsection 'Functions of Functions,'** in which the authors discuss how the tax you pay depends on how long you work.

[1]
After you work 8 hours, you earn _____ dollars, and pay _____ dollars in taxes.

---

[1] After you work 2 hours, you earn $16, and pay $0.80 in taxes.

📖 **Read Section 4.6 Example 1**, in which the authors find a formula for the tax you pay as a function of how long you work.

[2] *Confirm that after 2 hours of work, you pay $0.80 in taxes, using the formula for L. In function notation, we have, L(___) = ____.*

📖 **Read Section 4.6 Example 2**, in which the authors find a formula for the composition of two functions.

[3] *Review Section 1.5 Examples 1, 2 and 3. What rules do M(x) and N(x) represent? Answer in your own words:*

✏ **Section 4.6 Exercises 15 and 17** Two functions f and g are given.
   (a) Give a verbal description of the functions $M(x) = f(g(x))$ and $N(x) = g(f(x))$.
   (b) Find algebraic expressions for the functions $M(x)$ and $N(x)$.
   (c) Evaluate $M(3)$ and $N(-2)$.

   15. $f(x) = x + 2$, $g(x) = 3x$

   Solution

   17. $f(x) = 2x^2$, $g(x) = x - 1$

   Solution

---

[2] With the formula, we see $L(2) = 0.4(2) = 0.8$, so the formula agrees with the table of values in Table 1.

[3] $M(x)$ represents the rule, "Square the input, and then add 1." $N(x)$ represents the rule, "Add one to the input, and then square the result."

Reversing the Rule of a Function

In these exercises you discuss the inverse of a function, and find formulas for inverses by solving equations. Find the definition of the inverse of a function in a blue box and instructions for finding the formula for the inverse of a function in a red box in Section 4.6 the textbook.

📖 **Read Section 1.4 Example 9 again**, paying particular attention to the connections between the different representations of the function.

[4]*How are the independent and dependent variables represented in each of the four representations? Answer in your own words:*

📖 **Read the introduction to the subsection, 'Reversion the Rule of a Function,'** in which the authors define the inverse of a function.

[5]*In context, the independent variable for the function P represents _____, and the dependent variable for the function P represents _____. Verbally, the independent variable for the function P^{-1} represents _____, and the dependent variable for the function P^{-1} represents _____.*

📖 **Read Section 1.7 Example 2 again**, paying particular attention to the connection between the height of the graph and the values of the function.

[6]*To find the values of x for which T(x) = 25, the authors look for points on the graph for which the _____ (x-coordinate / y-coordinate) equals 25. In contrast, to find T(1), the authors look for points on the graph for which the _____ (x-coordinate / y-coordinate) equals 1.*

📖 **Read Section 4.6 Example 3**, in which the authors read the values of f^{-1} from the graph of f.

[7]*To find $f^{-1}(3)$, the authors look for points on the graph for which the _____ (x-coordinate / y-coordinate) equals 3. In contrast, to find f (3), you look for the points on the graph for which the _____ (x-coordinate / y-coordinate) equals 3.*

---

[4] Verbally, the independent variable is the number of toppings, and the dependent variable is the price of the pizza. Symbolically, the independent variable is x and the dependent variable is y. Numerically, the independent variable is the first row of the table, consisting of 0 toppings, 1 topping, 2 toppings, etc.; the dependent variable is the second row, consisting of the prices $10.00, $11.25, $12.50, and so on. Graphically, the values of the independent variable are on the x-axis; the values of the dependent variable are on the y-axis.

[5] The independent variable for P represents the number of toppings, and the dependent variable for P represents the price of the pizza. The independent variable for P^{-1} represents the price of the pizza, and the dependent variable for P^{-1} represents the number of toppings.

[6] To find the values of x for which $T(x) = 25$, the authors look for points on the graph with y-coordinate (or height) equal to 25. To find $T(1)$, the authors look for the points on the graph with the x-coordinate equal to 1.

[7] To find $f^{-1}(3)$, the authors look for points on the graph for which the y-coordinate (height of the graph) equals 3. To find $f(3)$ you would look for points on the graph for which the x-coordinate equals 3.

✐ **Section 4.6 Exercise 35** The graph of a function is given. Use the graph to find the indicated values.

(a) $f^{-1}(2)$,

(b) $f^{-1}(4)$

(c) $f^{-1}(6)$

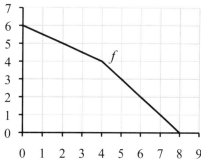

📖 **Read Section 4.6 Example 4**, in which the authors find a formula for the inverse of a linear function.

[8]*What rules do P(x) and P^{-1}(x) represent? Note that the verbal description makes it clear that inverse reverses the rule of the function.*
Answer in your own words:

📖 **Read Section 4.6 Example 5**, in which the authors find a formula for the inverse of a polynomial function.

[9]*What rules do f(x) and f^{-1}(x) represent? Note that the verbal description makes it clear that inverse reverses the rule of the function.*
Answer in your own words:

📖 **Read Section 4.6 Example 6**, in which the authors find a formula for the inverse of a rational function.

[10]*What rules do f(x) and f^{-1}(x) represent? Can you tell that the inverse reverses the rule of the function from this description?*
Answer in your own words:

---

[8] The function P is the rule, "Multiply by 2, and then add 12." The function P^{-1} is the rule, "Subtract 12, and then divide by 2."

[9] The function f is the rule, "Cube x, multiply by 2, and then add 1." The function f^{-1} is the rule, "Subtract 1, divide by 2, and then take the cube root."

[10] The function f is the rule, "Divide 3 times the input by the input plus 2." The function f^{-1} is the rule, "Divide 2 times the input by 3 minus the input." It is not clear from the rules that these functions are indeed inverses.

✎ **Section 4.6 Exercises 57, 61 and 63** Find the inverse function of f.

57. $f(x) = 4x + 7$

Solution

61. $f(x) = x^3 - 4$

Solution

63. $f(x) = \dfrac{x-2}{x+2}$

Solution

📖 **Read the paragraph after Section 4.6 Example 6**, in which the authors explain that in some cases, it is not apparent from the formulas that f^{-1} reverses the rule of f.

[11] *If f is a function with domain A and range B, and g is the inverse of f, then two equation hold:*

_____ *and* _____ .

📖 **Read Section 4.6 Example 7**, in which the authors show that one function is the inverse of another.

[12] *How do you evaluate f(5x +2)?*
Answer in your own words:

✎ **Section 4.6 Exercise 37** Show that $f(x) = 2x - 5$ and $g(x) = \dfrac{x+5}{2}$ are inverses of each other.

Solution

Which Functions Have Inverses?

In these exercises, you determine whether or not functions are one-to-one using a table, a formula or a graph. Find the definition of one-to-one and the horizontal line test in blue boxes in Section 4.6 of the textbook.

📖 **Read the introduction to the subsection, 'Which Functions Have Inverses?'** in which the authors show that a certain price scheme makes it impossible to determine how many toppings are on a pizza from the pizza's price.

[13] *Saying a function is one-to-one is the same as saying the function _____ (has / does not have) an inverse.*

---

[11] If f is a function with domain A and range B, and g is the inverse of f, then $g(f(x)) = x$ for all x in A, and $f(g(x)) = x$ for all x in B.

[12] To evaluate $f(5x + 2)$, replace x in the formula for f by $(5x + 2)$. Tip: to avoid making mistakes, include the parentheses when you make the substitution.

[13] A function is one-to-one if and only if the function has an inverse.

📖 **Read Section 4.6 Example 8**, in which the authors show a function is not one-to-one.

[14] *The function $f(x) = x^2$ _____ (has / does not have) an inverse. The authors show that f is not one-to-one (and does not have an inverse) by explaining that the output _____ corresponds to the inputs _____ and _____.*

> Refer back to page 23
> **Hints and Tips 1.4: Showing a relation is not a function**

When you show a relation is not a function, you name specific inputs that correspond to more than one output. When you show a function does not have an inverse, or equivalently is not one-to-one, you name specific outputs that correspond to more than one input. The existence of outputs corresponding to more than one input means that the inverse relation cannot be a function.

⇨ In the introduction to the subsection, 'Which Functions Have Inverses?' the authors explain that the output of a function, $12, corresponds to the inputs of the function, 0, 1 or 2 toppings. So the function has no inverse function.

⇨ In Example 8, the authors explain that the output 4 of f corresponds to the inputs 2 and –2, so the function f is not one-to-one and does not have an inverse.

✎ **Section 4.6 Exercise 53 and 55** A function is given by a formula or a verbal description. Determine if the function is one-to-one.

53. $g(x) = x^2 - 2x$

Solution

55. $h(t)$ is the height of a basketball t seconds after it is thrown in a free-throw shot.

Solution

---

[14] The function f does not have an inverse. The output 4 corresponds to the inputs 2 and –2.

📖 **Read Section 4.6 Example 9**, in which the authors use the horizontal line test to determine whether functions are one-to-one.

[15] *What is the difference between the function in part (a) and the function in part (b)? Do either of them have inverses?*
Answer in your own words:

✎ **Section 4.6 Exercise 71** Draw the graph of $f(x) = x^3 - x$ and use it to determine whether the function is one-to-one.

Solution

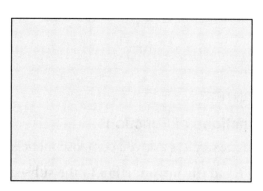

Exponential and Logarithmic Functions as Inverse Functions

These exercises ask you to apply the ideas of this section to connect the exponential and logarithmic functions you studied in the rest of this chapter. Find the connection between exponential functions and logarithmic functions in a blue box in Section 4.6 of the textbook.

📖 **Read Section 4.6 Example 10**, in which the authors show that one function is the inverse of another.

[16] *How do you evaluate $f(10^{x/2})$?*
Answer in your own words:

---

[15] In part (a) the domain of the function includes all real numbers. In part (b), only the nonnegative real numbers belong to the domain. Since being one-to-one is the same thing as having an inverse, the function in part (a) has no inverse, but the function in part (b) has an inverse.

[16] To evaluate $f(10^{x/2})$, replace x in the formula for f by $(10^{x/2})$. Tip: to avoid making mistakes, include the parentheses when you make the substitution.

🖉 **Section 4.6 Exercise 41** Show that $f(x) = 10 \cdot 4^x$ and $g(x) = \log_4\left(x/10\right)$ are inverses of each other.

Solution

CONTEXTS

Functions of Functions

These exercises ask you compose models of the real world.

📖 **Read the introduction to the subsection 'Functions of Functions' again,** paying particular attention to the units on the input and output of each function.

[17] *The input of P is _____ and the output of P is _____. The input of T is*
_____ and the output of T is _____. To find your income after 2 hours, you
calculate _____ (P(2) / T(2) / T(P(2))). To find your tax after 2 hours you calculate _____ (P(2) / T(2) /
T(P(2))). To find your tax on $16 dollars, you calculate _____ (P(16) / T(16) / T(P(16))).

> Refer back to page 11 (see also page 36).
> **Hints and Tips 1.2b: What does the algebra represent in context?**

📖 **Read Section 4.6 Example 1 again,** paying particular attention to what the functions represent in context.

[18] *What does L(2) represent in context?*
Answer in your own words:

---

[17] The input of P is the time you spend at work, and the output of P is your income. The input of T is your income and the output of T is the tax you pay. To find your income after 2 hours, you calculate $P(2)$. To find your tax after 2 hours, you calculate $T(P(2))$. To find your tax on $16, you calculate $T(16)$.

[18] We write the following sentences as scratch work. Algebraically, $L(2) = 0.80$ is the output when the input is 2. In context, $L(2) = \$0.80$ represents the tax you pay when how long you work is 2 hours. We reword this more eloquently for the final answer: $L(2) = \$0.80$ represents the tax you pay when you work 2 hours.

✎ **Section 4.6 Exercise 77 (extended) Multiple Discounts** You have a $50 coupon from the manufacturer, good for the purchase of a cell phone. The store where you are purchasing your cell phone is offering a 20% discount on all cell phones. Let x represent the sticker price of the cell phone.

(a) Suppose that only the 20% discount applies. Find a function f that models the purchase price of the cell phone as a function of the sticker price x.

[19] *What does f(x) represent in context?*
Answer in your own words:

Refer back to page 69
Hints and Tips 1.8a: Thinking about the problem

[20] *Following Hints and Tips 1.8a, think about the problem by calculating some examples. Fill out the table of values.*

Input x (sticker price)	Output f(x) (purchase price)	Input x (sticker price)	Output f(x) (purchase price)
$100	**$80**	$250	
$150		$300	
$200		$350	

(b) Suppose that only the $50 coupon applies. Find a function g that models the purchase price of the cell phone as a function of the sticker price x.

[21] *What does g(x) represent in context?*
Answer in your own words:

---

[19] Following Hints and Tips 1.2b, we write the following sentences as scratch work. Algebraically, $f(x)$ is the output when x is the input. In context, $f(x)$ dollars is the purchase price (with the 20% discount) when the sticker price is x dollars. Here is our more eloquent final answer. The number $f(x)$ represents the purchase price when the sticker price of x dollars is discounted by 20%.

[20] To find the output, we reduce the sticker price by 20%. The amounts in the table are (all in dollars) (100, 80), (150, 120), (200, 160), (250, 200), (300, 240), and (350, 280).

[21] Following Hints and Tips 1.2b, we write the following sentences as scratch work. Algebraically, $g(x)$ is the output when x is the input. In context, $g(x)$ dollars is the purchase price (with the $50 discount) when the sticker price is x dollars. Here is our more eloquent final answer. The number $g(x)$ represents the purchase price when the sticker price of x dollars is discounted by $50.

^{22}*Following Hints and Tips 1.8a, think about the problem by calculating some examples. Fill out the table of values.*

Input x (sticker price)	Output f(x) (purchase price)	Input x (sticker price)	Output f(x) (purchase price)
$100	$50	$250	
$150		$300	
$200		$350	

(c) If you can use the coupon and the discount, then the purchase price is either $f(g(x))$ or $g(f(x))$, depending on the order in which they are applied to the sticker price. Find both $f(g(x))$ and $g(f(x))$. Which composition gives a lower output?

^{23}*What does f(g(x)) represent in context?*
Answer in your own words:

(d) Do you get a lower price if you first apply the $50 discount and then apply the 20% discount, or if you first apply the 20% discount, and then apply the $50 discount?

Solution

---

^{22} To find the output, we reduce the sticker price by $50. The amounts in the table are (all in dollars) (100, 50), (150, 100), (200, 150), (250, 200), (300, 250), and (350, 300).

^{23} Following Hints and Tips 1.2b, we write the following sentences as scratch work. Algebraically, f(g(x)) is the output when x is the input. In context, $f(g(x))$ dollars is the purchase price (with the $50 discount taken first, and then followed by the 20% discount) when the sticker price is x dollars. Here is our more eloquent final answer. The number $f(g(x))$ represents the purchase price when the sticker price of x dollars is reduced by $50, and then by an additional 20%.

Reversing the Rule of a Function

✎ **Section 4.6 Exercise 83 Population of India** The function $f(x) = 846.3(1.031)^x$ is a model for the population of India, where x is the number of years since 1990 and the population is measured in millions (see Exercise 45 in Section 3.2, included in this study guide on page 172).
(a) Find f^{-1}. What does $f^{-1}(x)$ represent?
(b) Use the inverse function f^{-1} to predict when the population of India reaches 1.1 billion.

Solution

CONCEPTS

✓ **4.6 Exercises – Concepts – Fundamentals** Complete the Fundamentals section of the exercises at the end of Section 4.6. Compare your answers to those at the back of the textbook, and make corrections as necessary.

📖 **Read the Concept Check for Section 4.6**, in the Chapter Review at the end of Chapter 4.

4.6 Solutions to ✐ Exercises

✐ Section 4.6 Exercise 15

(a) The function $M(x) = f(g(x))$ first applies the rule of g and then applies the rule of f. So M is the rule "Multiply by 3, then add 2." The function $N(x) = g(f(x))$ first applies the rule of f and then applies the rule of g. So N is the rule "Add 2, then multiply by 3."

(b) In symbols, we have

$$M(x) = f(g(x)) \qquad \text{Definition of } M$$
$$= f(3x) \qquad \text{Definition of } g$$
$$= 3x + 2 \qquad \text{Definition of } f$$

So $M(x) = 3x + 2$. Similarly, we have

$$N(x) = g(f(x)) \qquad \text{Definition of } N$$
$$= g(x + 2) \qquad \text{Definition of } f$$
$$= 3(x + 2) \qquad \text{Definition of } g$$

So $N(x) = 3(x + 2)$.

(c) $M(3) = 3(3) + 2 = 11$, and $N(-2) = 3(-2 + 2) = 0$.

✐ Section 4.6 Exercise 17

(a) The function $M(x) = f(g(x))$ first applies the rule of g and then applies the rule of f. So M is the rule "Subtract 1, then square, then multiply by 2." The function $N(x) = g(f(x))$ first applies the rule of f and then applies the rule of g. So N is the rule "Square, then multiply by 2, and then subtract 1."

(b) In symbols, we have:

$$M(x) = f(g(x)) \qquad \text{Definition of } M$$
$$= f(x - 1) \qquad \text{Definition of } g$$
$$= 2(x - 1)^2 \qquad \text{Definition of } f$$

So $M(x) = 2(x - 1)^2$. Similarly, we have

$$N(x) = g(f(x)) \qquad \text{Definition of } N$$
$$= g(2x^2) \qquad \text{Definition of } f$$
$$= 2x^2 - 1 \qquad \text{Definition of } g$$

So $N(x) = 2x^2 - 1$.

(c) $M(3) = 2(3 - 1)^2 = 2(2)^2 = 8$ and $N(-2) = 2(-2)^2 - 1 = 8 - 1 = 7$.

✎ Section 4.6 Exercise 35

From the graph we see that $f(6) = 2$, $f(4) = 4$, and $f(0) = 6$. So we have:

(a) $f^{-1}(2) = 6$ (b) $f^{-1}(4) = 4$ (c) $f^{-1}(6) = 0$

✎ Section 4.6 Exercise 37

We to check that $f(g(x)) = x$ and $g(f(x)) = x$. We have

$$f(g(x)) = f\left(\frac{x+5}{2}\right) \qquad \text{Definition of } g$$

$$= 2\left(\frac{x+5}{2}\right) - 5 \qquad \text{Definition of } f$$

$$= (x+5) - 5 \qquad \text{Cancel 2}$$

$$= x \qquad \text{Simplify}$$

Also,

$$g(f(x)) = g(2x-5) \qquad \text{Definition of } f$$

$$= \frac{(2x-5)+5}{2} \qquad \text{Definition of } g$$

$$= \frac{2x}{2} \qquad \text{Simplify}$$

$$= x \qquad \text{Cancel 2}$$

Since $f(g(x)) = x$ and $g(f(x)) = x$, the functions f and g are inverses.

✎ Section 4.6 Exercise 41

We check that $f(g(x)) = x$ and $g(f(x)) = x$. We have

$$f(g(x)) = f\left(\log_4(x/10)\right) \qquad \text{Definition of } g$$

$$= 10 \cdot 4^{\log_4(x/10)} \qquad \text{Definition of } f$$

$$= 10 \cdot \left(\frac{x}{10}\right) \qquad \log_4 x \text{ and } 4^x \text{ are inverse functions}$$

$$= x \qquad \text{Cancel 10}$$

Also,

$$g(f(x)) = g(10 \cdot 4^x) \qquad \text{Definition of } f$$

$$= \log_4\left(\frac{10 \cdot 4^x}{10}\right) \qquad \text{Definition of } g$$

$$= \log_4(4^x) \qquad \text{Cancel 10}$$

$$= x \qquad \log_4 x \text{ and } 4^x \text{ are inverse functions}$$

Since $f(g(x)) = x$ and $g(f(x)) = x$, the functions f and g are inverses of each other.

✎ Section 4.6 Exercise 53

Following **Hints and Tips 1.4**, we explain that g is not one-to-one by giving an example of an output that corresponds to more than one input. The function g is not one-to-one because, for example, $g(0) = 0$ and $g(2) = 2^2 - 2(2) = 0$. So the inputs 0 and 2 have the same output, namely 0.

✎ Section 4.6 Exercise 55

Following **Hints and Tips 1.4**, we explain that h is not one-to-one by giving an example of an output that corresponds to more than one input. The function h is not one-to-one because what goes up must come down. For example, if the ball leaves the shooter's hands at 5 feet, then $h(0) = 5$. At some time t^*, the ball will return to 5 feet above the grown, so $h(t^*) = 5$. So the inputs 0 and t^* have the same output 5.

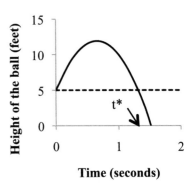

✎ Section 4.6 Exercise 57

First we write $f(x)$ in equation form: $y = 4x + 7$

Then we solve this equation for x:

$$y = 4x + 7 \qquad \text{Equation}$$

$$y - 7 = 4x \qquad \text{Subtract 7 from each side}$$

$$x = \frac{y - 7}{4} \qquad \text{Divide by 4 and switch sides}$$

Finally, we interchange x and y:

$$y = \frac{x - 7}{4}$$

So the inverse function is $f^{-1}(x) = \dfrac{x - 7}{4}$.

✐ Section 4.6 Exercise 61

First we write $f(x)$ in equation form: $y = x^3 - 4$.

Then we solve this equation for x:

$$y = x^3 - 4 \qquad \text{Equation}$$

$$y + 4 = x^3 \qquad \text{Add 4 to each side}$$

$$x = \sqrt[3]{y + 4} \qquad \text{Take the cube root and switch sides}$$

Finally, we interchange x and y:

$$y = \sqrt[3]{x + 4}$$

So the inverse function is $f^{-1}(x) = \sqrt[3]{x + 4}$.

✐ Section 4.6 Exercise 63

First we write $f(x)$ in equation form: $y = \dfrac{x - 2}{x + 2}$.

Then we solve this equation for x:

$$y = \frac{x - 2}{x + 2} \qquad \text{Equation}$$

$$y(x + 2) = x - 2 \qquad \text{Multiply each side by } x + 2$$

$$xy + 2y = x - 2 \qquad \text{Distributive Property}$$

$$2y + 2 = x - xy \qquad \text{Subtract } xy \text{ and add 2 to each side}$$

$$2y + 2 = x(1 - y) \qquad \text{Factor } x$$

$$x = \frac{2y + 2}{1 - y} \qquad \text{Divide by } 1 - y \text{ and switch sides}$$

Finally, we interchange x and y:

$$y = \frac{2x + 2}{1 - x}$$

So the inverse function is $f^{-1}(x) = \dfrac{2x + 2}{1 - x}$.

✐ Section 4.6 Exercise 71

From the figure we see that there are horizontal lines, such as the line $y = -0.3$, shown here as a dotted line, that intersect the graph of $f(x) = x^3 - x$ more than once. So by the Horizontal Line Test, f is not one-to-one.

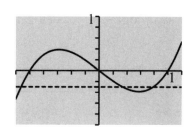

✎ **Section 4.6 Exercise 77 (extended) Multiple Discounts**

(a) If only the 20% discount applies, then the purchase price is the sticker price minus 20% of the sticker price. In symbols this means $f(x) = x - 0.2x$. Further simplifying gives us $f(x) = 0.8x$.

(b) If only the $50 coupon applies, then the purchase price is $50 less than the sticker price. In symbols this means $g(x) = x - 50$.

(c) We find $f(g(x))$ and $g(f(x))$.

$$\begin{aligned} f(g(x)) &= f(x - 50) & &\text{Definition of } g \\ &= 0.8(x - 50) & &\text{Definition of } f \\ &= 0.8x - 40 & &\text{Distributive Property} \end{aligned}$$

So $f(g(x)) = 0.8x - 40$.

$$\begin{aligned} g(f(x)) &= g(0.8x) & &\text{Definition of } f \\ &= 0.8x - 50 & &\text{Definition of } g \end{aligned}$$

So $g(f(x)) = 0.8x - 50$.

Since $f(g(x)) = 0.8x - 40$ and $g(f(x)) = 0.8x - 50$, we see that the output of $f(g(x))$ is 10 more than the output of $g(f(x))$.

(d) We find in part (c) that $g(f(x))$ yields the lower price, so we determine which order the discounts are applied in $g(f(x))$. When calculating $g(f(x))$, we first apply the function $f(x)$, which models the selling price when we receive a 20% discount. So the function $g(f(x))$ represents first applying the 20% discount, and then applying the $50 coupon, and using the discounts in this order gives a lower price.

✎ **Section 4.6 Exercise 83 Population of India**

(a) First we write $f(x)$ in equation form: $y = 846.3(1.031)^x$.

Then we solve this equation for x:

$$\begin{aligned} y &= 846.3(1.031)^x & &\text{Equation} \\[2mm] \frac{y}{846.3} &= (1.031)^x & &\text{Divide by 846.3 on each side} \\[2mm] \log\left(\frac{y}{846.3}\right) &= \log(1.031)^x & &\text{Take the logarithm of each side} \\[2mm] \log\left(\frac{y}{846.3}\right) &= x\log(1.031) & &\text{Law 3: ``Bring the exponent down''} \\[2mm] x &= \frac{\log(y/846.3)}{\log(1.031)} & &\text{Divide by } \log(1.031) \text{ and switch sides} \end{aligned}$$

Finally, we interchange x and y:

$$y = \frac{\log(x / 846.3)}{\log(1.031)}$$

So the inverse function is $f^{-1}(x) = \dfrac{\log(x / 846.3)}{\log(1.031)}$.

Following **Hints and Tips 1.2b**, we write the following sentences as scratch work. Algebraically, when the input is x, the output is $f^{-1}(x)$. In context, when the population of India is x million people, the number of years since 1990 is $f^{-1}(x)$ years. Here is our more eloquent final answer: The function $f^{-1}(x)$ represents the number of years since 1990 after which the population of India reaches x million people.

(b) We wish to know when the population of India reaches 1.1 billion = 1100 million. Since $f^{-1}(x)$ is a model for the number of years since 1990 after which the population reaches x million people, we replace x with 1100 in the formula for $f^{-1}(x)$.

$$f^{-1}(x) = \frac{\log(x / 846.3)}{\log(1.031)} \qquad \text{Formula}$$

$$f^{-1}(1100) = \frac{\log(1100 / 846.3)}{\log(1.031)} \qquad \text{Replace } x \text{ with } 1100$$

$$\approx 8.6 \qquad \text{Calculator}$$

So it takes almost 9 years for the population of India to reach 1.1 billion. This happens in the year $1990 + 9 = 1999$.

5.1 Working with Functions: Shifting and Stretching

OBJECTIVES

☑ *Check the box when you can do the exercises addressing the objective, such as those included in this study guide, which are listed here.*

Exercises

Shifting Graphs Up and Down

☐ Recognize from its formula when the graph of a function is an upward or downward shift of the graph of a basic function. **7, 13, 37, 45**

☐ Construct a formula for a function whose graph is the graph of a basic function after an upward or downward shift. **57, 63**

☐ Connect upward and downward shifts in the graph of a model to the real world context that the model represents. **65**

Shifting Graphs Left and Right

☐ Recognize from its formula when the graph of a function is the graph of a basic function after a shift to the left or right. **9, 13**

☐ Construct a formula for a function whose graph is the graph of a basic function after shift to the left or right. **57**

☐ Connect shifts to the left and right in the graph of a model to the real world context that the model represents. **67**

Stretching and Shrinking Graphs Vertically

☐ Recognize from its formula when the graph of a function is the graph of a basic function after it has been stretched or compressed vertically. **33, 35**

☐ Connect vertically stretching and shrinking the graph of a model to the real world context that the model represents. **69**

Reflecting Graphs

☐ Recognize from its formula that the graph of a function is the graph of a basic function after a reflection across the x- or y-axis. **37, 39, 49**

☐ Construct a formula for a function whose graph is the graph of a basic function after a reflection across the x- or y-axis. **63**

GET READY...

Review function notation in Section 1.5. Look over the graphs of basic functions in Section 1.6, and review how to graph functions by first making a table, and plotting points. Also review strategies for graphing on your graphing calculator in Algebra Toolkit D.3.

SKILLS

Shifting Graphs Up and Down

These exercises ask you to connect changes in the formula for a function and changes in the function's graph, when the graph is shifted up and down. Find this relationship in a blue box in Section 5.1 of the textbook.

📖 **Read the introduction to the subsection, 'Shifting Graphs Up and Down,'** in which the authors describe the height of a child jumping on a trampoline.

[1]*What is the relationship between the graph of the height of the child jumping on the trampoline and the graph of the height of the child jumping on the raised trampoline?*
Answer in your own words:

📖 **Read Section 5.1 Example 1**, in which the authors shift the graphs of parabolas up and down.

[2]*Relate the graph of g to the graph of h by filling in the blanks: g(x) = h(x) + _____. So the y-coordinate of each point on the graph of g is _____ units above the corresponding point on the graph of h. This means that to graph g, you can shift the graph of h _____ (upward / downward) by _____ units.*

✎ **Section 5.1 Exercise 7** Use the graph of $f(x) = x^2$ to graph the following.

(a) $g(x) = x^2 - 4$

(b) $g(x) = x^2 + 2$

Solution

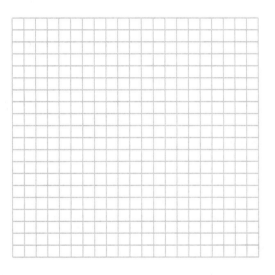

Shifting Graphs Left and Right

These exercises ask you to connect the changes in the formula for a function and changes in the function's graph, when the graph is shifted left and right. Find this relationship in a blue box in Section 5.1 of the textbook.

_____

[1] The graph of the height of the child jumping on the raised trampoline is the same as the graph for the original trampoline, just shifted 10 units upward.

[2] We have g(x) = h(x) + 5, so the y-coordinate of each point on the graph of g is 5 units above the corresponding point on the graph of h. To graph g you can shift the graph of h upward by 5 units.

📖 **Read Section 5.1 Example 4**, in which the authors shift the graphs of parabolas left and right.

[3]To graph f(x + c), move the graph of f(x) to the _____ (left /right) by ___ units. The function whose graph is the graph of f(x) shifted to the right by 5 units is f(_____).

✏ **Section 5.1 Exercise 9** Use the graph of $f(x) = x^2$ to graph the following.

(a) $g(x) = (x+2)^2$

(b) $g(x) = (x-4)^2$

Solution

x	x^2	$(x+2)^2$	$(x-4)^2$
−4	16	4	64
−3			
−2			
−1			
0			
1			
2			
3			
4			

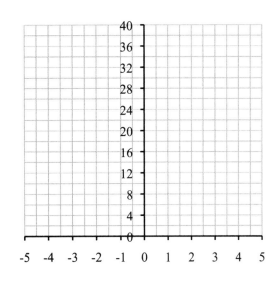

📖 **Read the subsection of Section 1.7, 'Domain and Range from a Graph,'** again, paying particular attention to the way the authors recognize the domain of a function from its graph.

[4]The domain of a function is a subset of the _____ (x-axis / y-axis).

📖 **Read Section 5.1 Example 5**, in which the authors shift the graph of the square root of x vertically and horizontally.

[5]You can obtain the graph of $f(x)=\sqrt{x-3}+4$ by shifting the graph of $y=\sqrt{x}$ _____ (to the left / to the right / upward / downward) by 3 units and _____ (to the left / to the right / upward / downward) by 4 units. The domain of $y=\sqrt{x}$ is _____. The domain of $f(x)=\sqrt{x-3}+4$ is _____.

---

[3] To graph $f(x + c)$, shift the graph of $f(x)$ to the left by c units. The function whose graph is the graph of f shifted to the right by 5 units is $f(x - 5)$.

[4] The domain is a subset of the x-axis.

[5] You can obtain the graph of f by shifting the graph of $y=\sqrt{x}$ to the right by 3 and upward by 4. The domain of $y=\sqrt{x}$ is $[0, +\infty)$, and the domain of f is $[3, +\infty)$, which is the domain of $y=\sqrt{x}$ shifted to the right by 3.

✎ **Section 5.1 Exercise 13 (modified)** Use the graph of $f(x) = \sqrt{x}$ to graph the following on the same coordinate axis.

(a) $g(x) = \sqrt{x+4}$ (b) $g(x) = \sqrt{x}+1$ (c) $g(x) = \sqrt{x+4}+1$

Solution

x	$\sqrt{x}$	$\sqrt{x}+1$	$\sqrt{x+4}$	$\sqrt{x+4}+1$
−4	−	−	**0**	**1**
−3	−	−	**1**	
−2	−	−	$\sqrt{2} \approx \mathbf{1.41}$	
−1	−	−		
0				
1				
2				
3				
4				

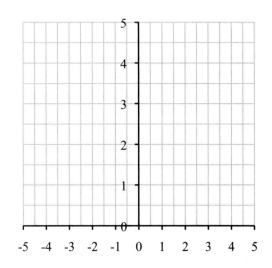

Stretching and Shrinking Graphs Vertically

These exercises ask you to connect the changes in the formula for a function and changes in the function's graph, when the graph is stretched or compressed. Find this relationship in a blue box in Section 5.1 of the textbook.

📖 **Read Section 5.1 Example 7**, in which the authors the authors stretch and compress the graph of a parabola.

[6] *At x = 2, the height of the graph of $f(x) = x^2$ is _____. So at x = 2, the height of the graph of*

$g(x) = \dfrac{1}{3}x^2$ *is _____, which is _____ times the height of the graph of f. Similarly at x = 2, the height of*

the graph of $h(x) = 3x^2$ is _____, which is _____ times the height of the graph of f.

---

[6] The height of the graph of f at $x = 2$ is $f(2) = 4$. The height of the graph of g at $x = 2$ is $g(2) = 4/3$, which is 1/3 times the height of the graph of f. The height of the graph of h at $x = 2$ is $h(2) = 12$, which is 3 times the height of the graph of f.

✎ **Section 5.1 Exercises 33 and 35** Sketch the graph of the function, not by plotting points, but by starting with the graph of a basic function and applying a vertical stretching or shrinking.

33. $f(x) = \dfrac{1}{4}x^2$

35. $f(x) = 2x^4$

[7]*For 33, start with the graph of y =* _____ . *For 35, start with the graph of y =* _____ .

Solution

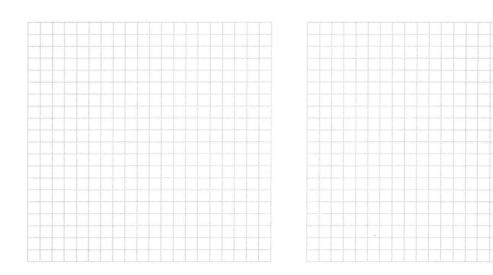

Reflecting Graphs

These exercises ask you to connect changes in the formula for a function and changes in the function's graph, when the graph is reflected across one of the axes. Find this relationship in a blue box in Section 5.1 of the textbook.

📖 **Read Section 5.1 Example 8**, in which the authors graph reflections of a parabola and a square root.

[8]*Complete the following table by looking at the graphs in Figures 9 and 10:*

Function	Domain	Range
$y = x^2$	$(-\infty, \infty)$	$[0, \infty)$
$f(x) = -x^2$		
$y = \sqrt{x}$		
$g(x) = \sqrt{-x}$		

---

[7] For 33, first graph $y = x^2$. For 35, first graph $y = x^4$.

[8] The domain of $f(x) = -x^2$ is $(-\infty,\infty)$, and the range is $(-\infty,0]$. The domain of $y=\sqrt{x}$ is $[0,\infty)$, and the range is $[0,\infty)$. The domain of $g(x)=\sqrt{-x}$ is $(-\infty,0]$, and the range is $[0,\infty)$.

✐ **Section 5.1 Exercise 37 and 39** Sketch the graph of the function not by plotting points, but by starting with the graph of a basic function and applying transformations.

37. $f(x) = 1 - x^2$ 39. $f(x) = -|x|$

[9] *For 37, start with the graph of y = _____. For 39, start with the graph of y = _____.*

Solution

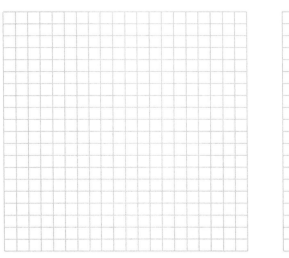   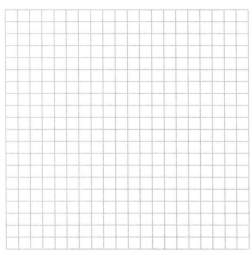

📖 **Read Section 5.1 Examples 9 and 10**, in which the authors transform exponential and logarithmic functions.

[10] *To graph $g(x) = 1 + 2^x$, transform the graph of y = _____, by _____ (shifting to the right / shifting to the left / shifting upwards / shifting downwards / reflecting across the x-axis / reflecting across the y-axis).*

To graph $g(x) = -2^x$, transform the graph of y = _____, by _____ (shifting to the right / shifting to the left / shifting upwards / shifting downwards / reflecting across the x-axis / reflecting across the y-axis).

To graph $g(x) = \log_{10}(x - 3)$, transform the graph of y = _____, by _____ (shifting to the right / shifting to the left / shifting upwards / shifting downwards / reflecting across the x-axis / reflecting across the y-axis).

To graph $h(x) = \log_{10}(-x)$, transform the graph of y =   , by   (shifting to the right / shifting to the left / shifting upwards / shifting downwards / reflecting across the x-axis / reflecting across the y-axis).

_____

[9] For 37, first graph $y = x^2$. For 39, first graph $y = |x|$.

[10] To graph $g(x) = 1 + 2^x$, transform the graph of $y = 2^x$, by shifting upwards. To graph $g(x) = -2^x$ transform the graph of $y = 2^x$, by reflecting across the x-axis. To graph $g(x) = \log_{10}(x - 3)$, transform the graph of $y = \log_{10} x$, by shifting to the right by 3 units. To graph $h(x) = \log_{10}(-x)$, transform the graph of $y = \log_{10} x$, by reflecting across the y-axis.

✎ **Section 5.1 Exercise 45 and 49** Sketch the graph of the function not by plotting points, but by starting with the graph of a basic function and applying transformations.

45. $f(x) = 2^x - 3$ 49. $f(x) = -\log(-x)$

[11] *For 45, start with the graph of $y = $ _____. For 49, start with the graph of $y = $_____.*

Solution

✎ **Section 5.1 Exercise 57** A function f is given, and the indicated transformations are applied to its graph (in the given order). Write the equation for the final transformed graph.

$$f(x) = |x| \text{; shift 3 units to the right and upward by 1 unit}$$

Solution

---

[11] For 45, first graph $y = 2^x$. For 39, first graph $y = \log x$.

✐ **Section 5.1 Exercise 63** The graphs of $f(x) = 2^x$ and g are given. Find a formula for the function g.

Solution

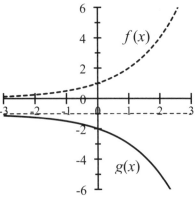

CONTEXTS

Shifting Graphs Up and Down

These exercises ask you to connect upward and downward shifts in the graph of a model to the real world context that the model represents.

📖 **Read Section 5.1 Example 2**, in which the authors consider a man walking in a moving train, and determine the speed he is traveling relative to the ground.

[12] *The formula for the man's speed relative to the ground is V(t) = _____, and the formula for the man's speed relative to the train is v(t) = _____, so you can get the graph of V by shifting the graph of v _____ (upward / downward) by _____ units (feet).*

✐ **Section 5.1 Exercise 65 Bungee Jumping** Luisa goes Bungee jumping from a 500-foot-high bridge. The graph shows Luisa's height $h(t)$ (in feet) after t minutes.

(a) Describe in words what the graph indicates about Luisa's Bungee jump.
(b) Suppose Luisa goes Bungee jumping from a 400-foot-high bridge. Sketch a new graph that shows Luisa's height $H(t)$ after t seconds.
(c) What transformation must be performed on the function h to obtain the function H? Express the function H in terms of h.

Solution

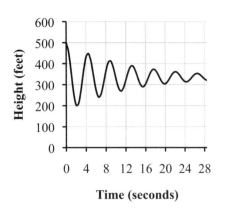

Time (seconds)

_____

[12] The speed relative to the ground is $V(t)=20\sqrt{t}+1320$, and the speed relative to the train is $v(t)=20\sqrt{t}$, so you can get the graph of V from the graph of v by shifting the graph of V upward by 1320 units.

Shifting Graphs Left and Right

These exercises ask you to connect shifts to the left and right in the graph of a model to the real world context that the model represents.

📖 **Read Section 5.1 Example 3**, in which the authors compare the distances two drivers travel, starting from the same place, but leaving at different times.

[13] *In the function j(t) from part (a), the input represents _____*
and the output represents _____.

In the function j(t) from part (a), the input represents _____
and the output represents _____.

✏️ **Section 5.1 Exercise 67 Distance** Alessandra leaves home 30 minutes after her brother Alberto and drives at a constant speed of 90 km/h on a straight road.

(a) Find a function d that models Alessandra's distance from home t hours after she started her trip.

(b) Find a function D that models Alessandra's distance from home t hours after Alberto started his trip.

(c) Graph the functions d and D. How are the graphs related?

Solution

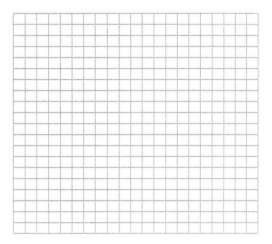

Stretching and Shrinking Graphs Vertically

These exercises ask you to connect vertically stretching and shrinking the graph of a model to the real world context that the model represents.

_____

[13] In part (a), the input t represents the time (in hours) since Jennifer started her trip, and the output $j(t)$ represents the distance (in miles) that Jennifer has traveled from home. In part (b), the input t represents the time (in hours) since Madeline started her trip, and the output $J(t)$ represents the distance (in miles) that Jennifer has traveled from home.

📖 **Read Section 2.1 Example 3 again,** paying particular attention to how the authors obtain the cycler's average speed from the graph of the distance the cycler travels.

[14]*To calculate the average speed Jodi traveled between t = 2 hours and t = 4 hours, the authors calculate the ratio $\frac{38-10}{4-2}$. How do the authors obtain the numbers 38 and 10 from the graph?*

Answer in your own words:

📖 **Read Section 5.1 Example 6,** in which the authors describe how far a driver gets from home.

[15]*If the function f(x) in Figure 6 describes Alan's distance from home, then Alan drives _____ miles in the first two hours. If the function f(2x) in Figure 7 describes Alan's distance from home, then Alan drives _____ miles in the first two hours.*

If f(x) in Figure 6 describes Alan's distance from home, then Alan's average speed over the first two hours is _____ miles per hour. If 2f(x) in Figure 7 describes Alan's distance from home, then Alan's average speed over the first two hours is _____ miles per hour.

✏ **Section 5.1 Exercise 69** Miyuki practices swimming laps with her team. The function $y = f(t)$ graphed below gives her distance (in meters) from the starting edge of the pool t seconds after she starts her laps.

(a) Describe in words Miyuki's swim practice. What is her average speed for the first thirty seconds?

(b) Graph the function $y = 1.2\,f(t)$ on the axes with the given graph. How is the graph of the new function related to the graph of the original function?

(c) What is Miyuki's new average speed for the first thirty seconds?

Solution

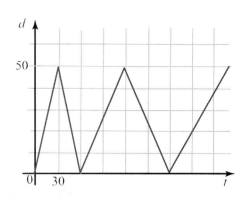

---

[14] In the ratio, 38 miles is the distance that Jodi travels after 4 hours, which is the height of the graph at $t = 4$. Similarly, 10 miles is the distance Jodi travels after 2 hours, which is the height of the graph at $t = 2$.

[15] From the graph of $f(x)$ in Figure 6, Alan drives 40 miles in the first two hours. From the graph of $2f(x)$ in Figure 7, Alan drives 80 miles in the first two hours. From the graph of $f(x)$ in Figure 6, Alan's average speed in the first two hours is $\frac{40-0}{2-0}=20$ miles per hour. For the graph of $2f(x)$ in Figure 7, Alan's average speed in the first two hours is $\frac{80-0}{2-0}=40$ miles per hour.

CONCEPTS

✓ **5.1 Exercises – Concepts – Fundamentals** Complete the Fundamentals section of the exercises at the end of Section 5.1. Compare your answers to those at the back of the textbook, and make corrections as necessary.

📖 **Read the Concept Check for Section 5.1**, in the Chapter Review at the end of Chapter 5.

5.1 Solutions to ✎ Exercises

✎ Section 5.1 Exercise 7

(a) Observe that $g(x) = x^2 - 4 = f(x) - 4$, so the y-coordinate of each point on the graph of g is 4 units below the y-coordinate of the corresponding point on the graph of f. This means that to graph g, we shift the graph of f downward 4 units.

$y = x^2 + 2$
$y = x^2$
$y = x^2 - 4$

(b) Similarly, here $g(x) = x^2 + 2 = f(x) + 2$ so to graph g, we shift the graph of f upward 2 units.

✎ Section 5.1 Exercise 9

(a) To graph g, we shift the graph of f to the left 2 units.

(b) To graph g, we shift the graph of f to the right 4 units.

x	x^2	$(x+2)^2$	$(x-4)^2$
-4	16	4	64
-3	9	1	49
-2	4	0	36
-1	1	1	25
0	0	4	16
1	1	9	9
2	4	16	4
3	9	25	1
4	16	36	0

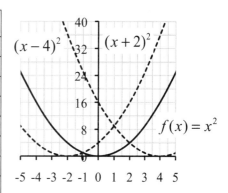

✐ Section 5.1 Exercise 13 (modified)

(a) We start with the graph of $f(x) = \sqrt{x}$ and shift it to the left 4 units to obtain the graph of $g(x) = \sqrt{x+4}$.

(b) We start with the graph $f(x) = \sqrt{x}$ and shift it upward 1 unit to obtain the graph of $g(x) = \sqrt{x}+1$.

(c) We start with the graph of $g(x) = \sqrt{x+4}$ from part (a) and shift it upward 1 unit to obtain the graph of $g(x) = \sqrt{x+4}+1$.

x	$\sqrt{x}$	$\sqrt{x}+1$	$\sqrt{x+4}$	$\sqrt{x+4}+1$
-4	$-$	$-$	0	1
-3	$-$	$-$	1	2
-2	$-$	$-$	$\sqrt{2} \approx 1.41$	$\sqrt{2}+1 \approx 2.41$
-1	$-$	$-$	$\sqrt{3} \approx 1.73$	$\sqrt{3}+1 \approx 2.73$
0	0	1	2	3
1	1	2	$\sqrt{5} \approx 2.24$	$\sqrt{5}+1 \approx 3.24$
2	$\sqrt{2} \approx 1.41$	$\sqrt{2}+1 \approx 2.41$	$\sqrt{6} \approx 2.45$	$\sqrt{6}+1 \approx 3.45$
3	$\sqrt{3} \approx 1.73$	$\sqrt{3}+1 \approx 2.73$	$\sqrt{7} \approx 2.65$	$\sqrt{7}+1 \approx 3.65$
4	2	3	$\sqrt{8} \approx 2.83$	$\sqrt{8}+1 \approx 3.83$

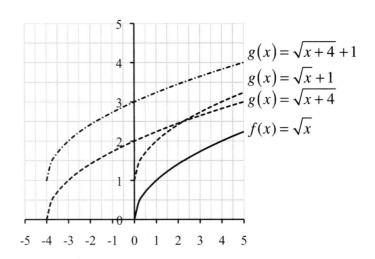

✐ Section 5.1 Exercise 33

We start with the graph of $g(x) = x^2$. The graph of f is obtained by multiplying the y-coordinate of each point on the graph of g by $\frac{1}{4}$. That is, to obtain the graph of f, we shrink the graph of g vertically by a factor of $\frac{1}{4}$. The result is the wider parabola in the graph.

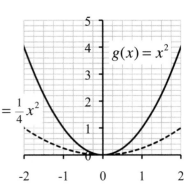

✐ Section 5.1 Exercise 35

We start with the graph of $g(x) = x^4$. The graph of f is obtained by multiplying the y-coordinate of each point on the graph of g by 2. That is, to obtain the graph of f, we stretch the graph of g vertically by a factor of 2. The result is the narrower parabola in the graph.

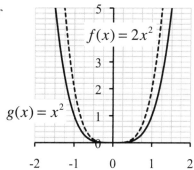

✐ Section 5.1 Exercise 37

We start with the graph of $y = x^2$. The graph of $y = -x^2$ is the graph of $y = x^2$ reflected in the x-axis. The graph of $f(x) = 1 - x^2$ is the graph of $y = -x^2$ shifted upward by 1 unit.

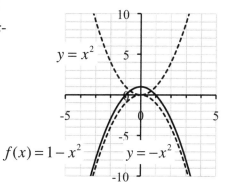

✐ Section 5.1 Exercise 39

We start with the graph of $y = |x|$. The graph of $f(x) = -|x|$ is the graph of $y = |x|$ reflected in the x-axis.

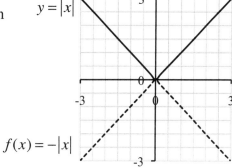

✐ Section 5.1 Exercise 45

To obtain the graph of $f(x) = 2^x - 3$, we start with the graph of $y = 2^x$ and shift it downward 3 units. Notice that the line $y = -3$ is now the horizontal asymptote.

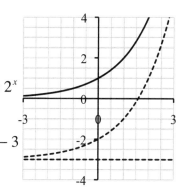

✎ Section 5.1 Exercise 49

To obtain the graph of $f(x) = -\log(-x)$, we start with the graph of $y = \log(x)$ and reflect it in the y-axis to get the graph $y = \log(-x)$. From here we reflect the graph of $y = \log(-x)$ in the x-axis to get the graph $f(x) = -\log(-x)$.

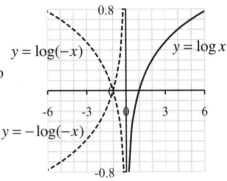

✎ Section 5.1 Exercise 57

First, we shift the graph of f by 3 units to the right. The formula becomes $y = |x - 3|$. Now we shift the graph upward by 1 unit. The equation for the final transformed graph is $y = |x - 3| + 1$.

✎ Section 5.1 Exercise 63

To obtain the graph of g, we reflect the graph of $f(x) = 2^x$ in the x-axis, giving us the graph of $y = -2^x$. Then we shift the graph of $y = -2^x$ downward by 1 unit, so that the resulting function has a horizontal asymptote at $y = -1$. This gives us the formula $g(x) = -2^x - 1$.

✎ Section 5.1 Exercise 65 Bungee Jumping

(a) Luisa drops to 200 feet, bounces up and down a bit, and then settles at 350 feet.

(b) Since Luisa jumps from 100 feet below her original jump, the graph of $H(t)$ is the same as the graph of $h(t)$, but shifted down by 100 units.

(c) When we shift the graph of $h(t)$ downward by 100 units, we get the graph of the function $H(t) = h(t) - 100$.

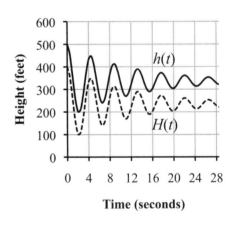

✎ Section 5.1 Exercise 67 Distance

(a) Since Alessandra drives at 90 kilometers per hour, her distance t hours after she starts her trip is given by $d(t) = 90t$.

(b) If we take time 0 to be the time when Alberto started his trip, then Alessandra should just be starting her trip when t is $\frac{1}{2}$ hour. In other words, in

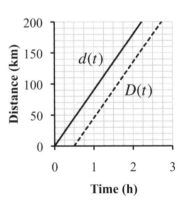

terms of this new starting time, Alessandra's distance at time t is given by

$$D(t) = 90\left(t - \frac{1}{2}\right) \text{ (for } t \geq \frac{1}{2}\text{).}$$

(c) We see that the graph of D (shown as a dashed line) is the same as the graph of d (shown as a solid line), but shifted half a unit to the right.

✎ **Section 5.1 Exercise 69**

(a) Here, the pool is 50 meters long. Miyuki swam two and a half laps, slowing down with each successive lap. Over the first 30 seconds Miyuki swam 50 meters, thus her average speed is $\dfrac{50 \text{ meters}}{30 \text{ seconds}} \approx 1.67$ meters per second.

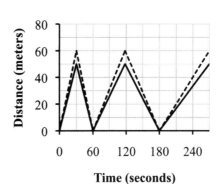

(b) The graph of $y = 1.2 f(t)$ (shown here as a dashed line) is a vertical stretching of the original function $f(t)$ (shown here as a solid line) by a factor of 1.2.

(c) Here, the pool is $50 \times 1.2 = 60$ meters long. Miyuki swam one 60-meter length in the first thirty seconds, so her average speed is $\dfrac{60 \text{ meters}}{30 \text{ seconds}} = 2$ meters per second.

5.2 Quadratic Functions and Their Graphs

OBJECTIVES

☑ Check the box when you can do the exercises addressing the objective, such as those included in this study guide, which are listed here.

Exercises

The Squaring Function

☐ Compare the values, graphs and rates of change of a quadratic function and a linear function.
43

Quadratic Functions in General Form

☐ Determine if a function is quadratic by comparing it to the general form for a quadratic function.
7

☐ Express a quadratic function in general form.
9, 13

Quadratic Functions in Standard Form

☐ Express a quadratic function in standard form by completing the square.
21, 23

Graphing Using the Standard Form

☐ Determine the coordinates of the vertex and the *y*-intercept of the graph of a quadratic function from its standard form. Also determine if the parabola opens upwards or downwards.
29, 33

☐ Graph quadratic functions by first expressing them in standard form.
29, 33

☐ Find the formula for a quadratic function from its graph.
39

GET READY...

This section uses the technique of completing the square, to rewrite quadratic expressions in standard form. Review this technique in Algebra Toolkit C.2. You also multiply linear expressions to write a quadratic function in general form. Review this in Algebra Toolkit B.1. For all of this, you must recognize the form of an algebraic expression. For practice comparing algebraic expressions, review Algebra toolkit B.2.

✓ **Algebra Checkpoint** Test yourself by completing the Algebra Checkpoint at the end of this section of the textbook, and comparing your answers to those in the back of the textbook. Refer to Algebra Toolkit B.1, B.2 and C.2 as necessary.

SKILLS

The Squaring Function

These exercises ask you to model the real world using the squaring function. Find its definition in bold typeface in Section 5.2 of the textbook.

📖 **Read Section 5.2 Example 1 and the paragraph preceding it**, in which the authors model the fall of skydivers, including one with superpowers.

[16] *The squaring function D(t) grows _____ (faster / more slowly) than the linear function H(t). The average rate of change of the squaring function D(t) is _____ (increasing / constant / decreasing). The average rate of change of the linear function H(t) is _____ (increasing / constant / decreasing).*

✏ **Section 5.2 Exercise 43** A linear function f and a quadratic function g are given.

(a) Complete the table, and find the average rate of change of each function on intervals of length 1 for x between 0 and 5.

(b) Sketch a graph of each function.

Solution

Sketch the graph of each function on the coordinate axis provided.

x	$f(x) = 3x$	Rate of change
0	*0*	–
1	*3*	*3*
2		
3		
4		
5		

x	$g(x) = 3x^2$	Rate of change
0		
1		
2		
3		
4		
5		

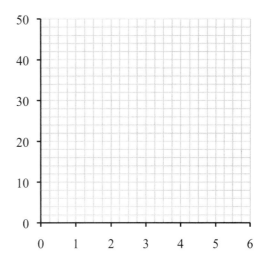

---

[16] The squaring function grows faster than the linear function. The average rate of change of the squaring function is increasing, while the average rate of change of the linear function is constant.

Quadratic Functions in General Form

These exercises ask you to write formulas for quadratic functions in general form. Find the general form for a quadratic function in a blue box in Section 5.2 of the textbook.

[17] *If f has the form,* $f(x) = ax^2 + bx + c$, *but a = 0, then f* _____ *(is / is not) quadratic.*

📖 **Read Section 5.2 Example 2**, in which the authors determine if a given function is quadratic by comparing it to general form.

[18] *This example demonstrates two ways in which a function can fail to be quadratic. What are they? Answer in your own words:*

✏ **Section 5.2 Exercise 7** Determine whether the given function is a quadratic function.

(a) $f(x) = \frac{1}{2}x^2 + 2x + 1$ (b) $g(x) = 3x^3 + 2x^2 + 1$ (c) $h(x) = 5^2 x + 1$

Solution

📖 **Read Section 5.2 Examples 3 and 4**, in which the authors rewrite quadratic functions in general form.

[19] *Why might you want to write these functions in general form? Answer in your own words:*

---

[17] If the coefficient a of x^2 is zero, then f is a linear function (with slope b and y-intercept c), and is hence not quadratic.

[18] The function in part (b) fails to be quadratic by having terms other than ax^2, bx or c (namely x^3). The function in part (c) fails to be quadratic since it does not have a nonzero term involving x^2.

[19] To determine whether a function is quadratic, you compare the function to the general form of a quadratic function, as in Example 2. When you are given a function, you make that comparison by trying to write your function in general form; in this case by multiplying the factors. Also in order to use the Quadratic Formula (see Algebra Toolkit C.2) to find the inputs at which a quadratic function equals zero, you must first write the equation in general form. Similarly, to use the formula for the maximum or minimum of a quadratic function in Section 5.3, you first write the function in general form.

✏ **Section 5.2 Exercises 9 and 13** Express the function in the general form of a quadratic function.

9. $f(x) = (2x+3)(4x-1)$ 13. $f(x) = (x-1)^2 + 5$

Solution   **Solution**

Quadratic Functions in Standard Form

These exercises ask you to write formulas for quadratic functions in standard form. Find the standard form for a quadratic function in a blue box in Section 5.2 of the textbook.

📖 **Read Section 5.2 Examples 5 and 6**, in which the authors rewrite quadratic functions in standard form by completing the square.

[20]*Why might you want to write these functions in standard form? Hint: Look back at Section 5.1. Answer in your own words:*

In the standard for the function in Example 5, a = _____*, h =* _____*, and k =* _____*. In the standard for the function in Example 6, a =* _____*, h =* _____*, and k =* _____*.*

✏ **Section 5.2 Exercises 21 and 23** Express the function in the standard form of a quadratic function.

21. $f(x) = x^2 + 2x - 5$

Solution

---

[20] By putting a quadratic function in standard form, you make it easy to use the ideas from Section 5.1 to relate the graph of the function to the graph of $y = x^2$. In the next section, you will use this to graph quadratic functions. In Example 5, $a = 1$, $h = -8$, and $k = -40$. In Example 6, $a = 2$, $h = 3$, and $k = 5$.

23. $f(x) = 2x^2 + 20x + 1$

Solution

Graphing Using the Standard Form

These exercises ask you to graph parabolas by transforming the graph of $y = x^2$. You will use the coefficients in the standard form of the quadratic function to find features of the graph such as the vertex. Find a summary of these ideas in a blue box, and the words parabola and vertex in bold typeface in Section 5.2 of the textbook.

📖 **Read the introduction to the subsection, 'Graphing Using the Standard Form,'** in which the authors relate the ideas about graphing from Section 5.1 to the standard form of a quadratic function.

[21] *Write the transformations that you do to the graph of $y = x^2$ in order to get the graph of the function* $f(x) = 2(x-1)^2 + 3$ *in words.*
Answer in your own words:

📖 **Read Section 5.2 Example 7 and 8**, in which the authors write a quadratic function in standard form in order to create its graph.

[22] *In the graph of f in Example 7, the authors label the vertex at (_____, _____) and the y-intercept at (_____, _____). While the sketch of a graph tends to be slightly inaccurate, it is important to get these feature in the right place (and label them to be sure your reader can tell they are in the right place), since it is visible and misleading if they are incorrect.*

What are the steps the authors follow to graph a quadratic function in these two examples?
Answer in your own words:

---

[21] Shift the graph of $y = x^2$ to the right by 1 unit, stretch the graph vertically by a factor of 2, and then shift the graph upward by 3 units.

[22] The vertex is (3, 5) and the y-intercept is (0, 25). The authors label the y-intercept on the y-axis, rather than providing the coordinates. First the authors put the expression in standard form. Then they find the coordinates for the vertex. Next, they determine if the parabola opens upward or downward. Last the authors find the y-intercept. Now they sketch the parabola passing through the y-intercept and the vertex.

✎ **Section 5.2 Exercises 29 and 33 (modified)** A quadratic function is given.

(a) Express the quadratic function in standard form.

(b) Find its vertex and y-intercept, and determine if the parabola opens upward or downward.

(c) Sketch its graph.

29. $f(x) = 2x^2 + 4x + 3$

Solution

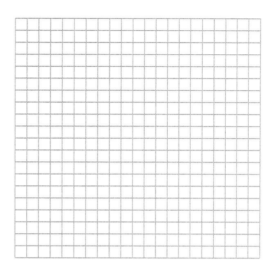

33. $f(x) = -x^2 - 3x + 3$

Solution

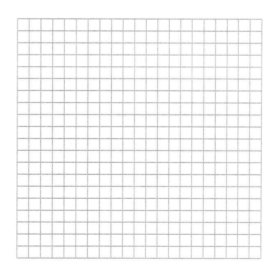

📖 **Read Section 5.2 Example 9**, in which the authors find the equation for a quadratic function given its graph.

[23]*To find the equation for the quadratic function in standard form, the authors use two points on the graph. they are _____ and _____. How do the authors find the value of the coefficient a?*
Answer in your own words:

✎ **Section 5.2 Exercise 39** Find the quadratic function whose graph is shown.

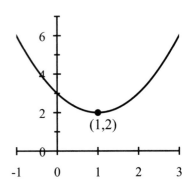

CONCEPTS

✓ **5.2 Exercises – Concepts – Fundamentals** Complete the Fundamentals section of the exercises at the end of Section 5.2. Compare your answers to those at the back of the textbook, and make corrections as necessary.

📖 **Read the Concept Check for Section 5.2**, in the Chapter Review at the end of Chapter 5.

5.2 Solutions to ✎ Exercises

✎ **Section 5.2 Exercise 7**

(a) The function f is a quadratic function with $a = \frac{1}{2}$, $b = 2$, and $c = 1$.

(b) The function g is not a quadratic function, since g has a term involving x^3.

(c) The function h is not a quadratic function, since h does not have a nonzero term involving x^2.

_____

[23] The authors use the vertex and the y-intercept of the graph. After reading the values of the coefficients $h = -1$ and $k = -2$, off the graph (from the coordinates of the vertex), the authors create the formula $f(x) = a(x + 1)^2 - 2$. From the graph, they know the y-intercept is –4, so $\underline{f(0) = -4.}$ They substitute 0 for x, and –4 for $f(0)$, and solve for a.

✎ Section 5.2 Exercise 9

$$f(x) = (2x + 3)(4x - 1) \qquad \text{Given function}$$

$$= 8x^2 + 12x - 2x - 3 \qquad \text{Distributive property}$$

$$= 8x^2 + 10x - 3 \qquad \text{Combine like terms}$$

The general form is $f(x) = 8x^2 + 10x - 3$.

✎ Section 5.2 Exercise 13

See Algebra Toolkit B.1 for Special Product Formula 1, used below.

$$f(x) = (x - 1)^2 + 5 \qquad \text{Given function}$$

$$= (x^2 - 2x + 1) + 5 \qquad \text{Special Product Formula 1}$$

$$= x^2 - 2x + 6 \qquad \text{Combine constant terms}$$

The general form is $f(x) = x^2 - 2x + 6$.

✎ Section 5.2 Exercise 21

The coefficient of x is 2. To get f in standard form we "complete the square" by first taking half of the coefficient of x and squaring it: $\left(\frac{2}{2}\right)^2 = 1$. We then add and subtract that result.

$$f(x) = x^2 + 2x - 5 \qquad \text{Given function}$$

$$= (x^2 + 2x + 1) - 1 - 5 \qquad \begin{array}{l} \text{Complete the square: Add 1} \\ \text{inside the parentheses, and} \\ \text{subtract 1 outside} \end{array}$$

$$= (x + 1)^2 - 6 \qquad \text{Factor and simplify}$$

The standard form is $f(x) = (x + 1)^2 - 6$.

✎ Section 5.2 Exercise 23

Since the coefficient of x^2 is not 1, we must factor this coefficient from the terms involving x before completing the square.

$$f(x) = 2x^2 + 20x + 1 \qquad \text{Given function}$$

$$= 2(x^2 + 10x) + 1 \qquad \text{Factor 2 from the } x\text{-terms}$$

$$= 2(x^2 + 10x + 25) - 2 \cdot 25 + 1 \qquad \begin{array}{l} \text{Complete the square: Add 25} \\ \text{inside the parentheses, and} \\ \text{subtract } 2 \cdot 25 \text{ outside} \end{array}$$

$$= 2(x + 5)^2 - 49 \qquad \text{Factor and simplify}$$

The standard form is $f(x) = 2(x + 5)^2 - 49$.

✎ Section 5.2 Exercise 29 (modified)

(a) Since the coefficient of x^2 is not 1, we must factor this coefficient from the terms involving x before completing the square.

$$f(x) = 2x^2 + 4x + 3 \qquad \text{Given function}$$

$$= 2(x^2 + 2x) + 3 \qquad \text{Factor 2 from the } x\text{-terms}$$

$$= 2(x^2 + 2x + 1) - 2 \cdot 1 + 3 \qquad \text{Complete the square: Add 1 inside the parentheses, and subtract } 2 \cdot 1 \text{ outside}$$

$$= 2(x + 1)^2 + 1 \qquad \text{Factor and simplify}$$

The standard form is $f(x) = 2(x + 1)^2 + 1$.

(b) The standard form tells us that we get the graph of f by taking the parabola $y = x^2$ and performing the following transformations: Shift to the left 1 unit, stretch by a factor of 2, then shift upward 1 unit. The vertex of the parabola is at $(-1, 1)$. The parabola opens upward because the coefficient 2 is positive.

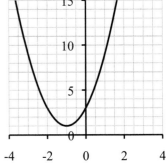

The y-intercept is obtained by evaluating f at 0. This is more easily done by using the general form of a quadratic function. In this case $f(0) = 2(0)^2 + 4(0) + 3 = 3$. So the y-intercept is 3.

(c) To create the graph, we plot the vertex $(-1, 1)$, and the y-intercept $(0, 3)$, and sketch the curve accordingly.

✎ Section 5.2 Exercise 33 (modified)

(a) Since the coefficient of x^2 is not 1, we must factor this coefficient from the terms involving x before completing the square.

$$f(x) = -x^2 - 3x + 3 \qquad \text{Given function}$$

$$= -(x^2 + 3x) + 3 \qquad \text{Factor } -1 \text{ from the } x\text{-terms}$$

$$= -\left(x^2 + 3x + \frac{9}{4}\right) - \left((-1)\frac{9}{4}\right) + 3 \qquad \text{Complete the square: Add } \frac{9}{4} \text{ inside the parentheses, and subtract } (-1)\frac{9}{4} \text{ outside}$$

$$= -\left(x + \frac{3}{2}\right)^2 + \frac{21}{4} \qquad \text{Factor and simplify}$$

The standard form is $f(x) = -\left(x + \frac{3}{2}\right)^2 + \frac{21}{4}$.

(b) From the standard form we see that the graph is a parabola with vertex $\left(-\dfrac{3}{2}, \dfrac{21}{4}\right)$.

The parabola opens downward because the coefficient -1 is negative.

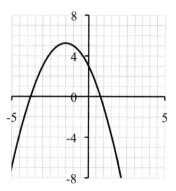

The y-intercept is obtained by evaluating f at 0. This is more easily done by using the general form of a quadratic function. In this case $f(0) = -(0)^2 - 3(0) + 3 = 3$. So the y-intercept is 3.

(c) To create the graph, we plot the vertex $\left(-\dfrac{3}{2}, \dfrac{21}{4}\right)$, and the y-intercept $(0, 3)$, and sketch the curve accordingly.

✏ Section 5.2 Exercise 39

We observe from the graph that the vertex is $(1, 2)$, so replacing h by 1 and k by 2 in the standard form of the quadratic function, we have

$$
\begin{aligned}
f(x) &= a(x - h)^2 + k && \text{Standard form} \\
&= a(x - 1)^2 + 2 && \text{Replace } h \text{ by 1, and } k \text{ by 2}
\end{aligned}
$$

We also observe from the graph that the y-intercept is 3, and hence $f(0)$ is 3. We can use this to find the value of a:

$$
\begin{aligned}
f(0) &= a(0 - 1)^2 + 2 && \text{Replace } x \text{ by 0} \\
3 &= a + 2 && \text{Replace } f(0) \text{ by 3} \\
a &= 1 && \text{Subtract 2 from each side and switch sides}
\end{aligned}
$$

So the function we seek is $f(x) = (x - 1)^2 + 2$.

✏ Section 5.2 Exercise 43

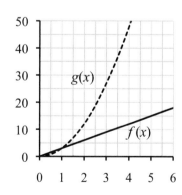

x	$f(x) = 3x$	Rate of change
0	0	–
1	3	3
2	6	3
3	9	3
4	12	3
5	15	3

x	$g(x) = 3x^2$	Rate of change
0	0	–
1	3	3
2	12	9
3	27	15
4	48	21
5	75	27

5.3 Maxima and Minima: Getting Information from a Model

OBJECTIVES

☑ *Check the box when you can do the exercises addressing the objective, such as those included in this study guide, which are listed here.*

Exercises

Finding Maximum and Minimum Values

☐ Find the maximum or minimum value of a quadratic function and the input at which it is achieved. — **9, 11**

☐ Write a quadratic function in general form in order to apply the formula to find its maximum or minimum value. — **25, 27, 33**

Modeling with Quadratic Functions

☐ Find the maximum or minimum value of a real world quantity modeled by a quadratic function. — **23, 25, 27, 33**

☐ Explain what the vertex of the graph of a quadratic model represents in context. — **23, 25, 27, 33**

☐ Construct quadratic models using geometry. — **27**

☐ Construct a quadratic model by multiplying two linear functions. — **33**

GET READY...

This section involves creating and analyzing quadratic functions that model the real world. Review Section 1.8 in which the authors construct models for a variety of contexts using ratios and geometry. Also review Section 2.3 in which the authors construct linear models using the point-slope form and the slope-intercept form.

SKILLS

Finding Maximum and Minimum Values

These exercises ask you to find the maximum or minimum value of a quadratic function and the input at which it is attained. Find the formula for these in a blue box in Section 5.3 of the textbook.

📖 **Read the introduction to the subsection, 'Finding Maximum and Minimum Values,'** in which the authors find the vertex for the quadratic function

$f(x) = ax^2 + bx + c$, by completing the square.

[24] *The coordinates of the vertex of the graph of this quadratic function are (_____, _____)*

📖 **Read Section 5.3 Example 1**, in which the authors find the maximum or minimum values of quadratic functions.

[25] *In part (a), the coordinates for the vertex are (___, ___). In part (b), the coordinates for the vertex are (___, ___).*

---

[24] The coordinates of the vertex of $f(x) = ax^2 + bx + c$ are $\left(-\dfrac{b}{2a}, c - \dfrac{b^2}{4a}\right)$.

[25] The coordinates for the vertex in part (a) are (–2, –4), and in part (b) are (1, –3).

✒ **Section 5.3 Exercises 9 and 11** Find the minimum or maximum value of the quadratic function, and find the *x*-value at which the minimum or maximum value occurs.

9. $f(x) = -\dfrac{x^2}{3} + 2x + 7$

Solution

11. $f(x) = x^2 + x + 1$

Solution

CONTEXTS

📖 **Read the introduction to Section 5.3**, in which the authors discuss the height reached by a model rocket.

[26] *Using a model for the height of the rocket as a function of time, you could find the time at which the rocket reaches a particular height, or the height of the rocket at a particular time. What else could you predict with this model?*
Answer in your own words:

Modeling with Quadratic Functions

These exercises ask you to find the maxima and minima of quadratic functions that model the real world.

Refer back to page 11
Hints and Tips 1.2b: What does the algebra represent in context?

In these exercises, you find a particular output (the maximum or minimum value) and its corresponding input. Your final answer should be in a sentence saying what that pair represents in context. Use Hints and Tips 1.2b to guide you as you write an eloquent conclusion to these exercises.

_____

[26] The input is the time since the rocket was launched, in seconds, and the output is the height of the rocket, in feet. The authors state that the goal for this model is to predict the maximum height of the rocket, and presumably, the time at which the rocket reaches that height.

📖 **Read Section 5.3 Example 2**, in which the authors find the maximum height of a rocket.

[27] *The input of the model is _____, and the units are _____. The output of the model is _____, and the units are _____. The authors first calculate the _____ (maximum height / time at which the rocket reaches the maximum height). How do they calculate the maximum height?*
Answer in your own words:

What does the point (25, 10,000) on the graph of h(t) represent in context?
Answer in your own words:

✏ **Section 5.3 Exercise 23 Height of a Ball** If a ball is thrown directly upward with a velocity of 12 m/s, its height (in meters) after t seconds is given by $y = 12t - 4.9t^2$. What is the maximum height attained by the ball, and after how many seconds is that height attained?

Solution

---

[27] The input of the model is time in seconds, and the output of the model is the height of the rocket in feet. The authors first calculate the time at which the rocket reaches the maximum height, using the formula. Then they calculate the maximum height by substituting the time they found into the model.

Following Hints and Tips 1.2b, we write the following sentences as scratch work: Algebraically, 25 is the t-value and 10,000 is the h-value. In context 25 seconds is the time since the rocket launch and 10,000 feet is the rocket's height. The authors conclude this example by writing the final answer more eloquently, including the additional information about the importance of that particular point on the graph: The rocket reaches a maximum height of 10,000 feet after 25 seconds.

📖 **Read Section 5.3 Example 3**, in which the authors find the speed at which a car obtains the best gas mileage.

[28] *The input of the model is _____, and the units are _____. The output of the model is _____, and the units are _____. The authors first calculate the _____ (maximum mileage / speed at which the mileage reaches its maximum). How do they calculate the maximum mileage?*
Answer in your own words:

What does the point (42, 32) on the graph of M(s) represent in context?
Answer in your own words:

✎ **Section 5.3 Exercise 25 Agriculture** The number of apples produced by each tree in an apple orchard depends on how densely the trees are planted. If n trees are planted on an acre of land, then each tree produces $900 - 9n$ apples. So the number of apples produced per acre is

$$A(n) = n(900 - 9n)$$

What is the maximum yield of the trees, and how many trees should be planted per acre to obtain the maximum yield of apples?

Solution

---

[28] The input of the model is speed in miles per hour, and the output of the model is the car's gas mileage, in miles per gallon. The authors first calculate the speed at which the mileage reaches its maximum, using the formula. Then they find the maximum mileage by substituting the speed they found into the model. Following Hints and Tips 1.2b, we write the following sentences as scratch work: Algebraically, 42 is the s-value and 32 is the M-value. In context 42 mi/h is the speed and 32 mi/gal is the mileage. The authors conclude this example by writing the final answer more eloquently, including the additional information about the importance of that particular point on the graph: The car's best gas mileage is 32 mi/gal, when it is traveling at 42 mi/h.

📖 **Read Section 1.8 Example 5 again**, paying particular attention to the table of values that the authors create while thinking about the problem.

[29]*From the table the authors construct while thinking about the problem the width of the garden with the largest area is approximately _____ How do the authors answer part (d)? Answer in your own words:*

✎ **Section 1.8 Exercise 29** *Find this exercise on page 72 of this Study Guide.*

📖 **Read Section 5.3 Example 4**, in which the authors find the largest vegetable garden that a gardener can enclose with 140 feet of fencing.

[30]*The input of the model is _____, and the units are _____. The output of the model is _____, and the units are _____. The authors first calculate the _____ (maximum area / width of the garden with the maximum area). How do they calculate the maximum area? Answer in your own words:*

What does the point (35, 1,225) on the graph of A(x) represent in context? Answer in your own words:

---

[29] The largest area on the table is 1200 ft^2, attained when the width is 30 ft and when the width is 40ft. So we guess that the maximum area occurs between these, perhaps at 35 ft. The authors find the dimensions of the garden with the largest area by graphing the model for the area on a graphing calculator, and tracing the parabola to its vertex.

[30] The input of the model is the width of the garden in feet, and the output of the model is the area of the garden in square feet. The authors first calculate the width of the garden with the maximal area, using the formula. Then they find the area by substituting the width they found into the model.

Following Hints and Tips 1.2b, we write the following sentences as scratch work: Algebraically, 35 is the x-value and 1,225 is the A-value. In context 35 ft is the width and 1,225 ft^2 is the area. The authors conclude this example by writing the final answer more eloquently, including the additional information about the importance of that particular point on the graph: The largest area she can fence is 1,225 square feet and the dimensions are 35 ft by 35 ft.

Section 5.3 Exercise 27 Fencing a Field A farmer has 2400 feet of fencing with which he wants to fence off a rectangular field that borders a straight river. He does not need a fence along the river (see the figure).

(a) Find a function A that models the area of the field in terms of one of its sides x.
(b) What is the largest area that he can fence, and what are the dimensions of that area?

Solution

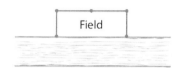

> Refer back to page 69
> **Hints and Tips 1.8a: Thinking about the problem**

📖 **Read Section 1.8 Example 3 again,** paying particular attention to how the authors think about the problem.

[31] *Before beginning the solution, the authors discuss thinking about the problem. What analysis do they do to help you understand the situation?*
Answer in your own words:

📖 **Read Section 5.3 Example 5,** in which the authors find the price per ticket that maximizes the revenue that a hockey team receives from ticket sales.

[32] *The input of the model is _____, and its unit are _____. The output of the model is _____, and its units are _____. Before beginning the solution, what analysis can you do to help understand the situation?*
Answer in your own words:

Do this analysis by using this sentence from the example to complete the table:

> *A market survey indicates that for each dollar the ticket price is lowered, the average attendance increases by 1000 people.*

Decreasing the ticket price from $14 to $12 dollars, lowers the price by 14 – 12 = _____ dollars. In this case, the number of tickets they sell increases by _____ tickets, so when the price is $12, they sell _____ tickets.

Decreasing the ticket price from $14 to x dollars, lowers the price by 14 – x dollars. In this case, the number of

Price (dollars per ticket)	Attendance (tickets)	Revenue (dollars)
14	9500	14(9500) = 133,000
13	10,500	13(10,500) = 136,500
12		
11		
10		

tickets they sell increases by _____ tickets, so when the price is $12, they sell _____ tickets.

---

[31] Following Hints and Tips 1.8a, the authors think about the problem by making up an example. In particular, they find the output of the model (the amount of water in the tank), when the input (days since the tank was filled) is equal to 10 days.

[32] The input of the model is the price of a hockey ticket, in dollars, and the output is the revenue obtained from ticket sales, also in dollars. Following Hints and Tips 1.8a and b, you can think about the problem by making up some example prices for the tickets. In particular, find the output (the revenue) for possible inputs (ticket prices).

Decreasing the price from $14 to $12 lowers the price by $2. The number of tickets they sell increases 2000 tickets (which is 2 times 1000 tickets). So when the price is $12, they sell 9500 + 2000 = 11,500 tickets. Decreasing the price from $14 to x dollars lowers the price by 14 – x dollars. The number of tickets they sell increases by 1000(14 – x). So when the price is x dollars, they sell 9500 + 1000(14 – x) tickets.

In the table, when the ticket price is $12, they sell 11,500 tickets, and earn $138,000 in revenue. When the ticket price is $11, they sell 12,500 tickets, and earn $137,500 in revenue. When the ticket price is $10, they sell 13,500 tickets, and earn $135,000 in revenue.

✎ **Section 5.3 Exercise 33** A baseball team plays in a stadium that holds 55,000 spectators. With the ticket price at $10, the average attendance at recent games has been 27,000. A market survey indicates that for every dollar the ticket price is lowered, the attendance increases by 3000.

(a) Find a function r that models the revenue in terms of ticket price.
(b) Find the price that maximizes revenue from ticket sales.

Solution

CONCEPTS

✓ **5.3 Exercises – Concepts – Fundamentals** Complete the Fundamentals section of the exercises at the end of Section 5.3. Compare your answers to those at the back of the textbook, and make corrections as necessary.

📖 **Read the Concept Check for Section 5.3**, in the Chapter Review at the end of Chapter 5.

5.3 Solutions to ✎ Exercises

✎ Section 5.3 Exercise 9

The function f is a quadratic function where a is $-\frac{1}{3}$ and b is 2. Thus the maximum or minimum value occurs at

$$x = -\frac{b}{2a} \qquad\qquad \text{Formula}$$

$$= -\frac{2}{2\left(-\frac{1}{3}\right)} \qquad\qquad \text{Replace } a \text{ by } -\frac{1}{3} \text{ and } b \text{ by 2}$$

$$= 3 \qquad\qquad \text{Simplify}$$

Since $a < 0$, the graph opens downward, and the function has a maximum value, namely $f(3) = -\frac{3^2}{3} + 2\cdot3 + 7 = 10$.

✎ Section 5.3 Exercise 11

The function f is a quadratic function where a is 1 and b is 1. Thus the maximum or minimum value occurs at

$$x = -\frac{b}{2a} \qquad\qquad \text{Formula}$$

$$= -\frac{1}{2\cdot1} \qquad\qquad \text{Replace } a \text{ by 1 and } b \text{ by 1}$$

$$= -\frac{1}{2} \qquad\qquad \text{Simplify}$$

Since $a > 0$, the graph opens upward, and the function has a minimum value, namely $f\left(-\frac{1}{2}\right) = \left(\frac{1}{2}\right)^2 + \frac{1}{2} + 1 = \frac{7}{4}$.

✎ Section 5.3 Exercise 23 Height of a Ball

The function y is a quadratic function where a is -4.9 and b is 12. Thus the maximum value occurs when

$$t = -\frac{b}{2a} \qquad\qquad \text{Formula}$$

$$= -\frac{12}{2\cdot(-4.9)} \qquad\qquad \text{Replace } a \text{ by } -4.9 \text{ and } b \text{ by 12}$$

$$\approx 1.224 \qquad\qquad \text{Calculator}$$

The maximum value is $y = 12(1.224) - 4.9(1.224)^2 \approx 7.35$. So the ball reaches a maximum height of approximately 7.35 meters after 1.224 seconds.

✐ Section 5.3 Exercise 25 Agriculture

Using the Distributive Property, we rewrite the function A as $A(n) = -9n^2 + 900n$, which is a quadratic function, where a is –9 and b is 900. Thus the maximum value occurs when

$$n = -\frac{b}{2a} \qquad \text{Formula}$$

$$= -\frac{900}{2(-9)} \qquad \text{Replace } a \text{ by } -9 \text{ and } b \text{ by } 900$$

$$= 50 \qquad \text{Simplify}$$

The maximum value is $A(50) = -9(50)^2 + 900(50) = 22,500$. So the maximum yield of the trees is 22,500 apples with 50 trees per acre.

✐ Section 5.3 Exercise 27 Fencing a Field

(a) In Section 1.8 Exercise 29, you find that when the width of the field is x, the length is $2400 - 2x$. Since the area is the length times the width, you find that the function

$$A(x) = x(2400 - 2x) = -2x^2 + 2400x$$

models the area of a field of width x that the farmer can enclose with 2,400 ft of fencing.

(b) The function A is a quadratic function where a is –2 and b is 2400. Thus the maximum value occurs when

$$x = -\frac{b}{2a} \qquad \text{Formula}$$

$$= -\frac{2400}{2(-2)} \qquad \text{Replace } a \text{ by } -2 \text{ and } b \text{ by } 2400$$

$$= 600 \qquad \text{Simplify}$$

When the width is 600, the length is $2400 - 2(600) = 1200$, and the maximum area is $A(600) = -2(600)^2 + 2400(600) = 720,000$. So the largest area the farmer can enclose is 720,000 square feet, and the dimensions are 600 by 1200 feet.

✎ Section 5.3 Exercise 33

(a) We want to find a function that gives the revenue for any ticket price.

$$\text{revenue} = \text{ticket price} \times \text{attendance}$$

There are two varying quantities: ticket price and attendance. Since the function we want is in terms of the price, we let x be the ticket price. Next, we express attendance in terms of x, as shown in the table.

In words	In algebra
Ticket Price	x
Amount ticket price is lowered	$10 - x$
Increase in attendance	$3000(10 - x)$
Attendance	$27{,}000 + 3000(10 - x)$

The model that we want is the function r that gives us the revenue for a given ticket price.

$$\text{revenue} = \text{ticket price} \times \text{attendance}$$

$$r(x) = x[27{,}000 + 3000(10 - x)] \qquad \text{Model}$$

$$r(x) = x(57{,}000 - 3000x) \qquad \text{Simplify}$$

$$r(x) = 57{,}000x - 3000x^2 \qquad \text{Distributive Property}$$

(b) Since r is a quadratic function where a is –3000 and b is 57,000, the maximum occurs at $x = -\dfrac{b}{2a} = -\dfrac{57000}{2(-3000)} = 9.5$. So a ticket price of \$9.50 gives the maximum revenue.

5.4 Quadratic Equations: Getting Information from a Model

OBJECTIVES

☑ *Check the box when you can do the exercises addressing the objective, such as those included in this study guide, which are listed here.*

Exercises

Solving Quadratic Equations: Factoring

☐ Use the Zero-Product Property of the real numbers in solving quadratic equations. — **7, 53, 63, 65, 67**

☐ Solve quadratic equations by rewriting the equation with one side factored into linear factors, and the other side equal to zero. — **7, 53, 63, 65, 67**

☐ Find the *x*-intercepts of a quadratic function from its formula or from its graph. — **17, 49, 63, 67**

☐ Find a formula of a quadratic function from its graph. — **17**

Solving Quadratic Equations: The Quadratic Formula

☐ Use the Quadratic Formula to find the solutions of a quadratic equation. — **27, 31, 33, 45, 47, 49, 63, 67**

The Discriminant

☐ Calculate the discriminant of a quadratic equation, and use it to determine the number of solutions. — **45, 46, 49**

☐ Determine the number of times the graph of a quadratic function *f* crosses the *x*-axis using the discriminant of the equation $f(x) = 0$. — **45, 47**

Modeling with Quadratic Functions

☐ Construct quadratic functions that model the real world using geometry. — **65**

☐ Construct quadratic functions that model the real world using ratios. — **67**

☐ Use a model to find inputs that correspond to particular outputs that are meaningful in the real world. — **63, 65, 67**

GET READY...

This section involves factoring quadratic expressions. Review techniques of factoring in Algebra Toolkit B.2.

✓ **Algebra Checkpoint** Test yourself by completing the Algebra Checkpoint at the end of this section of the textbook, and comparing your answers to those in the back of the textbook. Refer to Algebra Toolkit B.2 as necessary.

SKILLS

Solving Quadratic Equations: Factoring

These exercises ask you to solve quadratic equations by factoring and then using the zero-product property. Find the zero-product property in a blue box in Section 5.4 of the textbook. These exercises ask you use these techniques to find the *x*-intercepts of the graph of a quadratic function. Find the definitions of the *x*- and *y*-intercept of a function in a blue box in Algebra Toolkit B.2.

³³*If AB=4, then do you necessarily know that either A = 4 or B = 4?*
Answer in your own words:

📖 **Read Section 5.4 Example 1**, in which the authors solve a quadratic equation by factoring.

³⁴*What do the authors do first when solving this equation?*
Answer in your own words:

How do the authors check that their answers are correct?
Answer in your own words:

✎ **Section 5.4 Exercise 7** Solve the equation $x^2 + x = 12$ by factoring.

Solution

📖 **Read Section 5.4 Example 2**, in which the authors find the x-intercepts in order to label them on the graph of a quadratic function.

³⁵*The x-intercepts are the inputs at which the output equals _____. The authors label three features of the graph of the quadratic function. They are _____, _____ and the*
_____.
In factored form, f(x) = a(x – x₁)(x – x₂), where a = ___, x₁ = ___ and x₂ = ___.

---

³³ The Zero-Product Property says, if $AB = 0$, then either $A = 0$ or $B = 0$. The authors warn in red typeface after presenting this topic, that the method works only when the right-hand side of the equation is 0. Note that when $A = 2$ and $B = 2$, neither A nor B is 4, but $AB = 4$.

³⁴ The authors first rewrite the equation so that the right-hand side is 0. They check that $x = 3$ is a solution by verifying that the original equation is a true statement when $x = 3$, which they do by substituting $x = 3$ into the original equation (and similarly for $x = 8$).

³⁵ The x-intercepts are the inputs at which the output equals 0. The authors label the x-intercepts, the y-intercept and the vertex of the quadratic function on the graph. In factored form, $f(x) = (x – 3)(x + 8)$, which is of the form $f(x) = a(x – x_1)(x – x_2)$, where $a = 1$, $x_1 = 3$ and $x_2 = -8$.

✎ **Section 5.4 Exercise 53** Consider the quadratic function $f(x) = x^2 + 2x - 1$.

(a) Find the x-intercepts of the graph of f.

(b) Sketch the graph of f and label the x- and y-intercepts and the vertex.

Solution

📖 **Read Section 5.4 Example 3**, in which
the authors find an equation for a quadratic function from its graph.

[36]*Since the x-intercepts on the graph are –3 and 1, you know that the function can be expressed in the factored form $f(x) = a(x - x_1)(x - x_2)$, where a is unknown, $x_1 = $ ____ and $x_2 = $ ____. How do the authors solve for a?*
Answer in your own words:

✎ **Section 5.4 Exercise 17** A graph of a quadratic function f is given.

(a) Find the x-intercepts.

(b) Find an equation that represents the function f (as in Example 3).

Solution

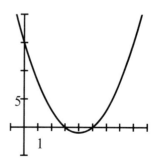

---

[36] In factored form, $f(x) = a(x + 3)(x - 1)$, which is of the form $f(x) = a(x - x_1)(x - x_2)$, where a is unknown, $x_1 = -3$ and $x_2 = 1$. To find a the authors see that the y-intercept is 6, so $f(0) = 6$. They substitute $x = 0$ and $f(0) = 6$, and solve for a.

Solving Quadratic Equations: The Quadratic Formula

These exercises ask you to solve quadratic equations using the quadratic formula. Find the Quadratic Formula in a blue box in Section 5.4 as well as in Algebra Toolkit C.3 in the textbook.

📖 **Section 5.4 Example 4**, in which the authors find the solutions to quadratic equations.

[37] In part (a), the number under the square root is _____ (positive / negative / zero), and the equation has ___ real solutions. In part (b), the number under the square root is _____ (positive / negative / zero), and the equation has ___ real solutions. In part (c), the number under the square root is _____ (positive / negative / zero), and the equation has ___ real solutions.

✎ **Section 5.4 Exercises 27, 31, 33** Find all real solutions of the equation.

27. $t^2 + 3t + 1 = 0$

Solution

31. $s^2 - \dfrac{3}{2}s + \dfrac{9}{16} = 0$

Solution

33. $w^2 = 3(w - 1)$

Solution

---

[37] In part (a), the number under the square root is positive, and the equation has two real solutions. In part (b), the number under the square root is zero, and the equation has one real solutions. In part (c), the number under the square root is negative, and the equation has no real solutions.

The Discriminant

These exercises ask you to find the discriminant of a quadratic equation in order to determine how many real solutions the equation has. Find the definition of discriminant and its relationship to the solutions of a quadratic equation in a blue box in Section 5.4 of the textbook.

📖 **Read Section 5.4 Example 5**, in which the authors determine how many times the graph of a quadratic equation crosses the x-axis.

38 *The graph of the function in part (a) has ___ x-intercepts, which means the discriminant of the equation f(x) = 0 is _____ (positive / negative / zero). The graph of the function in part (b) has ___ x-intercepts, which means the discriminant of the equation f(x) = 0 is _____ (positive / negative / zero). The graph of the function in part (c) has ___ x-intercepts, which means the discriminant of the equation f(x) = 0 is _____ (positive / negative / zero).*

✎ **Section 5.4 Exercise 49** A graph of a quadratic function $f(x) = ax^2 + bx + c$ is shown.

(a) Find the x-intercept(s), if there are any.

(b) Is the discriminant $D = b^2 - 4ac$ positive, negative, or 0?

(c) Find the solution(s) to the equation $ax^2 + bx + c = 0$.

(d) Find an equation that represents the function f.

Solution

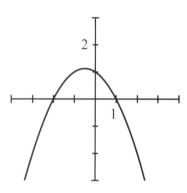

---

38 The graph in part (a) has two x-intercepts, which means the discriminant of the equation $f(x)$ is positive. The graph in part (b) has one x-intercepts, which means the discriminant of the equation $f(x)$ is zero. The graph in part (c) has zero x-intercepts, which means the discriminant of the equation $f(x)$ is negative.

✎ **Section 5.4 Exercises 43, 45 and 47** A quadratic function $f(x) = ax^2 + bx + c$ is given.

(a) Find the discriminant of the equation $ax^2 + bx + c = 0$. How many real solutions does this equation have?

(b) Use the answer to part (a) to determine the number of x-intercepts for the graph of the function $f(x) = ax^2 + bx + c$, and then graph the function to confirm your answer.

43. $f(x) = x^2 - 6x + 1$

Solution

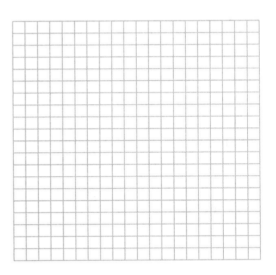

45. $f(x) = x^2 + 2.20x + 1.21$

Solution

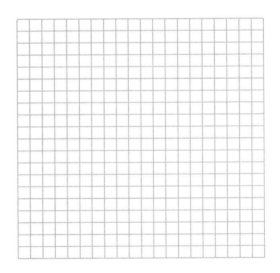

47. $f(x) = 32x^2 + 40x + 13$

Solution

CONTEXTS

Modeling with Quadratic Functions

These exercises ask you to use the techniques of solving quadratic equations to find inputs of a model that correspond to a particular output.

> Refer back to pages 69-70
> **Hints and Tips 1.8a and b: Thinking about the problem**

📖 **Read Section 1.8 Example 4 again**, paying particular attention to how the authors think about the problem.

[39]*Before beginning the solution, the authors discuss thinking about the problem. What analysis do they do to help you understand the situation?*
Answer in your own words:

📖 **Read Section 1.8 Example 5**, in which the authors find a model for the area of a garden that can be enclosed with a given length of fencing.

[40]*Before beginning the solution, the authors discuss thinking about the problem. What analysis do they do to help you understand the situation?*
Answer in your own words:

---

[39] Following Hints and Tips 1.8a and b, the authors think about the problem by making up some example dimensions of cereal boxes, and drawing diagrams. In particular, they find the output of the model (the volume of the cereal box), when the input (the depth of the cereal box) is equal to 1, 2, 3 or 4 inches.

[40] Following Hints and Tips 1.8a and b, the authors think about the problem by making up some example dimensions of gardens, and drawing diagrams. In particular, they find the output of the model (the area of the garden), when the input (width of the garden) equals to 10, 20, 30, 40, 50 or 60 feet.

📖 **Read Section 5.4 Example 6**, in which the authors model the area of a building lot.

[41]*Before beginning the solution, what analysis can you do to help you understand the situation?*
Answer in your own words:

Complete the table, and draw a diagram of each building lot represented on the table.

width (ft)	length (ft)	area (ft^2)	width (ft)	length (ft)	area (ft^2)
30	**38**	**1140**	50		
40			60		

In part (b), the authors substitute 2900 ft^2 for _____ (A(w) / w).

✎ **Section 5.4 Exercise 65 Dimensions of a Lot** A rectangular building lot is 40 ft longer than it is wide.

(a) Find a function A that models the area of the lot.

(b) If the lot has an area of 11,700 ft^2, then what are the dimensions of the lot?

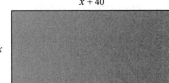

Solution

📖 **Read Section 5.3 Example 5 again**, paying particular attention to the input and output of the model.

The input of the model is _____, and the output of the model is _____.

---

[41] Following Hints and Tips 1.8a and b, you can think about the problem by making up some example dimensions for the building lot, and drawing diagrams. In the table, you find the output of the model (the area of the lot), when the input (the width of the lot) is 30, 40, 50 and 60. When the width is 40 ft, the lengths is 48 ft, and the area is 1920 ft^2; when the width is 50 ft, the lengths is 58 ft, and the area is 2900 ft^2; and, when the width is 60 ft, the lengths is 68 ft, and the area is 4080 ft^2.

In part (b), the authors substitute 2900 ft^2 for $A(w)$, since 2900 ft^2 is an area, not a width.

📖 **Read Section 5.4 Example 7**, in which the authors model the revenue from sales of tickets to a hockey game.

[42]*What does the point (0,0) on the graph of R(x) represent in context?*
Answer in your own words:

What does the point (23.5, 0) on the graph of R(x) represent in context?
Answer in your own words:

✏ **Section 5.4 Exercise 67 Stadium Revenue** A baseball team plays in a stadium that holds 55,000 spectators. With the ticket price at $10, the average attendance at recent games has been 27,000. A market survey indicates that for every dollar the ticket price is lowered, attendance increases by 3000.

(a) Find a function that models the revenue in terms of ticket price.
(b) What ticket price is so high that no revenue is generated?
(c) The baseball team hopes to have a revenue of $250,000. What ticket price should they charge to meet this goal?

Solution

---

[42] Following hints and tips 1.2b, we write the following sentences as scratch work. Algebraically, $(0, 0)$ is the point at which x is 0 and $R(x)$ is 0. In context, $(0, 0)$ represents the case in which the price is $0 and the revenue is $0. We write our final more eloquent answer: When the tickets are free, they earn no revenue.

Following hints and tips 1.2b, we write the following sentences as scratch work. Algebraically, $(23.5, 0)$ is the point at which x is 23.5 and $R(x)$ is 0. In context, $(23.5, 0)$ represents the case in which the price is $23.50, but the revenue is $0. We write our final more eloquent answer: When the tickets are $23.50, they earn no revenue, since they sell no tickets.

📖 **Read Section 5.4 Example 8**, in which the authors model the height of a rocket as a function of time.

[43] *In part (a), the authors solve by _____ (factoring / using the Quadratic Formula). In part (b), the authors solve by _____ (factoring / using the Quadratic Formula). How do they decide which technique to use?*
Answer in your own words:

Before using the Quadratic Formula, you rewrite the equation so that one side equals _____ .

✎ **Section 5.4 Exercise 63 Height of a Rocket** A model rocket shot straight upward with an initial velocity of 25 m/s will reach a height of h meters in t seconds, where the height h is modeled by

$$h(t) = -4.9t^2 + 25t$$

(a) When does the rocket reach a height of 20 m?
(b) When does the rocket reach the highest point of its path?
(c) When does the rocket hit the ground?
(d) Graph the function h and identify the points on the graph that correspond to your solutions to parts (a), (b), and (c).

Solution
Copy the graph from part (d) into the space provided.

---

[43] In part (a), they solve by using the Quadratic Formula. In part (b), they solve by factoring. They factor when they can do so easily, and use the Quadratic Formula when they do not see a way to factor the expression. To use the Quadratic Formula, you rewrite the equation so that one side equals 0.

CONCEPTS

✓ **5.4 Exercises – Concepts – Fundamentals** Complete the Fundamentals section of the exercises at the end of Section 5.4. Compare your answers to those at the back of the textbook, and make corrections as necessary.

📖 **Read the Concept Check for Section 5.4**, in the Chapter Review at the end of Chapter 5.

5.4 Solutions to ✎ Exercises

✎ Section 5.4 Exercise 7

We must first rewrite the equation so that the right hand side is 0.

$$x^2 + x = 12 \qquad \text{Given equation}$$

$$x^2 + x - 12 = 0 \qquad \text{Subtract 12 on each side}$$

$$(x+4)(x-3) = 0 \qquad \text{Factor}$$

$$x+4 = 0 \ \text{ or } \ x-3 = 0 \qquad \text{Zero-Product Property}$$

The solutions are $x = -4$ and $x = 3$.

✓ **Check** We substitute $x = -4$ into the left had side of original equation: $(-4)^2 + (-4) = 16 - 14 = 12$, which is equal to the right hand side, as desired.

We substitute $x = 3$ into the left had side of original equation:

$(3)^2 + (3) = 9 + 3 = 12$, which is equal to the right hand side, as desired.

✎ Section 5.4 Exercise 17

(a) We observe from the graph that the x-intercepts are 3 and 5.

(b) Because the x-intercepts are 3 and 5, we can express the function f in factored form as $f(x) = a(x-3)(x-5)$. We need to find a. From the graph we see that the y-intercept is 15, so $f(0) = 15$. We have

$$f(0) = a(0-3)(0-5) \qquad \text{Replace } x \text{ by } 0$$

$$15 = a(0-3)(0-5) \qquad y\text{-intercept is 15}$$

$$15 = 15a \qquad \text{Simplify}$$

$$a = 1 \qquad \text{Divide each side by 15 and switch sides}$$

It follows that $f(x) = (x-3)(x-5) = x^2 - 8x + 15$.

✎ Section 5.4 Exercise 27

In this quadratic equation a is 1, b is 3, and c is 1. By the Quadratic Formula, we have

$$x = \frac{-3 \pm \sqrt{3^2 - 4(1)(1)}}{2(1)} = \frac{-3 \pm \sqrt{5}}{2}$$

Thus the real solutions are $x = \frac{-3+\sqrt{5}}{2}$ and $x = \frac{-3-\sqrt{5}}{2}$.

✎ Section 5.4 Exercise 31

In this quadratic equation a is 1, b is $-\frac{3}{2}$, and c is $\frac{9}{16}$.

By the Quadratic Formula

$$x = \frac{-\left(-\frac{3}{2}\right) \pm \sqrt{\left(-\frac{3}{2}\right)^2 - 4(1)\left(\frac{9}{16}\right)}}{2(1)} = \frac{3}{4}$$

Thus the only real solution is $x = \frac{3}{4}$.

✎ Section 5.4 Exercise 33

To use the Quadratic Formula, our equation must have the general form of a quadratic function on the left hand side, and 0 on the right hand side. We subtract $3(w-1)$ from each side, and use the Distributive Property to get the equation $w^2 - 3w + 3 = 0$. In this quadratic equation a is 1, b is -3, and c is 3. By the Quadratic Formula

$$x = \frac{-(-3) \pm \sqrt{(-3)^2 - 4(1)(3)}}{2(1)} = \frac{3 \pm \sqrt{-3}}{2}$$

Since the square of any real number is nonnegative, $\sqrt{-3}$ is not a real number, and the equation has no real solution.

✎ Section 5.4 Exercise 43

(a) The equation is $x^2 - 6x + 1 = 0$, so a is 1, b is -6, and c is 1. The discriminant is
$D = b^2 - 4ac = (-6)^2 - 4(1)(1) = 32$. Since the discriminant is positive, there are two distinct real solutions.

(b) Since there are two distinct real solutions to the equation $x^2 - 6x + 1 = 0$, there are two x-intercepts for the graph of f.

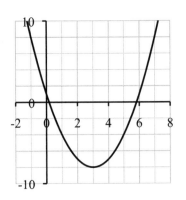

✐ Section 5.4 Exercise 45

(a) The equation is $x^2 + 2.20x + 1.21 = 0$, so a is 1, b is 2.20, and c is 1.21. The discriminant is $D = b^2 - 4ac = (2.2)^2 - 4(1)(1.21) = 0$. Since the discriminant is 0, there is exactly one real solution.

(b) Since there is exactly one real solution to the equation $x^2 + 2.20x + 1.21 = 0$, there is exactly one x-intercept for the graph of f.

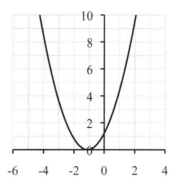

✐ Section 5.4 Exercise 47

(a) The equation is $32x^2 + 40x + 13 = 0$, so a is 32, b is 40, and c is 13. The discriminant is $D = b^2 - 4ac = (40)^2 - 4(32)(13) = -64$. Since the discriminant is negative, there are no solutions.

(b) Since there are no solutions to the equation $32x^2 + 40x + 13 = 0$, there are no x-intercepts for the graph of f.

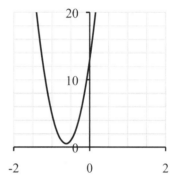

✐ Section 5.4 Exercise 49

(a) We observe from the graph that the x-intercepts are –2 and 1.

(b) Since there are two x-intercepts, there are two distinct real solutions to the equation $ax^2 + bx + c = 0$. This means that D must be positive.

(c) The x-intercepts by definition satisfy $f(-2) = 0$ and $f(1) = 0$. Thus both –2 and 1 are solutions to the equation $ax^2 + bx + c = 0$.

(d) Because the x-intercepts are –2 and 1, we can express the function f in factored form as $f(x) = a(x + 2)(x - 1)$.

We need to find a. From the graph we see that the y-intercept is 1, so $f(0) = 1$. We have

$$f(0) = a(0 + 2)(0 - 1) \qquad \text{Replace } x \text{ by } 0$$

$$1 = a(0 + 2)(0 - 1) \qquad y\text{-intercept is } 1$$

$$1 = -2a \qquad \text{Simplify}$$

$$a = -\frac{1}{2} \qquad \text{Divide each side by } -2 \text{ and switch sides}$$

It follows that $f(x) = -\frac{1}{2}(x + 2)(x - 1) = -\frac{1}{2}x^2 - \frac{1}{2}x + 1$.

✐ **Section 5.4 Exercise 53**

(a) To find the x-intercepts, we use the Quadratic Formula to solve the equation $x^2 + 2x - 1 = 0$. In this quadratic equation a is 1, b is 2, and c is -1. By the Quadratic Formula,

$$x = \frac{-2 \pm \sqrt{2^2 - 4(1)(-1)}}{2(1)} = \frac{-2 \pm \sqrt{8}}{2} = \frac{2(-1 \pm \sqrt{2})}{2} = -1 \pm \sqrt{2}$$

Thus the x-intercepts are $x = -1 + \sqrt{2}$ and $x = -1 - \sqrt{2}$.

(b) To find the y-intercept, we set x equal to 0. We get $f(0) = (0)^2 + 2(0) - 1 = -1$. So the y-intercept is -1.

The function f is a quadratic function where a is 1 and b is 2. So the x-coordinate of the vertex is

$$x = -\frac{b}{2a} \qquad \text{Formula}$$

$$= -\frac{2}{2 \cdot 1} \qquad \text{Replace } a \text{ by 1 and } b \text{ by 2}$$

$$= -1 \qquad \text{Simplify}$$

The y-coordinate of the vertex is

$$y = f\left(-\frac{b}{2a}\right) \qquad \text{Formula}$$

$$= f(-1) \qquad -\frac{b}{2a} = -1 \text{ from above}$$

$$= (-1)^2 + 2(-1) - 1 \qquad \text{Definition of } f$$

$$= -2 \qquad \text{Simplify}$$

The parabola opens upward because $a > 0$. Thus the graph is a parabola, opening upward, with vertex at $(-1, -2)$, x-intercepts at $x = -1 + \sqrt{2}$ and $x = -1 - \sqrt{2}$, and y-intercept at $y = -1$.

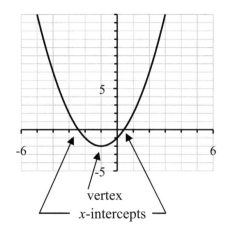

vertex

x-intercepts

✎ Section 5.4 Exercise 63 Height of a Rocket

(a) We want to find the time t such that $h = 20$.

$$h(t) = -4.9t^2 + 25t \qquad \text{Model}$$

$$20 = -4.9t^2 + 25t \qquad \text{Height is 20}$$

$$4.9t^2 - 25t + 20 = 0 \qquad \text{Subtract } -4.9t^2 + 25t \text{ from each side}$$

The expression on the left-hand side does not factor easily, so we use the Quadratic Formula where a is 4.9, b is –25, and c is 20.

$$t = \frac{-(-25) \pm \sqrt{(-25)^2 - 4(4.9)(20)}}{2(4.9)} = \frac{25 \pm \sqrt{233}}{9.8}$$

Using a calculator we find that $t \approx 0.99$ or $t \approx 4.1$. So the rocket is at a height of 20 meters after about 0.99 seconds and again after about 4.1 seconds.

(b) The function h is a quadratic function where a is –4.9 and b is 25. Thus the maximum value occurs when

$$x = -\frac{b}{2a} \qquad \text{Formula}$$

$$= -\frac{25}{2(-4.9)} \qquad \text{Replace } a \text{ by } -4.9 \text{ and } b \text{ by } 25$$

$$\approx 2.55 \qquad \text{Calculator}$$

So the rocket reaches the highest point of its path after about 2.55 seconds.

(c) At ground level the height is 0, so we must solve the equation

$$h(t) = -4.9t^2 + 25t \qquad \text{Model}$$

$$0 = -4.9t^2 + 25t \qquad \text{Height is 0}$$

$$0 = 4.9t\left(-t + \frac{25}{4.9}\right) \qquad \text{Factor}$$

$$t = 0 \text{ or } t = \frac{25}{4.9} \approx 5.1 \qquad \text{Zero-Product Property}$$

According to this model, the rocket hits the ground after 5.1 seconds. (Of course the rocket is also at ground level at time 0.)

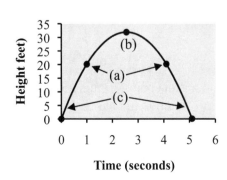

✎ Section 5.4 Exercise 65 Dimensions of a Lot

(a) We want a function A that models the area of the lot in terms of its width, so let x be the width of the lot. We translate the information given in the problem to the language of algebra in this table.

In Words	*In Algebra*
Width of lot	x
Length of lot	$x + 40$

Now we set up the model:

area of lot = width of lot × length of lot

$$A(x) = x(x + 40) = x^2 + 40x$$

(b) Suppose that the lot has an area of 11,700 square feet. Then

$A(x) = x^2 + 40x$	Model
$11{,}700 = x^2 + 40x$	Area of the lot is 11,700 ft^2
$0 = x^2 + 40x - 11{,}700$	Subtract 11,700 from each side
$0 = (x - 90)(x + 130)$	Factor
$x = 90$ or $x = -130$	Zero-Product Property

Since the width of a lot must be a positive number, we conclude that the width is 90 feet. The length of the lot is $x + 40 = 90 + 40 = 130$ feet. So the dimensions of the lot with area 11,700 ft^2 are 90 ft by 130 ft.

✎ Section 5.4 Exercise 67 Stadium Revenue

(a) In Section 5.3 Exercise 33, we find the following function, which models the revenue r in terms of the ticket price x.

$$r(x) = 57{,}000x - 3000x^2$$

(b) We want to find the ticket price for which $r(x) = 0$.

$r(x) = 57{,}000x - 3000x^2$	Model
$0 = 57{,}000x - 3000x^2$	Revenue is \$0
$0 = 19x - x^2$	Divide each side by 3000
$0 = x(19 - x)$	Factor
$x = 0$ or $x = 19$	Zero-Product Property

According to this model, a ticket price of \$19 is just too high; at that price, no one attends the baseball game. (Of course the revenue is also zero if the ticket price is \$0.)

(c) We want to find the ticket price for which $r(x) = 250,000$.

$$r(x) = 57,000x - 3000x^2 \quad \text{Model}$$

$$250,000 = 57,000x - 3000x^2 \quad \text{Revenue is \$250,000}$$

$$250 = 57x - 3x^2 \quad \text{Divide each side by 1000}$$

$$3x^2 - 57x + 250 = 0 \quad \text{Subtract } 57x - 3x^2$$

The expression on the left-hand side of this equation does not factor easily, so we use Quadratic Formula, where a is 3, b is -57, and c is 250.

$$x = \frac{-(-57) \pm \sqrt{(-57)^2 - 4(3)(250)}}{2(3)} = \frac{57 \pm \sqrt{249}}{6}$$

Using a calculator we find that $x \approx 6.87$ or $x \approx 12.13$. So the revenue is \$250,000 when the ticket price is \$12.13 or \$6.87.

5.5 Fitting Quadratic Curves to Data

OBJECTIVES

☑ *Check the box when you can do the exercises addressing the objective, such as those included in this study guide, which are listed here.*

Exercises

Modeling Data with Quadratic Functions

☐ Determine a quadratic function is appropriate to model a collection of data, by looking at a scatter plot of the data.   **5, 9**

☐ Use a graphing calculator to create a quadratic model for a collection of data.   **5, 9**

☐ Use a model to determine an output that corresponds to a particular input, or an input that corresponds to an output.   **5, 9**

☐ Use real-world data to create a model, and then use the model to make predictions about the real world.   **9**

SKILLS

Modeling Data with Quadratic Functions

📖 **Read the introduction to the subsection, 'Modeling Data with Quadratic Functions,'** in which the authors show data that is better modeled by a quadratic function than the linear functions you studied in Section 2.5.

[44]*It is appropriate to model a collection of data with a quadratic function when the data exhibits what trend?*
Answer in your own words:

📖 **Read Section 5.5 Example 1**, in which the authors create a quadratic model using a graphing calculator.

[45]*Why might you want to construct a model for data?*
Answer in your own words:

---

[44] When the data appears to decrease and then increase, or increase and then decrease, then a quadratic model is appropriate.
[45] We construct models to predict outputs for inputs that do not belong to the domain of the data. Also we use models to determine inputs that correspond to particular outputs. For example, you could use the model to find the data's maximum value.

Section 5.5 Exercise 5 A set of data is given.

(a) Make a scatter plot of the data. Is it appropriate to model the data by a quadratic function?

(b) Use a graphing calculator to find the quadratic model that best fits the data. Draw a graph of the model.

(c) Use the model to predict the value of y when x is 7.

x	0.1	0.8	0.8	1.2	1.7	2.3	2.6
y	101.2	106.7	105.2	110.1	112.7	114.6	113.3

x	3.1	3.3	3.4	3.9	4.1	5.2	5.9
y	113.1	109.1	110.4	109.2	107.1	97.6	95.5

Solution

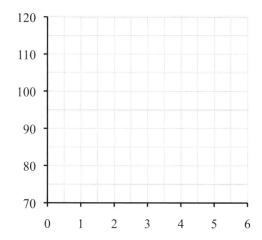

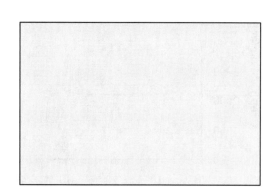

CONTEXTS

📖 **Read Section 5.5 Example 2**, in which the authors model the duration of the life of a tire with respect to its tire pressure.

[46] *The input of the model is _____ and its units are _____. The output of the model is _____ and its units are _____. What does the point (36, 80.4) on the graph of the model represent in context? Answer in your own words:*

---

[46] The input of the model is the tire pressure (in pounds per cubic inch), and the output of the model is the life of the tire (in thousands of miles). Following Hints and Tips 1.2b, we write the following two sentences as scratch work: Algebraically, (36, 80.4) is the point with x-value 36, and y-value 80.4. In context, (36, 80.4) is the point at which the pressure is 36 lbs/in³, and the tire life is 80.4 thousand miles. Our final answer is a more eloquent version: The point (36, 80.4) represents a tire with the maximum life of 80,400 miles, which is attained by maintaining a tire pressure of 36 lbs/in³.

✎ **Section 5.5 Exercise 9 Rainfall and Crop Yield** Rain is essential for crops to grow, but too much rain can diminish crop yields. The data give rainfall and cotton yield per acre for several seasons in a certain county.

(a) Make a scatter plot of the data. Does a quadratic function seem appropriate for modeling the data?
(b) Use a graphing calculator to find the quadratic model that best fits the data. Draw a graph of the model.
(c) Use the model that you found to estimate the yield if there are 25 in. of rainfall.

Solution

Season	Rainfall (in)	Yield (kg/acre)
1	23.3	5311
2	20.1	4382
3	18.1	3950
4	12.5	3137
5	30.9	5113
6	33.6	4814
7	35.8	3540
8	15.5	3850
9	27.6	5071
10	34.5	3881

CONCEPTS

✓ **5.5 Exercises – Concepts – Fundamentals** Complete the Fundamentals section of the exercises at the end of Section 5.5. Compare your answers to those at the back of the textbook, and make corrections as necessary.

📖 **Read the Concept Check for Section 5.5**, in the Chapter Review at the end of Chapter 5.

5.5 Solutions to ✐ Exercises

✐ Section 5.5 Exercise 5

(a) From the data we see that the y-values appear to increase and then decrease, so a quadratic model is appropriate.

(b) Using the QuadReg command on a graphing calculator, we obtain the quadratic function that best fits the data: $y = -1.7529x^2 + 8.983x + 100.898$.

(c) When x is 7, the model predicts the following value for y.

$$y = -1.7529(7)^2 + 8.983(7) + 100.898 \approx 77.9$$

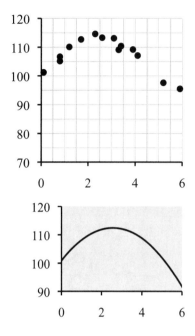

✐ Section 5.5 Exercise 9

(a) From the data we see that the y-values appear to increase and then decrease, so a quadratic model is appropriate.

(b) Using the QuadReg command on a graphing calculator, we obtain the quadratic function that best fits the data:
$y = -12.627x^2 + 651.547x - 3283.157$.

(c) When x is 25, the model predicts the following value for y.

$$y = -12.627(25)^2 + 651.547(25) - 3283.157$$
$$\approx 5114$$

Thus the crop yield is approximately 5114 kilograms per acre when there is 25 inches of rain.

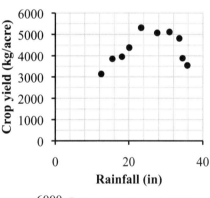

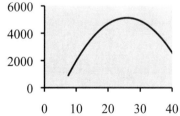

6.1 Working with Functions: Algebraic Operations

OBJECTIVES

☑ *Check the box when you can do the exercises addressing the objective, such as those included in this study guide, which are listed here.*

		Exercises
Adding and Subtracting Functions		
☐	Find the sum or difference of two functions using their formulas.	**13, 15**
☐	Determine the domain of the sum or difference of two functions.	**15, 21,**
☐	Evaluate the sum or difference of two functions at a point in its domain.	**15**
☐	Relate the graph of the sum or difference of two functions to the graphs of the functions.	**13, 21,**
☐	Model the real world with sums or difference of models.	**35**
Multiplying and Dividing Functions		
☐	Find the product or quotient of two functions using their formulas.	**33**
☐	Find the product or quotient of two functions using a table of values.	**39**
☐	Determine the domain of the product or quotient of two functions.	**33**
☐	Interpret products and quotients of models of the real world.	**39**
☐	Describe trends in two-variable data in context.	**39**

GET READY...

This section includes a discussion of the domain of a function. Review this vocabulary in Section 1.7 of the textbook. This section also includes exercises in which you add and subtract graphically. Reinforce your understanding of the connection between a function and its graph by reading Section 1.7 Examples 2 and 3.

SKILLS

Adding and Subtracting Functions

These exercises ask you to add and subtract functions, and find the formulas, domains and graphs of the resulting functions. Find the definitions of the sum and the difference of functions in a blue box in Section 6.1 of the textbook.

📖 **Read Section 6.1 Example 1**, in which the authors find the sums and differences of functions and look at their graphs.

[1] *If the domain of f(x) is A and the domain of g(x) is B, then the domain of f + g and f − g is _____. In the second step of calculating the difference between f and g, the authors substitute _____ for f(x) and _____ for g(x).*

---

[1] The domain of $f + g$ and $f - g$ is $A \cap B$. The authors substitute $(x^2 + 4)$ for $f(x)$ and $(2x + 8)$ for $g(x)$. It is a good idea to include the parentheses when substituting complicated expressions. The expression $x^2 + 4 - 2x + 8$ is incorrect, as it is not the same as $(x^2 + 4) - (2x + 8)$.

✎ **Section 6.1 Exercises 13** Let $f(x) = 2x^2 + 1$ and $g(x) = 3x + 2$.

(a) Find the function $f + g$. Draw the graphs of f, g and $f + g$ in the same viewing rectangle.
(b) Find the function $f - g$. Draw the graphs of f, g and $f - g$ in the same viewing rectangle.

Solution

Copy the graphs from your calculator into the spaces provided.

✎ **Section 6.1 Exercise 15** Let $f(x) = x^2$ and let $g(x) = x - 3$.

(a) Find the functions $f + g$ and $f - g$ and their domains.
(b) Evaluate the functions $f + g$ and $f - g$ at the point $x = 1$, if it is defined.

Solution

📖 **Read Section 6.1 Example 2**, in which the authors find the graph of $f + g$ using the graphs of f and g.

[2] *How do you find the height of the graph of $f + g$ at a point $x = c$ graphically?*
Answer in your own words:

---

[2] The height of the graph of $f + g$ at $x = c$ is the height of the graph of f at $x = c$ plus the height of the graph of g at $x = c$. This is because the height of the graph of $f + g$ at $x = c$ is equal to $f(c) + g(c)$.

📖 **Read Section 6.1 Example 3**, in which the authors model the height of a ball as a difference of two functions.

[3] *How do you find the height of the graph of $f - g$ at a point $x = c$ graphically?*
Answer in your own words:

✐ **Section 6.1 Exercise 21 (modified)** Use graphical addition to sketch the graphs of $f + g$ and $f - g$ on the given coordinate axis provided.

Solution

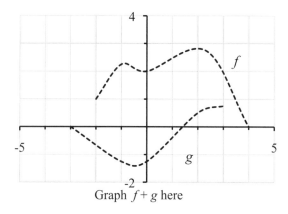

Graph $f + g$ here

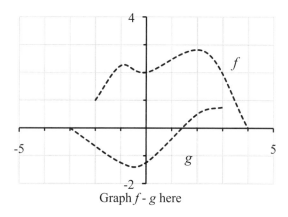

Graph $f - g$ here

[4] *If $f(c) = g(c)$, then $(f - g)(c) =$ _____ . If $g(c) = 0$, then $(f - g)(c) =$ _____ , and $(f + g)(c) =$ _____ .*

---

[3] The height of the graph of $f - g$ at $x = c$ is the height of the graph of f at $x = c$ minus the height of the graph of g at $x = c$. This is because the height of the graph of $f - g$ at $x = c$ is equal to $f(c) - g(c)$. To construct the graph of $f - g$, begin at a point $(c, f(c))$ on the graph of f and go down (if $g(c)$ is positive) or up (if $g(c)$ is negative) by the distance $|g(c)|$ (which is the distance from $(c, g(c))$ to the x-axis).

[4] If $f(c) = g(c)$, then $(f - g)(c) = f(c) - g(c) = f(c)$. If $g(c) = 0$, then $(f - g)(c) = f(c) - g(c) = f(c)$, and $(f + g)(c) = f(c)$.

Multiplying and Dividing Functions

These exercises ask you to multiply and divide functions, finding the formulas, domains and graphs of the resulting functions. Find the definitions of the sum and difference of functions in a blue box in Section 6.1 of the textbook.

📖 **Read Section 6.1 Example 5**, in which the authors find the product and quotient of functions and calculate their domains.

[5]*The domain of f is _____, and the domain of g is _____. So, the domain of fg is _____.*
How do the authors obtain the domain of f/g?
Answer in your own words:

✐ **Section 6.1 Exercise 33** Let $f(x) = \dfrac{2}{x}$, and let $g(x) = \dfrac{4}{x+4}$.

(a) Find the functions fg and f/g and their domains.
(b) Evaluate the functions fg and f/g at $x = 4$, if defined.

Solution

---

[5] The domain of f is all real numbers, and the domain of g is the set of all nonnegative real numbers $\{x \mid x \geq 0\}$. The domain of fg is the intersection of these which is the set, $\{x \mid x \geq 0\}$, of all real numbers greater than or equal to zero. To find the domain of f/g, the authors find the intersection of the domain of f and the domain of g, and then remove any values of x for which g(x) = 0. In this case, g(x) = 0 when x = 0, so we eliminate x = 0 point from the intersection $\{x \mid x \geq 0\}$, to get the domain of f/g, which is the set $\{x \mid x > 0\}$ of all real numbers strictly greater than zero.

CONTEXTS

Adding and Subtracting Functions

These exercises ask you to model profit. Find the formula for profit highlighted in yellow in Section 6.1 of the textbook.

📖 **Read Section 6.1 Example 4**, in which the authors model a donut shop's profit for making and selling bear claws.

[6] *In the second step of calculating the difference between R(x) and C(x), the authors substitute _____ for R(x), and _____ for C(x).*

How do the authors find the number of bear claws the shop has to sell in order for revenues to exceed costs?
Answer in your own words:

✎ **Section 6.1 Exercise 35 Revenue, Cost, and Profit** A store sells a certain type of digital camera. The revenue from selling x cameras is modeled by $R(x) = 450x - 0.5x^2$.

The cost of producing x cameras is modeled by $C(x) = 9300 + 150x - 0.1x^2$. These models are valid for x in the interval [0, 450].

(a) Find a function that models the profit from producing and selling x cameras.
(b) Sketch a graph of the revenue, cost and profit functions. How many cameras must be sold in one week before revenue exceeds cost?

Solution
Copy the graph from your calculator into the space provided.

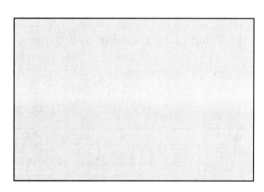

---

[6] The authors substitute $0.80x$ for $R(x)$ and $(50 + 0.40x - 0.0002x^2)$ for $C(x)$. It is a good idea to include the parentheses when substituting complicated expressions. The expression $0.80x - 50 + 0.40x - 0.0002x^2$ is incorrect, as it is not the same as $0.80x - (50 + 0.40x - 0.0002x^2)$. To find the number of bear claws that the shop must sell to break even, the authors look on the diagram for the x-value at which the graph of the revenue crosses to lie above the graph of the costs. This is also the x-value at which the graph of $P(x)$ crosses the x axis.

Multiplying and Dividing Functions

These exercises ask you to interpret products and quotients of functions that model the real world.

> Refer back to page 11 (see also page 36)
> **Hints and Tips 1.2b: What does the algebra represent in context?**

📖 **Read Section 6.1 Example 6**, in which the authors find the amount that an employee invests in stock each paycheck.

[7]*The authors explain that R(1) = 180. What does this represent in context?*
Answer in your own words:

> Refer back to page 12
> **Hints and Tips 1.2c: Describing trends in relations**

The authors follow Hints and Tips 1.2c when drawing a conclusion from this data.

⇨ **Section 6.1 Example 6(b)** What do you conclude?

Solution

The authors describe how the data changes as time increases. They say, this employee is investing more money every month in his company's stock.

✎ **Section 6.1 Exercise 39 Home Sales** Let $f(t)$ be the number of existing homes sold in the United States in month t of 2007, and let $g(t)$ be the same function as f but for 2008.

(a) What is the meaning of the function $h(t) = (f - g)(t)$?

(b) What is the meaning of the function $R(t) = \dfrac{h(t)}{f(t)}$?

(c) Calculate the values of $R(t)$ for each month. What do you conclude?

Month t	1	2	3	4	5	6	7	8	9	10	11	12
2007 sales: $f(t)$ (thousands)	532	550	509	494	494	479	480	458	426	422	418	409
2008 sales: $g(t)$ (thousands)	408	419	412	408	416	404	418	409	428	409	370	396

---

[7] Following Hints and Tips 1.2b, we write the following sentences as scratch work: Algebraically, $R(1) = 180$ means when the input is 1, the output is 180. In context, $R(1) = 180$ means when the time is 1 month, the amount he invests is $180. We express our final answer more eloquently: In the first month, the employee spends $180 on his company's stock.

Solution
Enter your answers to part (c) in the table provided.

Month t	1	2	3	4	5	6	7	8	9	10	11	12
$R(t)$												

CONCEPTS

✓ **6.1 Exercises – Concepts – Fundamentals** Complete the Fundamentals section of the exercises at the end of Section 6.1. Compare your answers to those at the back of the textbook, and make corrections as necessary.

📖 **Read the Concept Check for Section 6.1**, in the Chapter Review at the end of Chapter 6.

6.1 Solutions to ✐ Exercises

✐ Section 6.1 Exercises 13

(a) $(f+g)(x) = f(x) + g(x)$ Definition of $f + g$

$\quad\quad\quad\quad = \left(2x^2 + 1\right) + \left(3x + 2\right)$ Definitions of f and g

$\quad\quad\quad\quad = 2x^2 + 3x + 3$ Simplify

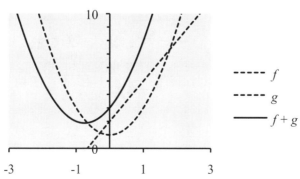

(b) $(f - g)(x) = f(x) - g(x)$ Definition of $f - g$

$\qquad\qquad = \left(2x^2 + 1\right) - \left(3x + 2\right)$ Definitions of f and g

$\qquad\qquad = 2x^2 - 3x - 1$ Simplify

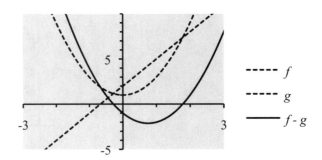

$$\cdots\cdots\ f$$
$$\cdots\cdots\ g$$
$$\text{——}\ f\text{-}g$$

✎ **Section 6.1 Exercise 15**

(a) The functions f and g have domain $\mathbb{R}$, so the domain of $f + g$ is also $\mathbb{R}$.

$(f + g)(x) = f(x) + g(x)$ Definition of $f + g$

$\qquad\qquad = \left(x^2\right) + \left(x - 3\right)$ Definitions of f and g

$\qquad\qquad = x^2 + x - 3$ Simplify

The domain of $f - g$ is also $\mathbb{R}$. We have

$(f - g)(x) = f(x) - g(x)$ Definition of $f - g$

$\qquad\qquad = \left(x^2\right) - \left(x - 3\right)$ Definitions of f and g

$\qquad\qquad = x^2 - x + 3$ Simplify

(b) We can calculate each of these values, because $x = 1$ is in the domain of each function. From part (a) we have

$$(f + g)(1) = (1)^2 + (1) - 3 = -1$$
$$(f - g)(1) = (1)^2 - (1) + 3 = 3$$

✎ **Section 6.1 Exercise 21 (modified)**

Since the domain of f is $[-2, 4]$, and the domain of g is $[-3, 3]$, the domain of $f + g$ is $[-2, 4] \cap [-3, 3] = [-2, 3]$.

We obtain the graph of $f + g$ by "graphically adding" the values of $f(x)$ and $g(x)$.

Note that the length of segment PQ is $g(3) \approx 0.75$. The length of segment PR is $f(3) \approx 2$. We create segment PS by copying segment PQ on top of PR. Thus the length of segment PS is the sum of the lengths of PR and PQ, approximately 2.75. So, we can plot the point $S(3, 2.75)$ on the graph of $f + g$.

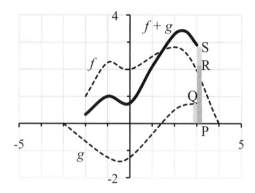

We obtain the graph of $f - g$ by "graphically subtracting" the values of $f(x)$ and $g(x)$. The height of the graph of $f - g$ at $x = c$ is the height of the graph of f at $x = c$ minus the height of the graph of g at $x = c$. To construct the graph of $f - g$, begin at a point $(c, f(c))$ on the graph of f and go down (if $g(c)$ is positive) or up (if $g(c)$ is negative) by the distance $|g(c)|$ (which is the distance from $(c, g(c))$ to the x-axis).

For example, the length of the segment PQ is $g(3) \approx 0.75$. The length of segment PR is $f(3) \approx 2$. We create segment PS by dragging the segment PQ so that Q moves to R. The point at the bottom of the segment PQ moves to S. So segment PS has height $f(3) - g(3) \approx 1.25$, and we can plot the point $S(3, 1.25)$ on the graph of $f - g$.

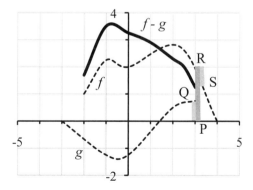

✎ **Section 6.1 Exercise 33**

(a) The domain of f is all real numbers except zero, $\{x \mid x \neq 0\}$, and the domain of g is all real numbers except -4, $\{x \mid x \neq -4\}$. Thus, the intersection of the domains is all real numbers except 0 and -4, $\{x \mid x \neq 0, -4\}$. We have

$$
\begin{aligned}
(fg)(x) &= f(x) \cdot g(x) && \text{Definition of } fg \\[2mm]
&= \left(\frac{2}{x}\right)\left(\frac{4}{x+4}\right) && \text{Definitions of } f \text{ and } g \\[2mm]
&= \frac{8}{x(x+4)} && \text{Simplify}
\end{aligned}
$$

The domain of f/g is $\{x \mid x \neq 0, -4\}$. Note that $g(x) = \dfrac{4}{x+4}$ can never equal 0, so dividing by g does not further restrict the domain.

$$
\begin{aligned}
\left(\frac{f}{g}\right)(x) &= \frac{f(x)}{g(x)} && \text{Definition of } \frac{f}{g} \\[2mm]
&= \frac{\left(\dfrac{2}{x}\right)}{\left(\dfrac{4}{x+4}\right)} && \text{Definitions of } f \text{ and } g \\[2mm]
&= \left(\frac{2}{x}\right)\left(\frac{x+4}{4}\right) && \text{Property of division of fractions} \\[2mm]
&= \frac{x+4}{2x} && \text{Simplify}
\end{aligned}
$$

(b) We can calculate each of these values because 4 is in the domain of each function. Using the results of part (a), we have

$$
(fg)(4) = \frac{8}{4(4+4)} = \frac{8}{4(8)} = \frac{1}{4}, \text{ and}
$$

$$
\left(\frac{f}{g}\right)(4) = \frac{4+4}{2 \cdot 4} = \frac{8}{8} = 1
$$

✎ **Section 6.1 Exercise 35 Revenue, Cost, and Profit**

(a) The profit function is the difference between the revenue and cost functions.

$$
\begin{aligned}
P(x) &= R(x) - C(x) && \text{Definition of profit} \\[2mm]
&= \left(450x - 0.5x^2\right) - \left(9300 + 150x - 0.1x^2\right) && \text{Definitions of } R \text{ and } C \\[2mm]
&= 450x - 0.5x^2 - 9300 - 150x + 0.1x^2 && \text{Distributive Property} \\[2mm]
&= -0.4x^2 + 300x - 9300 && \text{Simplify}
\end{aligned}
$$

(b) Since these models are valid in the interval [0, 450], we sketch the graphs of R, C, and P on that interval.

By zooming in with a graphing calculator, we see that revenue exceeds cost when x is greater than about 32. So about 32 cameras must be sold in one week before the company makes a profit. (Note from the graph that the values of P are positive when x is greater than 32.)

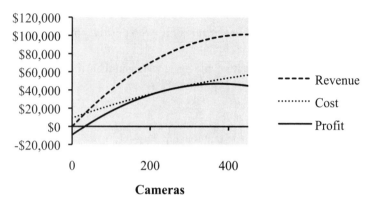

✎ Section 6.1 Exercise 39 Home Sales

(a) Following **Hints and Tips 1.2b**, we write the following sentences as scratch work. Algebraically, the input is t and the output is $f(t) - g(t)$. In context, the input is the month, and the output is the difference between 2007 sales and 2008 sales. We write our final answer more eloquently: The function $h(t)$ gives the change in number of existing homes sold from 2007 to 2008, in the tth month.

(b) Following **Hints and Tips 1.2b**, we write the following sentences as scratch work. Algebraically, the input is t and the output is $h/f = (f - g)/f$. In context, the input is the month, and the output is the ratio of the difference in sales, between 2007 and 2008, to the 2007 sales, which we recognize as the percentage change in sales in the tth month. We give our final answer more eloquently: The function $R(t)$ gives the percentage change in number of homes sold from 2007 to 2008, in the tth month.

(c) The value of $R(t)$ at any t is the quotient of $f(t) - g(t)$ and $f(t)$. Therefore $R(1) = \frac{f(1) - g(1)}{f(1)} = \frac{532 - 408}{532} \approx 0.233$. The other entries in the table below are calculated similarly.

Month t	1	2	3	4	5	6	7	8	9	10	11	12
Percentage Change $R(t)$	0.23	0.24	0.19	0.17	0.16	0.16	0.13	0.11	0	0.03	0.12	0.03

Following **Hints and Tips 1.2c**, we describe the trend in the data as time increases. It appears that the percentage change in home sales decreases throughout the year, leveling off in the last months.

6.2 Power Functions: Positive Powers

OBJECTIVES

☑ Check the box when you can do the exercises addressing the objective, such as those included in this study guide, which are listed here.

Exercises

Power Functions with Positive Integer Powers

☐ Compare the rates of growth of power functions with positive integer powers to the growth rates of exponential functions.

17

☐ Graph power functions with positive integer powers and their transformations.

7, 8, 17

Direct Proportionality

☐ Write equations using proportionality.

23, 35

☐ Use proportionality to describe the relationship between two variables.

37

☐ When a variable is proportional to a power of another variable, relate changes in one variable to changes in the other.

35

Fractional Positive Powers

☐ Graph transformations of power functions with positive fractional powers.

9

☐ Find inverses of power functions.

37

Modeling with Power Functions

☐ Use power functions to model the real world.

35, 37, 39

GET READY...

This section involves power functions, and includes exercises in which you create new functions from power functions by stretching or shrinking, shifting, and reflecting the graph. Review transformations of functions in Section 5.1 of the textbook. This section also discusses the inverses of power functions. Review inverse functions in Section 4.6 of the textbook. All of the exercises in this section involve manipulating and solving expressions with exponents. Review the rules of exponents in Algebra Toolkit A.3 and A.4 and solving power equations in Algebra toolkit C.1.

✓ **Algebra Checkpoint** Test yourself by completing the Algebra Checkpoint at the end of this section of the textbook, and comparing your answers to those in the back of the textbook. Refer to Algebra Toolkit A.3, A.4 and C.1 as necessary.

SKILLS

Power Functions with Positive Integer Powers

These exercises ask you to compare power functions and use the techniques of Section 5.1 to create new functions by translating, stretching or shrinking, and reflecting power functions. Find the definition of a power function in a blue box in Section 6.2 of the textbook.

📖 **Read the introduction to the subsection, 'Power Functions with Positive Integer Powers,'** in which the authors compare the graphs of power functions with positive integer powers.

[8]
Consider the power function f(x) = x^n. The function f has a minimum at x =_____ when n is _____ (even / odd). The function f is increasing on $(-\infty,\infty)$ when n is _____ (even / odd). The function f is decreasing on $(-\infty,0]$ and increasing on $[0,\infty)$ when n is _____ (even / odd).

What are the differences between the graphs of x^2 and x^4?
Answer in your own words:

📖 **Read Section 6.2 Example 1**, in which the authors compare the graphs of a power function and an exponential function.

[9]
The graphs of f(x) = 2^x and g(x) = x^2 intersect at x = _____ and x = _____. Eventually, for large values of x, the graph of the _____ (power / exponential) function is higher.

✏ **Section 6.2 Exercise 17** Compare the rates of growth of the functions $f(x) = 2^x$ and $g(x) = x^5$ by drawing the graphs of both functions in each of the following viewing rectangles.

(a) [0, 5] by [0, 20]
(b) [0, 25] by [0, 10^7]
(c) [0, 50] by [0, 10^8]

Solution

---

[8]
The graph of f has a minimum at x = 0 when n is even. The function f is increasing on $(-\infty,\infty)$ when n is odd. The function f is decreasing on $(-\infty,0]$ and increasing on $[0,\infty)$ when n is even. The graph of x^4 is flatter than the graph of x^2 near x = 0, and steeper than the graph of x^2 when x < −1 and when x > 1.
[9]
The graph of f and g intersect at x = 2 and x = 4. For large values of x, the exponential function is higher.

📖 **Read Section 6.2 Example 2**, in which the authors transform power functions by translating, reflecting and stretching.

[10]*The graph of $P(x) = -x^3$ is the reflection of the graph of $y = x^3$ across the _____ (x-axis / y-axis). The graph of $Q(x) = (x-2)^4$ is the graph of $y = x^4$ shifted to the _____ (left / right) by 2 units. To find the graph of $R(x) = -2x^5 + 4$, first _____ (stretch and reflect / shift) the graph of $y = x^5$.*

✏️ **Section 6.2 Exercises 7 and 8** Sketch the graph of each function by transforming the graph of an appropriate function of the form $y = x^p$ from Figure 1, page 494 of the textbook. Indicate all x- and y-intercepts.

7. (a) $P(x) = x^3 - 8$ (b) $Q(x) = -x^3 + 27$

Solution

   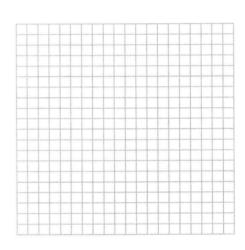

8. (a) $R(x) = -(x+2)^3$ (b) $S(x) = \frac{1}{2}(x-1)^3 + 27$

Solution

   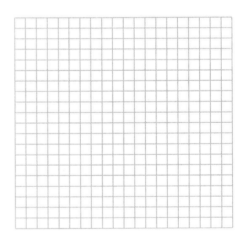

---

[10] The graph of $P(x) = -x^3$ is the reflection of the graph of $y = x^3$ across the x-axis. The graph of $Q(x) = (x-2)^4$ is the graph of $y = x^4$ shifted to the right by 2 units. To find the graph of $R(x) = -2x^5 + 4$, first stretch (by a factor of 2) and reflect (across the x-axis) the graph of $y = x^5$. Next shift the resulting graph upwards by 4 units.

Direct Proportionality

These exercises ask you to describe the relationships between variables using proportionality. Find the definition of directly proportional variables in a blue box in Section 6.2 of the textbook.

📖 **Read Section 6.2 Example 3**, in which the authors use the formula for proportionality.

[11]If you know that L is proportional to w^5, then you can write the equation $L = kw^5$. What additional information do you need to find the value of k?
Answer in your own words:

✎ **Section 6.2 Exercise 23** Express the statement as an equation. Use the given information to find the constant of proportionality.

A is directly proportional to the third power of x. If $x = 2$, then $A = 40$.

Solution

Fractional Positive Powers

These exercises ask you to compare and graph root functions. Find the definition of a root function in bold typeface in Section 6.2 of the textbook.

📖 **Read the introduction to the subsection, 'Fractional Positive Powers,'** in which the authors graph and compare root functions.

[12]Consider the root function $f(x) = x^{1/n}$. The function f has domain $[0, \infty)$ when n is (even / odd), and domain $(-\infty, \infty)$ when n is _____ (even / odd). The root function is _____ (decreasing / increasing) on its domain.

---

[11] To find the constant of proportionality k, you need a value of L and a corresponding value of w. For example, in part (b) you are given $L = 80$ when $w = 2$. Then you substitute these into the equation, and solve for k.

[12] The function f has domain $[0, \infty)$ when n is even, and domain $(-\infty, \infty)$ when n is odd. A root function is increasing on its domain.

✎ **Section 6.2 Exercise 9 (modified)** Sketch the graph of each function by transforming the graph of an appropriate function of the form $y = x^p$ from Figure 5, page 497 of the textbook. Indicate all x- and y-intercepts.

(a) $P(x) = x^{1/4} + 2$ (b) $Q(x) = (x+3)^{1/4}$

Solution

CONTEXTS

Direct Proportionality

These exercises ask you to discuss relationships in the real world using proportionality.

📖 **Read Section 6.2 Example 4**, in which the authors relate the radius of a ball to its volume.

[13] *How do the authors determine what happens to the volume when the radius is doubled?*
Answer in your own words:

---

[13] The authors find a formula for $V(2r)$ in terms of $V(r)$. This means that they substitute $2r$ into the formula for V, and group together the terms $\frac{4}{3}\pi r^3$ from the original formula. Then they substitute $V(r)$ for $\frac{4}{3}\pi r^3$, to get a formula for $V(2r)$ in terms of $V(r)$.

Section 6.2 Exercise 36 Law of the Pendulum The period of a pendulum (the time elapsed during one complete swing of the pendulum) is directly proportional to the square root of the length of the pendulum.

(a) Express the relationship by writing an equation.
(b) To double the period, how would we have to change the length l?

Solution

Fractional Positive Powers

These exercises ask you to use power functions with fractional powers to express the inverses of power functions with integer powers.

📖 **Read Section 6.2 Example 5**, in which the authors model the height of a stone dropped from a cliff.

[14] *Express the proportionality of $d = 16t^2$ and $t = \frac{1}{4}\sqrt{d}$ in words.*

Answer in your own words:

✎ **Section 6.2 Exercise 37 Stopping Distance** The stopping distance D (in feet) of a car after the brakes have been applied is directly proportional to the square of the speed s. The stopping distance of a car on dry asphalt is given by the function $f(s) = 0.033s^2$, where s is measured in mi/h.

(a) Find the inverse function of f.
(b) Interpret the inverse function.
(c) If a car skids 50 ft on dry asphalt, how fast was it moving when the brakes were applied?

Solution

---

[14] The distance the stone falls is proportional to the square of the time that the stone has been falling. The time that the stone has been falling is proportional to the square root of the distance that the stone falls.

Modeling with Power Functions

These exercises ask you to use power functions to model the real world.

📖 **Read Section 6.2 Example 6**, in which the authors model the number of bat-species living in central Mexico.

[15] In part (a), you are given _____ (a number of species / an area), and are asked to find _____ (a number of species / an area). In part (b), you are given _____ (a number of species / an area), and are asked to find _____ (a number of species / an area).

✏ **Section 6.2 Exercise 39 Ostrich Flight?** The weight W (in pounds) of a bird (that can fly) has been related to the wingspan L (in inches) of the bird by the equation $L = 30.6 \cdot W^{0.3952}$. (In Exercise 19 of Section 6.4 this model will be derived from data.)

(a) The bald eagle has a wingspan of about 7.5 ft. Use the model to estimate the weight of the bald eagle.
(b) An ostrich weighs about 300 pounds. Use the model to estimate what the wingspan of an ostrich should be in order for it to be able to fly.
(c) The wingspan of an ostrich is about 72 inches. Use your answer to part (b) to explain why ostriches can't fly.

Solution

CONCEPTS

✓ **6.2 Exercises – Concepts – Fundamentals** Complete the Fundamentals section of the exercises at the end of Section 6.2. Compare your answers to those at the back of the textbook, and make corrections as necessary.

📖 **Read the Concept Check for Section 6.2**, in the Chapter Review at the end of Chapter 6.

_____

[15] In part (a), you are given the surface area of a cave, namely 60 m^2, and you are asked to find the approximate number of bat species present. In part (b), you are given the number of bat species present in a cave, namely 4 species, and you are asked to find the approximate surface area of the cave.

6.2 Solutions to ✐ Exercises

✐ Section 6.2 Exercise 7

(a) The graph of $P(x) = x^3 - 8$ is the graph of $y = x^3$ shifted downward 8 units. The x-intercept of P is 2. The y-intercept of P is –8.

(b) The graph of $Q(x) = -x^3 + 27$ is the graph of $y = x^3$ reflected in the x-axis, and then shifted upward 27 units. The x-intercept of Q is 3. The y-intercept of Q is 27.

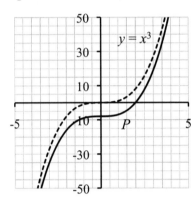
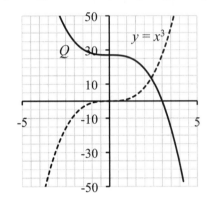

✐ Section 6.2 Exercise 8

(a) The graph of $R(x) = -(x+2)^3$ is the graph of $y = x^3$ shifted to the left 2 units, and then reflected in the x-axis. The x-intercept of R is –2. The y-intercept of R is –8.

(b) We begin with the graph of $y = x^3$. We obtain the graph of $y = \frac{1}{2}(x-1)^3$ by shifting the graph of $y = x^3$ to the right by one unit, and then shrinking the graph vertically by a factor of $\frac{1}{2}$. Finally, we shift the resulting graph upward by 27 units to get the graph of $S(x) = \frac{1}{2}(x-1)^3 + 27$. The x-intercept of S is $1 - 3\sqrt[3]{2} \approx -2.78$. The y-intercept of S is 26.5.

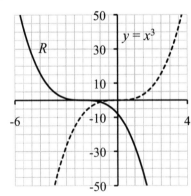
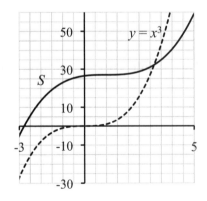

✐ **Section 6.2 Exercise 9 (modified)**

(a) The graph of $P(x) = x^{1/4} + 2$ is the graph of $y = x^{1/4}$ shifted upward by 2 units. The graph of P has no x-intercepts. The y-intercept is 2.

(b) The graph of $Q(x) = (x+3)^{1/4}$ is the graph of $y = x^{1/4}$ shifted to the left by 3 units. The x-intercept of Q is -3. The y-intercept is $(3)^{1/4} = \sqrt[4]{3}$.

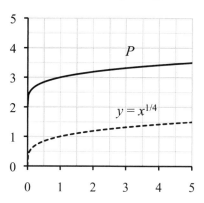
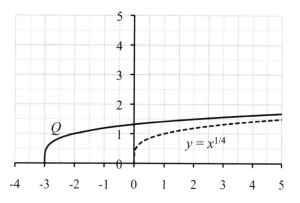

✐ **Section 6.2 Exercise 17**

(a) This figure shows that the graph of $g(x) = x^5$ catches up with, and becomes higher than, the graph of $f(x) = 2^x$ at $x = 1.177$.

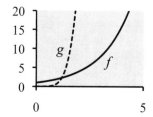

(b) The larger viewing rectangle shows that the graph of $f(x) = 2^x$ overtakes that of $g(x) = x^5$ when $x = 22.44$.

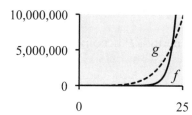

(c) The global view shows that when x is large, $f(x) = 2^x$ is much larger than $g(x) = x^5$.

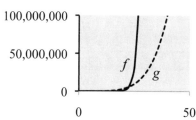

🖊 Section 6.2 Exercise 23

We are given that A is proportional to x^3, so we can write $A = kx^3$, where k is a constant and $k \neq 0$. To find k, we use that $A = 40$ when $x = 2$. So

$$A = kx^3 \qquad\qquad A \text{ is proportional to } x^3$$

$$40 = k \cdot 2^3 \qquad\qquad \text{Replace } A \text{ by 40 and } x \text{ by 2}$$

$$k = \frac{40}{2^3} \qquad\qquad \text{Divide by } (2)^3, \text{ and switch sides}$$

$$k = 5 \qquad\qquad \text{Simplify}$$

The proportionality constant is $k = 5$, so $A = 5x^3$.

🖊 Section 6.2 Exercise 36 Law of the Pendulum

(a) Let P be the period of the pendulum, and l be its length. We are given that P is proportional to $\sqrt{l}$, so we can write $P = k\sqrt{l}$, where k is a constant and $k \neq 0$.

(b) This question asks about how the length l changes as a function of the period P. First we solve the equation of proportionality for l.

$$P = k\sqrt{l} \qquad\qquad \text{Equation}$$

$$\frac{P}{k} = \sqrt{l} \qquad\qquad \text{Divide each side by } k$$

$$l = \left(\frac{P}{k}\right)^2 \qquad\qquad \text{Square each side and switch sides}$$

Thus the length as a function of period is given by $l(P) = \left(\frac{P}{k}\right)^2$. If we double the period, P becomes $2P$, and the corresponding length becomes

$$l = \left(\frac{2P}{k}\right)^2 \qquad\qquad \text{Replace } P \text{ by } 2P$$

$$l = 2^2\left(\frac{P}{k}\right)^2 \qquad\qquad \text{Rules of Exponents}$$

$$l = 4\left(\frac{P}{k}\right)^2 \qquad\qquad \text{Simplify}$$

$$l = 4I(P) \qquad\qquad \text{Replace } \left(\frac{P}{k}\right)^2 \text{ by } I(P)$$

So to double the period, you must make the length of the string 4 times longer.

✎ Section 6.2 Exercise 37 Stopping Distance

(a) First we write the function f in equation form, and then we solve for s.

$$D = 0.033s^2 \qquad \text{Equation form}$$

$$\frac{D}{0.033} = s^2 \qquad \text{Divide by 0.033}$$

$$s = \sqrt{\frac{D}{0.033}} \qquad \text{Take square roots and switch sides}$$

So the inverse function is $f^{-1}(D) = \sqrt{\frac{D}{0.033}}$.

(b) The input of $f^{-1}(D)$ is D, which is the stopping distance. The output of $f^{-1}(D)$ is s, which is the speed. So the speed of a car with stopping distance D is

$s = f^{-1}(D) = \sqrt{\frac{D}{0.033}}$. This says that the speed is proportional to the square root of the stopping distance.

(c) We want to know the speed of a car with stopping distance $D = 50$.

$$s = \sqrt{\frac{D}{0.033}} \qquad \text{Model}$$

$$= \sqrt{\frac{50}{0.033}} \qquad \text{Replace } D \text{ with 50}$$

$$\approx 38.9 \qquad \text{Calculator}$$

So if a car skids 50 ft on dry asphalt, it had been traveling about 38.9 miles per hour when the brakes were applied.

✎ Section 6.2 Exercise 39 Ostrich Flight?

(a) If the bald eagle has a wingspan of about 7.5 feet, then in inches its wingspan is $7.5(12) = 90$ inches. Thus we have

$$L = 30.6 \cdot W^{0.3952} \qquad \text{Model}$$

$$90 = 30.6 \cdot W^{0.3952} \qquad \text{Replace } L \text{ by 90}$$

$$W^{0.3952} = \frac{90}{30.6} \qquad \text{Divide by 30.6 and switch sides}$$

$$W = \left(\frac{90}{30.6}\right)^{1/0.3952} \qquad \text{Raise each side to the power 1/0.3952}$$

$$W \approx 15.3 \qquad \text{Calculator}$$

We predict that the bald eagle weighs approximately 15.3 pounds.

(b) According to the model, $L = 30.6 \cdot W^{0.3952}$, the wingspan of the ostrich should be $L = 30.6 \cdot (300)^{0.3952} \approx 291.5$. So for an ostrich to be able to fly it would need a wingspan of approximately 291.5 inches or 24.3 feet.

(c) The wingspan of an ostrich is about 72 inches which is $291.5 - 72 = 219.5$ inches too short for ostriches to be able to fly.

6.3 Polynomial Functions: Combining Power Functions

OBJECTIVES

☑ *Check the box when you can do the exercises addressing the objective, such as those included in this study guide, which are listed here.*

Exercises

Polynomial Functions

☐ Graph a polynomial by transforming the graphs of power functions, and then using graphical addition.   **7**

Graphing Polynomial Functions by Factoring

☐ Factor a given polynomial, or verify that a given factorization is correct.   **9, 19, 23**

☐ Determine the *x*-intercepts of a function from its formula.   **9, 19, 23**

☐ Determine the intervals on which a polynomial is positive or negative.   **9, 19, 23**

☐ Sketch the graph of a polynomial from a factored version of its formula.   **9, 19, 23**

End Behavior and the Leading Term

☐ Determine the end behavior of a polynomial by comparing it to a power function.   **27**

Modeling with Polynomial Functions

☐ Construct models using geometry.   **47**

☐ Use a graphing calculator to find the maximum of a polynomial function. Also, use a graphing calculator to solve equations graphically.   **47**

GET READY...

This section involves creating polynomials by adding and stretching, or shrinking power functions. Review transformations like stretching and shrinking in Section 5.1, and the addition and subtraction of functions in Section 6.1. When modeling with polynomial functions, you solve equations graphically, since it is often difficult or impossible to solve them algebraically. Review finding solutions using your graphing calculator in Section 1.7 Example 3. This section involves factoring polynomial expressions. Review this skill in Algebra Toolkit B.2.

✓ **Algebra Checkpoint** Test yourself by completing the Algebra Checkpoint at the end of this section of the textbook, and comparing your answers to those in the back of the textbook. Refer to Algebra Toolkit B.2 as necessary.

SKILLS

Polynomial Functions

These exercises ask you to graph polynomials by writing them as combinations of power functions. Find the definition of a polynomial in a blue box in Section 6.3 of the textbook.

📖 **Read Section 6.3 Example 1**, in which the authors use graphical addition and subtraction to graph a polynomial.

[16]To find the graph of $f(x) = x^3 - 3x + 1$, the authors first graph $y =$ _____ and $y =$ _____ and then use graphical addition to construct the graph of $y = x^3 - 3x$. Then they shift that graph _____ (upward / downward) by ___ units to obtain the graph of f.

✏ **Section 6.3 Exercise 7** Graph the polynomial function $f(x) = -x^3 + 2x^2 + 5$ by expressing it as a sum of power functions, and then using graphical addition.

Solution

Do the graphical addition to graph $y = -x^3 + 2x^2$ on the first coordinate axis, and then copy the result to the second coordinate axis and apply the appropriate transformation to graph f.

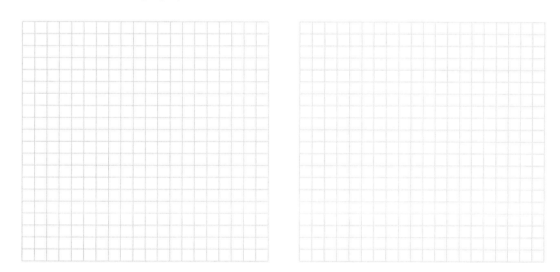

---

[16] The authors first graph $y = x^3$ and $y = -3x$, and then use graphical addition to construct the graph of $y = x^3 - 3x$, which they shift up by one unit to obtain the graph of f.

Graphing Polynomial Functions by Factoring

These exercises ask you to graph a polynomials by first finding the *x*-intercepts.

📖 **Read Section 6.3 Example 2**, in which the authors analyze the formula for a function and use the information they learn to construct a graph.

[17] *The x-intercepts of the graph of f are* ____, ____, *and* ____. *In part (c) the authors divide the x-axis into intervals. The endpoints of these intervals are* −∞, ____, ____, ____, *and* ∞. *How do the authors determine if the graph is above or below the x-axis in the interval* (−∞, −2)? *Answer in your own words:*

✎ **Section 6.3 Exercise 9** Consider the polynomial function $f(x) = x^3 + x^2 - x - 1$.

(a) Show that $f(x) = (x+1)^2(x-1)$.
(b) Find the *x*-intercepts of the graph of *f*.
(c) Find the sign of *f* on each of the intervals determined by the *x*-intercepts.
(d) Sketch a graph of *f*.

Solution
In part (c), plot and label the *x*-intercepts and your test points on the number line.

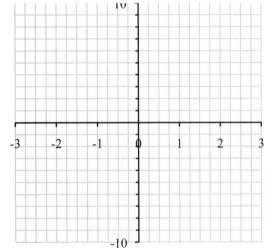

---

[17] The *x*-intercepts are −2, 2 and 3. The endpoints of the intervals are −∞, −2, 2, 3 and ∞. The authors determine if the graph of *f* is above or below the *x*-axis in (−∞, −2) by choosing a test point in that interval, namely −3, and evaluating the function at that point. If the function is positive at the test point, then the graph is above the axis in that interval. If the function is negative at the test point, then the graph is below the axis in that interval. In this case $f(-3) = -30$, so the graph of *f* is below the axis in the interval (−∞, −2).

📖 **Read Section 6.3 Example 3**, in which the authors analyze the formula for a function and use the information they learn to construct a graph.

[18] *List the steps that the authors use to find the graph of the polynomial f from its formula in Examples 2 and 3.*
Answer in your own words:

✎ **Section 6.2 Exercise 19** Consider the polynomial $P(x) = x^3 - x^2 - 6x$.

(a) Express the function P in factored form.
(b) Sketch a graph of P.

Solution
Plot and label the x-intercepts and your test points on the given number line.

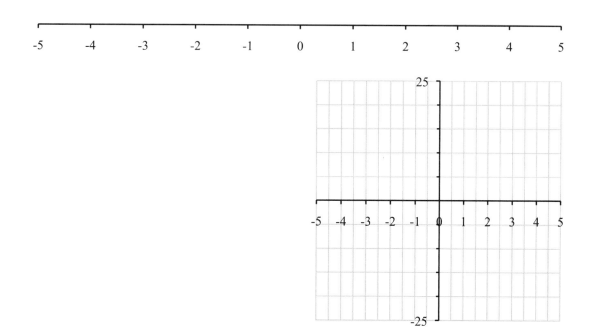

_____

[18] First the authors factor f. Then they use the factored form to find the x-intercepts. They plot each of the x-intercepts on a number line, dividing the number line into intervals, and choose a test point in each interval. They evaluate the function at each test point. If the function is positive at a test point, then the graph lies above the x-axis on that interval. They plot the x-intercepts and test points on the graph, and draw the curve through these points, making f positive or negative on each interval, as appropriate.

📖 **Read Section 6.3 Example 4**, in which the authors analyze the formula for a function and use the information they learn to construct a graph.

[19]*What do the graphs in Examples 3 and 4 have in common? What do the formulas for f have in common in those examples?*
Answer in your own words:

✎ **Section 6.2 Exercise 23** Consider the polynomial $P(x) = x^4 - 3x^3 + 2x^2$.

(a) Express the function P in factored form.
(b) Sketch a graph of P.

Solution
Plot and label the x-intercepts and your test points on the given number line.

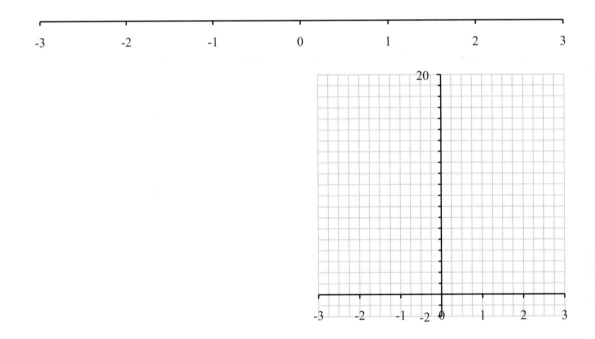

_____

[19] In Example 3, the graph of f does not cross the x-axis at the x-intercept at $x = 0$. On the number line, the test points -1 and 0.5 on either side of $x = 0$ both give positive answers, showing that the graph is above the axis to the left of $x = 0$, then on the axis at $x = 0$, and then above the axis again to the right of $x = 0$. The graph in Example 4 has the same feature at $x = 2$. In the formula for f in Example 3, the factor that is associated with the x-intercept $x = 0$ is squared (in particular, the factor is x^2). Similarly in Example 4, the factor associated with the x-intercept $x = 2$, is squared (in particular, the authors write the factored expression with the term $(x - 2)$ appearing twice).

End Behavior and the Leading Term

These exercises use arrow notation to describe the end behavior of a function. Find the definition of arrow notation in a blue box in Section 6.3 of the textbook.

📖 **Read the introduction to the subsection, 'End Behavior and the Leading Term,'** in which the authors describe how the graphs of polynomials look for large positive and negative values of x.

[20] *What does, "end behavior," mean?*
Answer in your own words:

What is the leading term of a polynomial? Give an example of a polynomial and its leading term to explain.
Answer in your own words:

📖 **Read Section 6.3 Example 5**, in which the authors find the end behavior of the leading term of a polynomial and compare it graphically to the end behavior of the polynomial.

[21] *Fill out the table to describe the end behavior of each power function. Use arrow notation.*

Power Function	End behavior as $x \to -\infty$	End behavior as $x \to \infty$
$y = 3x^5$	$y \to -\infty$	$y \to \infty$
$y = -3x^5$		

Power Function	End behavior as $x \to -\infty$	End behavior as $x \to \infty$
$y = 3x^4$		
$y = -3x^4$		

---

[20] The end behavior of a polynomial function is a description of what happens to the values of the function as x becomes large in the positive or negative direction. The leading term in a polynomial is the term with the highest power of x. For example, in the polynomial $p(x) = 3x^5 - 5x^3 + 2x$, the leading term is $3x^5$.

[21] In the table, for $y = -3x^5$, as $x \to -\infty$, we have $y \to +\infty$, and as $x \to \infty$, we have $y \to -\infty$. For $y = 3x^4$, as $x \to -\infty$, we have $y \to \infty$, and as $x \to \infty$, we have $y \to \infty$. $y = -3x^4$, as $x \to -\infty$, we have $y \to -\infty$, and as $x \to \infty$, we have $y \to -\infty$.

✐ **Section 6.3 Exercise 27** Let $P(x) = 3x^3 - x^2 + 5x + 1$ and let $Q(x) = 3x^3$. Determine the end behavior of P. Compare the graphs of P and Q on large and small viewing rectangles as Example 5.

Solution

Copy a large scale and a small scale graph in the spaces provided.

Contexts

Modeling with Polynomial Functions

When you construct a model from a description of the real world, you can help yourself understand the situation by following Hints and Tips 1.8a and b.

> Refer back to pages 69-70
> **Hints and Tips 1.8a and b: Thinking about the problem**

📖 **Read Section 1.8 Example 4 again**, paying particular attention to how the authors think about the problem and how they use a graphing calculator to get information from the model.

[22] *Before beginning the solution, the authors discuss thinking about the problem. What analysis do they do to help you understand the situation?*
Answer in your own words:

---

[22] Following Hints and Tips 1.8a and b, the authors think about the problem by making up some example dimensions of cereal boxes, and drawing diagrams. In particular, they find the output of the model (the volume of the cereal box), when the input (the depth of the cereal box) is equal to 1, 2, 3 or 4 inches. The authors graph $y = V(x) = 15x^2$, and $y = 60$. The authors find that $V(x) \geq 60$ when $x \geq 1.59$. (The number 1.59 is the x-coordinate of the point of intersection of the two graphs.)

📖 **Read Section 6.3 Example 6**, in which the authors find the largest volume of a box that can be mailed through the U.S. Postal Service.

[23]*What could you do before beginning the solution to help you understand the situation?*
Answer in your own words:

If the side of the base is 14 inches, then the girth is _____ *inches. The girth plus the length should be 108 in, so the length of the box is* _____ *inches. Fill in the table and draw a diagram of each box.*

Side of the base (in)	Girth (in)	Length (in)	Volume (in³)
14	*56*	*52*	*10,192*
16			
18			
20			

📖 **Read Section 6.3 Example 7**, in which the authors use the model in Example 6 to determine the dimensions of a box with a volume of 8000 in^3.

[24]*The authors graph y =* _____ *and y =* _____ *in the same viewing window to find side for which the volume is 8000 in^3.*

_____

[23] Following Hints and Tips 1.8a and b, you could think about the problem by making up some example dimensions of packages and drawing diagrams.

If the side length is 14 inches, then the girth is 4(14) = 56 inches. Since the length plus the girth must be 108, the length equals 108 – 56 =52 in. In that case the volume is 10,192 in^3. In the table, when the side length is 16 in, the girth is 64 in and the length is 44 in. In that case the volume is 11,264 in^3. When the side length is 18 in, the girth is 72 in and the length is 36 in. In that case the volume is 11,664 in^3. In the last row, when the side length is 20 in, the girth is 80 in and the length is 28 in. In that case the volume is 11,200 in^3.

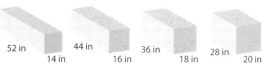

52 in   44 in   36 in   28 in
   14 in   16 in   18 in   20 in

[24] The authors graph $y = 108x^2 - 4x^3$ and $y = 8000$ in the same viewing window, and use the calculator's trace function to find the x-values at which the graphs intersect.

🖉 **Section 6.3 Exercise 47 (extended) Volume of a box** A cardboard box has a square base, with each edge of the base having length x inches, as shown in the figure. The total length of all 12 edges of the box is 144 in.

(a) Show that the volume of the open box is given by the function

$$V(x) = 2x^2(18 - x)$$

(b) What is the domain of V? (Use the fact that length and volume must be positive.)

(c) Draw a graph of the function V, and use it to estimate the maximum volume for such a box.

(d) What are the dimensions of the box with the greatest volume?

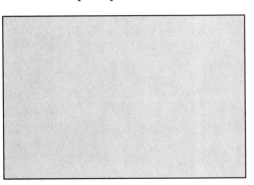

Solution

In part (c), copy the graph from your calculator into the space provided.

CONCEPTS

✓ **6.3 Exercises – Concepts – Fundamentals** Complete the Fundamentals section of the exercises at the end of Section 6.3. Compare your answers to those at the back of the textbook, and make corrections as necessary.

📖 **Read the Concept Check for Section 6.3**, in the Chapter Review at the end of Chapter 6.

6.3 Solutions to ✎ Exercises

✎ Section 6.3 Exercise

The polynomial function f is the sum of the three power functions $y = -x^3$, $y = 2x^2$, and $y = 5$. In the figure on the left, we first graph $y = -x^3$ and $y = 2x^2$; then we use graphical addition to graph their sum $y = -x^3 + 2x^2$. Then we shift this graph upward by 5 units to obtain the graph of $f(x) = -x^3 + 2x^2 + 5$, which we show in the figure on the right.

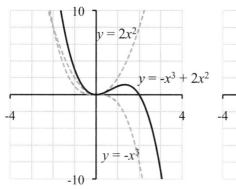 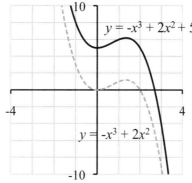

✎ Section 6.3 Exercise 9

(a) We expand the product to see whether it is the same expression as the expression defining f.

$$(x+1)^2(x-1) = (x^2 + 2x + 1)(x - 1) \qquad \text{Special Product Formula 1}$$

$$= (x^2 + 2x + 1)x - (x^2 + 2x + 1) \qquad \text{Distributive Property}$$

$$= x^3 + 2x^2 + x - x^2 - 2x - 1 \qquad \text{Distributive Property}$$

$$= x^3 + x^2 - x - 1 \qquad \text{Combine like terms}$$

The last expression is the same as the expression defining f.

(b) The x-intercepts are the solutions to the equation $f(x) = 0$. We have

$$x^3 + x^2 - x - 1 = 0 \qquad \text{Set } f(x) \text{ equal to } 0$$

$$(x+1)^2(x-1) = 0 \qquad \text{Factored form}$$

$$x + 1 = 0 \text{ or } x - 1 = 0 \qquad \text{Zero-Product Property}$$

$$x = -1 \text{ or } x = 1 \qquad \text{Solve}$$

So the x-intercepts are -1 and 1.

(c) The *x*-intercepts separate the real line into three intervals: $(-\infty,-1)$, $(-1,1)$, and

Interval	Test point	f (test point)	Sign of f in the interval
$(-\infty,-1)$	-2	$f(-2)=-3$	negative
$(-1,1)$	0	$f(0)=-1$	negative
$(1,\infty)$	2	$f(2)=9$	positive

$(1,\infty)$. The function *f* is either positive or negative on each of those intervals. To decide which it is, we pick a test point in the interval and evaluate *f* at that point.

First, in the interval $(-\infty,-1)$, let's pick the test point –2 (any point in the interval would do). Evaluating *f* at –2, we get $f(-2)=(-2+1)^2(-2-1)=-3$. Since $f(-2)$ is less than 0, *f* is negative

● x-intercepts   △ test points

negative   negative   positive

-3   -2   -1   0   1   2   3

at each point in the entire interval $(-\infty,-1)$. Similarly, in the interval $(-1,1)$, we choose the test point 0, and we find $f(0)=(0+1)^2(0-1)=-1$, which is negative. So *f* is negative in the interval $(-1,1)$. In the interval $(1,\infty)$, we choose the test point 2, and we find $f(2)=(2+1)^2(2-1)=9$, which is positive. So *f* is positive in the interval $(1,\infty)$. The results are summarized in the table, and displayed on the number line.

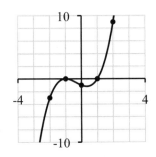

(d) The function *f* is negative on the intervals $(-\infty,-1)$ and $(-1,1)$, and is positive on the interval $(1,\infty)$. We plot our test points and *x*-intercepts to sketch the graph.

✎ Section 6.3 Exercise 19

(a) We factor the polynomial completely.

$$P(x) = x^3 - x^2 - 6x \qquad \text{Definition of } P$$

$$= x(x^2 - x - 6) \qquad \text{Factor } x$$

$$= x(x-3)(x+2) \qquad \text{Factor trinomial}$$

(b) The *x*-intercepts are the solutions to the equation $P(x)=0$. We use the factored form of *P* to solve this equation.

$$x(x-3)(x+2) = 0 \qquad \text{Set } P(x) \text{ equal to } 0$$

$$x=0 \text{ or } x-3=0 \text{ or } x+2=0 \qquad \text{Zero-Product Property}$$

$$x=0 \text{ or } x=3 \text{ or } x=-2 \qquad \text{Solve}$$

So the *x*-intercepts are –2, 0 and 3.

The x-intercepts separate the real line into four intervals: $(-\infty,-2)$, $(-2,0)$, $(0,3)$, and $(3,\infty)$.

We use test points to determine the sign of f in each interval. The results are summarized in the table, and displayed on the number line. We find that function P is negative on the intervals $(-\infty,-2)$ and $(0,3)$, and is positive on the intervals $(-2,0)$ and $(3,\infty)$.

Interval	Test point	P (test point)	Sign of P in the interval
$(-\infty,-2)$	-3	$P(-3) = -18$	negative
$(-2,0)$	-1	$P(-1) = 4$	positive
$(0,3)$	1	$P(1) = -6$	negative
$(3,\infty)$	4	$P(4) = 24$	positive

● x-intercepts   △ test points

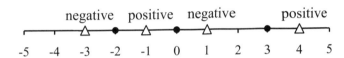

negative   positive   negative   positive

The graph of P is below the x-axis where P is negative and above the x-axis where P is positive. We plot the test points and x-intercepts to sketch the graph.

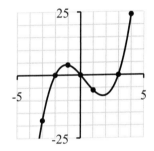

✐ Section 6.3 Exercise 23

(a)   We factor the polynomial completely.

$$P(x) = x^4 - 3x^3 + 2x^2 \qquad \text{Definition of } P$$

$$= x^2\left(x^2 - 3x + 2\right) \qquad \text{Factor } x^2$$

$$= x^2(x-2)(x-1) \qquad \text{Factor trinomial}$$

(b) The x-intercepts are the solutions to the equation $P(x)=0$. We use the factored form of P to solve this equation.

$$x^2(x-2)(x-1) = 0 \qquad \text{Set } P(x) \text{ equal to } 0$$

$$x=0 \ \text{ or } \ x-2=0 \ \text{ or } \ x-1=0 \qquad \text{Zero-Product Property}$$

$$x = 0 \ \text{ or } \ x = 2 \ \text{ or } x = 1 \qquad \text{Solve}$$

So the x-intercepts are 0, 1, and 2.

The x-intercepts separate the real line into four intervals: $(-\infty,0)$, $(0,1)$, $(1,2)$, and $(2,\infty)$. To determine the sign of P on each of these intervals, we use test points. We record the results in the table, and display them on the number line.

The function P is positive on the intervals $(-\infty,0)$, $(0,1)$ and $(2,\infty)$, and is negative on the interval $(1,2)$. The graph of P is below the x-axis where P is negative and above the x-axis where P is positive. We plot the x-intercepts and test points to sketch the graph.

Interval	Test point	P (test point)	Sign of P in the interval
$(-\infty,0)$	-1	$P(-1) = 6$	positive
$(0,1)$	$\dfrac{1}{2}$	$P\left(\dfrac{1}{2}\right) = \dfrac{3}{16}$	positive
$(1,2)$	$\dfrac{3}{2}$	$P\left(\dfrac{3}{2}\right) = -\dfrac{9}{16}$	negative
$(2,\infty)$	3	$P(3) = 18$	positive

● x-intercepts △ test points

Section 6.3 Exercise 27

The leading term of P is the function $Q(x) = 3x^3$. Since Q is a power function with an odd power, its end behavior is as follows:

$$y \to \infty \text{ as } x \to \infty \text{ and } y \to -\infty \text{ as } x \to -\infty$$

We make a table of values of P and Q for large values of x. We see that when x is large, P and Q have approximately the same value (in each case $y \to \infty$). This confirms that they have the same end behavior as $x \to \infty$.

x	$P(x) = 3x^3 - x^2 + 5x + 1$	$Q(x) = 3x^3$
15	9,976	10,125
30	80,251	81,000
50	372,751	375,000

The larger the viewing rectangle, the more the graphs look alike. This confirms that they have the same end behavior.

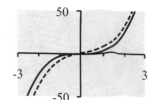
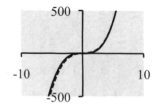

✐ **Section 6.3 Exercise 47 (extended) Volume of a box**

(a) Let x be the side of the base and let L be the length. The volume of the box is

$$V = x^2 L \qquad\qquad \text{Volume of the box}$$

To express V as a function of the variable x alone, we need to eliminate L. We know that the box has eight edges of length x, and four edges of length L. Since the lengths of all the edges is 144 in., we have $8x + 4L = 144$. Solving for L, we find $L = \dfrac{144 - 8x}{4} = 36 - 2x$. Using this expression for L, we can now express the function V as a function of x alone.

$$V = x^2 L \qquad\qquad \text{Volume of the box}$$

$$V = x^2(36 - 2x) \qquad\qquad \text{Replace } L \text{ by } 36 - 2x$$

$$V = 2x^2(18 - x) \qquad\qquad \text{Factor 2}$$

(b) Since the base length must be positive we have that $x > 0$. Similarly,

$$L = 36 - 2x > 0 \qquad\qquad \text{Length must be positive}$$

$$36 > 2x \qquad\qquad \text{Add } 2x$$

$$18 > x \qquad\qquad \text{Divide by 2}$$

$$x < 18 \qquad\qquad \text{Switch sides}$$

Also the volume must be positive, so we have

$$V > 0 \qquad\qquad \text{Volume must be positive}$$

$$2x^2(18 - x) > 0 \qquad\qquad \text{Replace } V \text{ by the model from part (a)}$$

For the product $2x^2(18 - x)$ to be positive, the terms $2x^2$ and $(18 - x)$ must have the same sign. We have one of two cases:

Case I: $2x^2 > 0$ and $18 - x > 0$

or

Case II: $2x^2 < 0$ and $18 - x < 0$

However, $2x^2$ is always positive, so Case II is impossible.

Therefore,

$$18 - x > 0 \qquad\qquad \text{Case I}$$

$$18 > x \qquad\qquad \text{Add } x$$

$$x < 18 \qquad\qquad \text{Switch sides.}$$

Thus, the domain of V is $(0,18)$.

(c) Zooming in on the graph on a graphing calculator, we see that the maximum volume of the box is 1728 square inches.

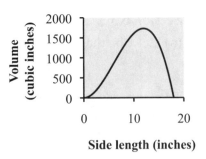

(d) The maximum volume (1728 in^2) occurs when x is 12. Thus the length of the box must be

$$L = 36 - 2x \qquad \text{From part (a)}$$

$$= 36 - 2(12) \qquad \text{Replace } x \text{ by } 12$$

$$= 12 \qquad \text{Calculate}$$

So the dimensions of the box with maximum volume are $12 \times 12 \times 12$ inches.

6.4 Fitting Power and Polynomial Curves to Data

OBJECTIVES

☑ *Check the box when you can do the exercises addressing the objective, such as those included in this study guide, which are listed here.*

Exercises

Fitting Power Curves to Data

☐ Use the PwrReg function to find power models on a graphing calculator.   **15,**

☐ Graph a model and the data on the same scatter plot to determine the quality of the fit.   **11, 15, 21**

☐ Use power functions to model the real world.   **15**

☐ Use real-world data to create a model, and then use the model to make predictions about the real world.   **15, 21**

A Linear, Power, or Exponential Model?

☐ Plot data on scatter, semi-log and log-log plots.   **11**

☐ Determine what type of model to use to model data.   **11**

Fitting Polynomial Curves to Data

☐ Use the CubicReg function to find power models on a graphing calculator.   **21**

☐ Model the real world with cubic polynomials.   **21**

GET READY...

This section involves fitting polynomial and power functions to data using the regression functions on a graphing calculator. Review the other regressions that you have studied in Section 2.5 (linear), Section 3.5 (exponential) and Section 5.5 (quadratic).

SKILLS

A Linear, Power, or Exponential Model?

These exercises ask you to use semi-log and log-log plots to determine if you should choose a linear, power, or exponential model to fit a set of data. Find semi-log and log-log plot in bold typeface, and find a scheme for determining which model is appropriate in an orange box in Section 6.4 of the textbook.

📖 **Read the introduction to the subsection, 'A Linear, Power, or Exponential Model?'** in which the authors derive the scheme to determine which model is appropriate for a set of data.

[25] *Suppose your data is a relation with input x and output y. To construct a scatter plot, plot the points (____, ____). To construct a semi-log plot, plot the points (____, ____). To construct a log-log plot, plot the points (____, ____).*

A linear model of the data is appropriate if the data lie approximately along a line when plotted on a _____ (scatter / semi-log / log-log) plot. A exponential model of the data is appropriate if the data lie approximately along a line when plotted on a _____ (scatter / semi-log / log-log) plot. A power model of the data is appropriate if the data lie approximately along a line when plotted on a _____ (scatter / semi-log / log-log) plot.

📖 **Section 6.4 Example 2**, in which the authors determine what type of function to use to model a given data set.

[26] *The authors chose to model the data with _____ (a linear / an exponential / a power) function because the data, plotted on a _____ (scatter / semi-log / log-log) plot, look like a line.*

---

[25] To construct a scatter plot, plot the points (x, y); to construct a semi-log plot, plot the points $(x, \log y)$; and to plot a log-log plot, plot the points $(\log x, \log y)$. A linear model is appropriate if the data lie along a line when plotted on a scatter plot; an exponential model is appropriate if the data lie along a line when plotted on a semi-log plot; and a power function is appropriate if the data lie along a line when plotted on a log-log plot.

[26] The authors model the data with a power function, since the log-log plot looks like a line.

✎ **Section 6.4 Exercise 11** Data points (x, y) are given.

(a) Draw a scatter plot of the data.

(b) Make semi-log and log-log plots of the data.

(c) Is a linear, power, or exponential function appropriate for modeling these data?

(d) Find an appropriate model for the data, and then graph the model on the axes with the scatter plot of the data.

Solution

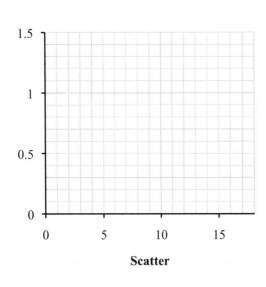

Scatter

	Complete the table to help you make the graphs in parts (b) and (d).			
x	y	$\log x$	$\log y$	Part (d) model output
2	0.08	*0.30103*	*−1.09691*	
4	0.12			
6	0.18			
8	0.25			
10	0.36			
12	0.52			
14	0.73			
16	1.06			

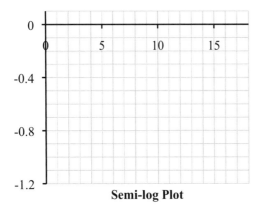

Semi-log Plot

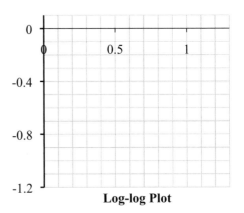

Log-log Plot

CONTEXTS

Fitting Power Curves to Data

These exercises ask you to find power functions that model data.

📖 **Read Section 6.4 Example 1**, in which the authors model the number of species of cave-bat in a cave as a function of the cave's surface area. You studied this model in Section 6.2 Example 6.

[27] *The PwrReg command produces two numbers, a =* _____ *and b =* _____ . *This means that in the model* $S(A) = CA^p$, $C =$ _____ *and* $A =$ _____ .

✎ **Section 6.4 Exercise 15 A Falling Ball** In a physics experiment a lead ball is dropped from a height of 5 m. The students record the distance the ball has fallen every one-tenth of a second. (This can be done by using a camera and a strobe light.)

(a) Make a scatter plot of the data.

(b) Use a graphing calculator to find a power function of the form $d = at^b$ that models the distance d that the ball has fallen after t seconds.

(c) Draw a graph of the function you found on the axes with the scatter plot. How well does the model fit the data?

(d) Use your model to predict how far a dropped ball would fall in 3 seconds.

Solution

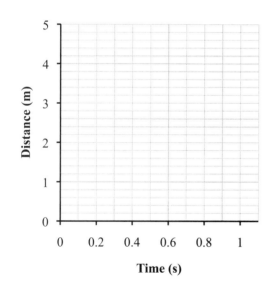

Time (s)	Distance (m)	Part (c) Model output
0.1	0.048	
0.2	0.197	
0.3	0.441	
0.4	0.882	
0.5	1.227	
0.6	1.765	
0.7	2.401	
0.8	3.136	
0.9	3.969	
1.0	4.902	

---

[27] The PwrReg command output $a = 0.140019$ and $b = 0.640512$. This means that in the model, $C = 0.140019$ and $p = 0.640512$. In the model the authors use, they round to two decimal places.

Fitting Polynomial Curves to Data

These exercises ask you to find polynomial functions to model data.

📖 **Read the introduction to the subsection, 'Fitting Polynomial Curves to Data,'** in which the authors introduce polynomial models.

[28] *Polynomial functions are ideal for modeling data for which the scatter plot has _____.*
The more peaks and valleys the data exhibits, the _____ (higher / lower) the degree of the polynomial.

📖 **Read Section 6.4 Example 3**, in which the authors model the length of a Rockfish as a function of its age.

[29] *In part (c), you are given the _____ (length / age) and asked to estimate the _____ (length / age)*
of the fish. In part (d), you are given the _____ (length / age) and asked to estimate the _____
(length / age) of the fish.

What does the data point (14, 26.9) represent in context?
Answer in your own words:

✏️ **Section 6.4 Exercise 21 How Fast Can You Name Your Favorite Things?** If you are asked to make a list of objects in a certain category, how fast you can list them follows a predictable pattern. For example, if you try to name as many vegetables as you can, you'll probably think of several right away— for example, carrots, peas, beans, corn, and so on. Then after a pause you might think of ones you eat less frequently—perhaps zucchini, eggplant, and asparagus. Finally, a few more exotic vegetables might come to mind—artichokes, jicama, bok choy, and the like. A psychologist performs this experiment on a number of subjects. The table below gives the average number of vegetables that the subjects named by a given number of seconds.

(a) Find the cubic polynomial that best fits the data.
(b) Draw a graph of the polynomial from part (a) together with a scatter plot of the data.
(c) Use your result from part (b) to estimate the number of vegetables that subjects would be able to name in 40 seconds.
(d) According to the model, how long (to the nearest 0.1 s) would it take a person to name five vegetables?

---

[28] Polynomial functions are idea to model data for which the scatter plot has peeks and valleys. The more peeks and valleys, the higher the degree of the polynomial.

[29] In part (c), you are given the age of a rockfish, namely 5 years old, and you are asked to estimate the fish's length. In part (d), you are given the length of a rockfish, namely 20 inches long, and you are asked to estimate the fish's age. Referring to Hints and Tips 1.2b, we write the following two sentences as scratch work. Algebraically, (14, 26.9) is a data point for which x-value is 14 and the y-value is 26.9. In context, (14, 26.9) represents a fish with age 14 years old and length 26.9 inches long. We write our final answer more eloquently: The data point (14, 26.9) represents a 26.9-inch long fish that is a 14 year old.

Solution

Seconds	Number of vegetables	Part (a) model output
1	2	
2	6	
5	10	
10	12	
15	14	
20	15	
25	18	
30	21	

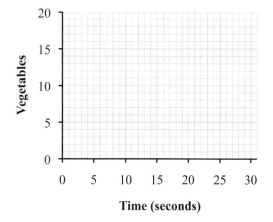

CONCEPTS

✓ **6.4 Exercises – Concepts – Fundamentals** Complete the Fundamentals section of the exercises at the end of Section 6.4. Compare your answers to those at the back of the textbook, and make corrections as necessary.

📖 **Read the Concept Check for Section 6.4**, in the Chapter Review at the end of Chapter 6.

6.4 Solutions to ✎ Exercises

✎ Section 6.4 Exercise 11

(a) For a scatter plot of the data we graph the points (x, y).

x	y	$\log x$	$\log y$
2	0.08	*0.30103*	*−1.09691*
4	0.12	*0.60206*	*−0.92082*
6	0.18	*0.77815*	*−0.74473*
8	0.25	*0.90309*	*−0.60206*
10	0.36	*1*	*−0.44370*
12	0.52	*1.07918*	*−0.28400*
14	0.73	*1.14613*	*−0.13668*
16	1.06	*1.20412*	*0.02531*

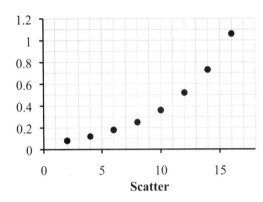

Scatter

(b) For the semi-log plot we graph the points $(x, \log y)$. For the log-log plot we graph the points $(\log x, \log y)$.

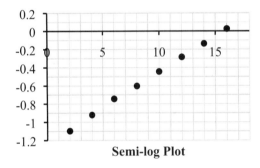

Semi-log Plot

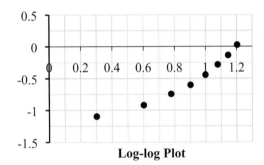

Log-log Plot

(c) From the graphs in parts (a) and (b) we see that the semi-log plot lies along a line much more closely than the other plots, so we conclude that an exponential model is more appropriate.

(d) Using the ExpReg command on a graphing calculator, we get the power model $y = 0.0577(1.2)^x$. We graph this function along with the original data.

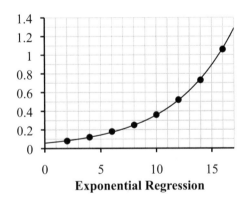

Exponential Regression

✐ Section 6.4 Exercise 15 A Falling Ball

(b) Using a graphing calculator and the PwrReg command, we get the power model
$d = 4.96t^{2.0027}$.

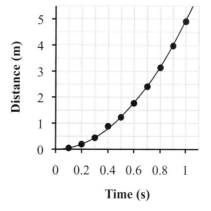

(a, c) This graph shows the scatter plot together with the graph of the model $d = 4.96t^{2.0027}$. The model appears to fit the data well.

(d) When the time t is 3 seconds, our model gives

$$d = 4.96t^{2.0027} \qquad \text{Model}$$

$$d = 4.96(3)^{2.0027} \qquad \text{Replace } t \text{ with } 3$$

$$d = 44.77$$

The ball drops 44.79 meters in 3 seconds.

✐ Section 6.4 Exercise 21 How Fast Can You Name Your Favorite Things?

(a) Using the CubicReg command on a graphing calculator, we find the cubic polynomial of best fit: $y = 0.002x^3 - 0.105x^2 + 1.97x + 1.46$

(c) We replace x by 40 in the model, to get

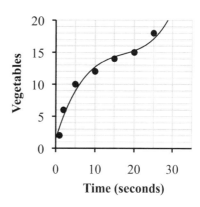

$$y = 0.002(40)^3 - 0.105(40)^2 + 1.97(40) + 1.46$$
$$= 40.26$$

So in 40 seconds subjects could name 40 vegetables.

(d) Tracing the graph of the polynomial on a graphing calculator, we find that y is 5 when x is approximately 2.0 seconds. So it takes a person approximately 2 seconds to name 5 vegetables.

6.5 Power Functions: Negative Powers

OBJECTIVES

☑ Check the box when you can do the exercises addressing the objective, such as those included in this study guide, which are listed here.   **Exercises**

The Reciprocal Function

☐ Analyze power functions with odd negative powers numerically.   **5**

☐ Use arrow notation to describe asymptotes.   **5, 7**

☐ Identify, graph and describe in words horizontal and vertical asymptotes.   **5, 7, 13, 15**

Inverse Proportionality

☐ Write equations using inverse proportionality.   **25, 31, 37**

☐ Use inverse proportionality to describe the relationship between two variables.   **25, 31, 37**

☐ Model the real world using inverse proportionality.   **31, 37**

Inverse Square Laws

☐ When a variable is inversely proportional to a power of another variable, relate changes in one variable to changes in the other.   **37**

☐ Analyze power functions with even negative powers numerically.   **7**

☐ Model the real world using the inverse square law.   **37**

GET READY...

This section involves analysis of graphs with asymptotes, and uses arrow notation to describe the features of a graph. Review arrow notation in Section 6.3 of the textbook. The functions in this section are reciprocals of power functions, so you use the laws of exponents extensively. Review the laws of exponents and negative exponents in Algebra Toolkits A.3 and A.4.

✓ **Algebra Checkpoint** Test yourself by completing the Algebra Checkpoint at the end of this section of the textbook, and comparing your answers to those in the back of the textbook. Refer to Algebra Toolkits A.3 and A.4 as necessary.

SKILLS

The Reciprocal Function

These exercises ask you to analyze the reciprocal function, and use arrow-notation to describe horizontal and vertical asymptotes. Find the definition of the reciprocal function and the asymptotes in bold typeface in Section 6.5 of the textbook.

📖 **Read Section 6.5 Example 1**, in which the authors analyze and graph the reciprocal function.

[30]*The authors explain that $y \to -\infty$ as $x \to 0^-$. What does that mean and how do the authors demonstrate this fact in the table?*
Answer in your own words:

Complete the table to summarize the analysis of the reciprocal function.

Behavior of x	Corresponding behavior of $y = 1/x$	Meaning in words
$x \to -\infty$	$y \to 0$	**As x approaches $-\infty$, y approaches 0**
$x \to 0^-$		
$x \to 0^+$		
$x \to \infty$		

The reciprocal function has a vertical asymptote at $x =$ _____, and a horizontal asymptote at $y =$ _____.

---

[30] The notation $y \to -\infty$ as $x \to 0^-$ means that y approaches negative infinity, as x approaches 0 from the left. This means that if you trace the graph of y, starting to the left of 0 and moving toward 0, the values of y become large negative numbers. In the table, the authors list points starting with $x = -0.1$ to the left of 0, and then choosing numbers closer and closer to 0, namely $x = -0.01$ and $x = -0.00001$. We estimate that $y \to -\infty$, since the corresponding y-values in the table are -10, -100, and $-100,000$, which are increasingly large negative numbers.

Behavior of x	Corresponding behavior of $y = 1/x$	Meaning in words
$x \to -\infty$	$y \to 0$	As x approaches $-\infty$, y approaches 0
$x \to 0^-$	$y \to -\infty$	As x approaches 0 from the left, y approaches $-\infty$
$x \to 0^+$	$y \to \infty$	As x approaches 0 from the right, y approaches ∞
$x \to \infty$	$y \to 0$	As x approaches ∞, y approaches 0

The reciprocal function has a vertical asymptote at $x = 0$, and a horizontal asymptote at $y = 0$.

✎ **Section 6.5 Exercise 5** Consider the function $f(x) = x^{-3}$.

(a) Complete each table with the values of the function.
(b) Describe the behavior of the function near its vertical asymptote, based on Tables 1 and 2.
(c) Determine the behavior of the function near its horizontal asymptote, based on tables 3 and 4.

Solution

x	$f(x)$
−0.1	*−1,000*
−0.01	
−0.001	
−0.00001	

x	$f(x)$
0.1	
0.01	
0.001	
0.00001	

x	$f(x)$
10	
50	
100	
100,000	

x	$f(x)$
−10	
−50	
−100	
−100,000	

✎ **Section 6.5 Exercise 13 (modified)** Sketch a graph of the function $f(x) = x^{-3}$. Use your work from Section 6.5 Exercise 5, above.

Solution

Inverse Proportionality

These exercises ask you to describe the relationship between variables in terms of inverse proportionality. Find the definition of inverse proportionality in a blue box in Section 6.5 of the textbook.

📖 **Read Section 6.5 Example 2**, in which the authors use Boyle's Law to relate the pressure and volume of a sample of gas under compression.

[31] *Since P is inversely proportional to V, you can write the equation $P = k/V$, where k is a constant and $k \neq 0$. What additional information do you need to find the value of k?*
Answer in your own words:

✎ **Section 6.5 Exercise 25** Express the statement as an equation Use the given information to find the constant of proportionality.

z is inversely proportional to t. If t is 3, then z is 2.

Solution

---

[31] To find the constant of proportionality, you need a value of P and a corresponding value of V. For example, in part (b) you are given $P = 50$ kilopascals when $V = 0.106$ cubic meters. Then you substitute these into the equation, and solve for k.

Inverse Square Laws

These exercises ask you to model quantities related by an inverse square law. Find the definition of an inverse square law in bold typeface in Section 6.5 of the textbook.

📖 **Read Section 6.5 Example 3**, in which the authors analyze and graph the reciprocal of x^2.

[32] *Complete the table to summarize the analysis of the reciprocal function.*

Behavior of x	Corresponding behavior of $y = 1/x^2$	Meaning in words
$x \to -\infty$	$y \to 0$	**As x approaches −∞, y approaches 0**
$x \to 0^-$		
$x \to 0^+$		
$x \to \infty$		

🖊 **Section 6.5 Exercise 7** Consider the function $f(x) = x^{-4}$.

(a) Complete each table with the values of the function.
(b) Describe the behavior of the function near its vertical asymptote, based on Tables 1 and 2.
(c) Determine the behavior of the function near its horizontal asymptote, based on tables 3 and 4.

Solution

x	$f(x)$
−0.1	**10,000**
−0.01	
−0.001	
−0.00001	

x	$f(x)$
0.1	
0.01	
0.001	
0.00001	

x	$f(x)$
10	
50	
100	
100,000	

x	$f(x)$
−10	
−50	
−100	
−100,000	

---

[32] The function $y = 1/x^2$ has a vertical asymptote at $x = 0$, and a horizontal asymptote at $y = 0$.

Behavior of x	Corresponding behavior of $y = 1/x^2$	Meaning in words
$x \to -\infty$	$y \to 0$	As x approaches −∞, y approaches 0
$x \to 0^-$	$y \to \infty$	As x approaches 0 from the left, y approaches ∞
$x \to 0^+$	$y \to \infty$	As x approaches 0 from the right, y approaches ∞
$x \to \infty$	$y \to 0$	As x approaches ∞, y approaches 0

✐ **Section 6.5 Exercise 15 (modified)** Sketch a graph of the function $f(x) = x^{-4}$. Use your work from Section 6.5 Exercise 7, above.

Solution

Contexts

Inverse Proportionality

These exercises ask you to model quantities that are inversely proportional.

📖 **Read Section 6.5 Example 2 again**, paying particular attention to which number in part (b) is the temperature, which is the pressure and which is the volume. Also note the graph in figure 2 on page 530.

[33] *In part (b), the volume of the gas is* _____, *the temperature is* _____, *and the pressure is* _____ *(include units). From the graph in Figure 2, as volume increases ($V \to \infty$) the pressure tends to* _____ *(which we write as* _____ *in arrow notation).*

---

[33] The volume is 0.106 m^3, the pressure is 50 kilopascals, and the temperature is 25°C. As volume increases, the pressure tends to 0 kilopascals.

✎ **Section 6.5 Exercise 31 Boyle's Law** The pressure P of a sample of oxygen gas that is compressed at a constant temperature is inversely proportional to the volume V of the gas.

(a) Find the constant of proportionality if a sample of oxygen gas that occupies 0.671 m^3 exerts a pressure of 39 kPa at a temperature of 293 K (absolute temperature measured on the Kelvin scale). Write the equation that expresses the inverse proportionality.

(b) If the sample expands to a volume of 0.916 m^3, find the new pressure.

Solution

Inverse Square Laws

These exercises ask you to model quantities related by the inverse square law.

📖 **Read Section 6.5 Example 4**, in which the authors model the intensity of light as a function of distance from the source.

[34] *How do you evaluate $I(\frac{1}{2}d)$?*

Answer in your own words:

How do the authors explain what happens to the intensity when the distance is halved?
Answer in your own words:

----

[34] To evaluate $I(\frac{1}{2}d)$, replace x in the formula for I by $(\frac{1}{2}d)$. Tip: to avoid making mistakes, include the parentheses when you make the substitution. The authors find a formula for $I(\frac{1}{2}d)$ in terms of $I(d)$. This means that they substitute $(\frac{1}{2}d)$ into the formula for I, and group together the terms $\dfrac{k}{d^2}$ from the original formula. Then they substitute $I(d)$ for $\dfrac{k}{d^2}$, to get a formula for $I(\frac{1}{2}d)$ in terms of $I(d)$.

✏ **Section 6.5 Exercise 37 Heat of Campfire** The heat experienced by a hiker at a campfire is inversely proportional to the cube of his distance from the fire.

(a) Write an equation that expresses the inverse proportionality.

(b) If the hiker is too cold and moves halfway closer to the fire, how much more heat does the hiker experience?

Solution

📖 **Read the end of the subsection, 'Inverse Square Laws,'** in which the authors explain why many of the laws of nature are inverse square laws.

[35] *At a distance r from the source, the strength of the source is spread out over the surface of a sphere of radius r, which has area A = _____. So the intensity at a distance r from the source equals the _____ divided by the _____. Thus the intensity is_____ (proportional / inversely proportional) to the area and hence to r^2.*

CONCEPTS

✓ **6.5 Exercises – Concepts – Fundamentals** Complete the Fundamentals section of the exercises at the end of Section 6.5. Compare your answers to those at the back of the textbook, and make corrections as necessary.

📖 **Read the Concept Check for Section 6.5**, in the Chapter Review at the end of Chapter 6.

_____

[35] At a distance r from the source, the strength of the source is spread out over the surface of a sphere of radius r, which has area $A = 4\pi r^2$. So the intensity at a distance r from the source equals the strength of the source divided by the surface area of the sphere. Thus the intensity is inversely proportional to r^2.

6.5 Solutions to ✎ Exercises

✎ **Section 6.5 Exercise 5**

(a) We calculate the values of $f(x) = x^{-3}$.

x	$f(x)$
−0.1	−1000
−0.01	−1,000,000
−0.001	−1,000,000,000
−0.00001	−1,000,000,000,000,000

x	$f(x)$
0.1	1000
0.01	1,000,000
0.001	1,000,000,000
0.00001	1,000,000,000,000,000

x	$f(x)$
10	0.001
50	0.000 008
100	0.000 001
100,000	0.000 000 000 000 001

x	$f(x)$
−10	−0.001
−50	−0.000 008
−100	−0.000 001
−100,000	−0.000 000 000 000 001

(b) The first table shows that as x approaches zero from the left, the values of $y = f(x)$ become large in the negative direction. The second table shows that as x approaches zero from the right, the values of $f(x) = x^{-3}$ become large in the positive direction. We describe this behavior in symbols and in words as follows:

$y \to -\infty$ as $x \to 0^-$ — y approaches negative infinity as x approaches zero from the left

$y \to \infty$ as $x \to 0^+$ — y approaches infinity as x approaches zero from the right

(c) The next two tables show that as x becomes large in the positive or negative direction, the values of $y = f(x)$ get closer and closer to zero. We describe this situation in symbols as follows:

$y \to 0$ as $x \to -\infty$ — y approaches zero as x approaches negative infinity

$y \to 0$ as $x \to \infty$ — y approaches zero as x approaches infinity

✎ Section 6.5 Exercise 7

(a) We calculate the values of $f(x) = x^{-4}$, noting that $f(-x) = f(x)$.

x	$f(x)$
−0.1	**10,000**
−0.01	**100,000,000**
−0.001	**1,000,000,000,000**
−0.00001	**1×10^{20}**

x	$f(x)$
0.1	**10,000**
0.01	**100,000,000**
0.001	**1,000,000,000,000**
0.00001	**1×10^{20}**

x	$f(x)$
10	**0.0001**
50	**0.00000016**
100	**0.00000001**
100,000	**1×10^{-20}**

x	$f(x)$
−10	**0.0001**
−50	**0.00000016**
−100	**0.00000001**
−100,000	**1×10^{-20}**

(b) The first table shows that as x approaches zero from the left, the values of $y = f(x)$ become large in the positive direction. Similarly, the second table shows that as x approaches zero from the right, the values of $f(x) = x^{-4}$ become large in the positive direction. We describe this behavior in symbols and in words as follows:

$y \to \infty$ as $x \to 0^-$ y approaches infinity as x approaches zero from the left

$y \to \infty$ as $x \to 0^+$ y approaches infinity as x approaches zero from the right

(c) The next two tables show that as x becomes large in the positive or negative direction, the values of $y = f(x)$ get closer and closer to zero. We describe this situation in symbols as follows:

$y \to 0$ as $x \to -\infty$ y approaches zero as x approaches negative infinity

$y \to 0$ as $x \to \infty$ y approaches zero as x approaches infinity

✐ Section 6.5 Exercise 13 (modified)

Using the end behavior from Exercise 5, and plotting a few additional points, we obtain the graph of $f(x) = x^{-3}$.

x	$f(x)$
-2	$-\dfrac{1}{8}$
-1	-1
$-\dfrac{1}{2}$	-8
$\dfrac{1}{2}$	8
1	1
2	$\dfrac{1}{8}$

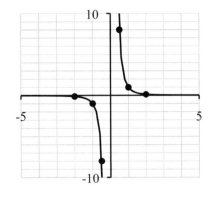

✐ Section 6.5 Exercise 15 (modified)

Using the end behavior from Exercise 7 and plotting a few additional points, we obtain the graph of $f(x) = x^{-4}$.

x	$f(x)$
-2	$\dfrac{1}{16}$
-1	1
$-\dfrac{1}{2}$	16
$\dfrac{1}{2}$	16
1	1
2	$\dfrac{1}{16}$

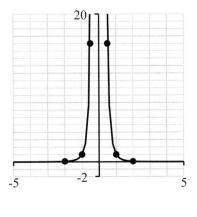

✐ Section 6.5 Exercise 25

By the definition of inverse proportionality we have $z = \dfrac{k}{t}$ where k is a constant and $k \neq 0$. To find k, we use the fact that when z is 2, t is 3.

$$z = \frac{k}{t} \qquad \text{Equation of proportionality}$$

$$2 = \frac{k}{3} \qquad \text{Replace } z \text{ by 2 and } t \text{ by 3}$$

$$k = 6 \qquad \text{Solve for } k$$

Putting this value of k back into the probability equation, we have $z = \dfrac{6}{t}$.

✎ Section 6.5 Exercise 31 Boyle's Law

(a) By the definition of inverse proportionality we have $P = \dfrac{k}{V}$ where k is a constant and $k \neq 0$. To find k, we use the fact that P is 39 when V is 0.671.

$$P = \frac{k}{V} \qquad\qquad \text{Equation of proportionality}$$

$$39 = \frac{k}{0.671} \qquad\qquad \text{Replace } P \text{ by 39 and } V \text{ by 0.671}$$

$$k = 26.169 \qquad\qquad \text{Solve for } k$$

Putting this value of k into the proportionality equation, we have $P = \dfrac{26.169}{V}$.

(b) We use the equation from part (a), replacing V by 0.916.

$$P = \frac{26.169}{V} \qquad\qquad \text{Equation of proportionality}$$

$$P = \frac{26.169}{0.916} \qquad\qquad \text{Replace } V \text{ by 0.916}$$

$$P \approx 28.57 \qquad\qquad \text{Calculator}$$

So the new pressure is 28.57 kPa.

✎ Section 6.5 Exercise 37 Heat of Campfire

(a) If $H(d)$ is the heat experience when the distance to the fire is d, then $H(d)$ is inversely proportional to the cube of d: $H(d) = \dfrac{k}{d^3}$, where k is a constant and $k \neq 0$.

(b) If the hiker is at a distance x from the fire, then the heat the hiker experiences is $H(x) = \dfrac{k}{x^3}$. If the hiker moves halfway closer to the fire, then the hiker's new distance to the fire is $\frac{x}{2}$, so the heat experienced is now

$$H(x/2) = \frac{k}{(x/2)^3} \qquad\qquad \text{Replace } x \text{ by } \tfrac{x}{2}$$

$$= \frac{k}{\frac{1}{8}x^3} \qquad\qquad \text{Properties of Exponents}$$

$$= 8\left(\frac{k}{x^3}\right) \qquad\qquad \text{Property of Fractions}$$

$$= 8\,H(x) \qquad\qquad \text{Replace } \frac{k}{x^3} \text{ by } H(x)$$

Thus, the heat experienced by the hiker is now eight times greater.

6.6 Rational Functions

OBJECTIVES

☑ *Check the box when you can do the exercises addressing the objective, such as those included in this study guide, which are listed here.*

Exercises

Graphing Quotients of Linear Functions

☐ Graph the quotient of linear functions by transforming the graph of the reciprocal function.

29, 31, 41

Graphing Rational Functions

☐ Write a rational function in standard form, factored form or compound fraction form.

13, 23, 37

☐ Find the x-intercepts and vertical asymptotes from the factored form of a rational function.

13, 23, 37

☐ Find the y-intercepts from the standard form of a rational function.

13, 37

☐ Find the end behavior of a rational function using compound fraction form.

23, 37

☐ Write a rational function in a form that is appropriate to determine the desired information about the rational function's graph.

13, 23, 29, 31, 37, 41

☐ Graph a rational function by getting information from its formula.

37

GET READY...

This section involves manipulating rational functions. Review rational expressions and long division of polynomials in Algebra Toolkit B.3. You will use the function's formula to find features of its graph, like x- and y-intercepts. Review finding intercepts in Algebra Toolkit D.2.

✓ **Algebra Checkpoint** Test yourself by completing the Algebra Checkpoint at the end of this section of the textbook, and comparing your answers to those in the back of the textbook. Refer to Algebra Toolkit B.3 as necessary.

SKILLS

Graphing Quotients of Linear Functions

These exercises ask you to graph rational functions. Find the definition of a rational function in bold typeface in Section 6.6 of the textbook.

📖 **Read Section 6.6 Example 1**, in which the authors graph quotients of linear functions.

[36]*The authors obtain the graph of the function r(x) = 2/(x – 3) from the graph of y = 1/x by shifting _____ (upwards / downwards / to the left / to the right) by _____ units, and _____ (stretching / shrinking) by a factor of _____. The authors obtain the graph of s(x) = –1/(x + 2) + 3 from the graph of y = 1/x by shifting _____(upwards / downwards / to the left / to the right) by _____ units, reflecting across the _____ (x-axis / y-axis), and shifting _____(upwards / downwards / to the left / to the right) by _____ units.*

What is the first step the authors do to rewrite s(x) in terms of the reciprocal function f(x) = 1/x? Answer in your own words:

_____

_____

✏ **Section 6.6 Exercise 29 and 31** Use transformations of the graph of the reciprocal function $y = \dfrac{1}{x}$ to graph the rational function, as in Example 1.

29. $s(x) = \dfrac{3}{x+1}$

Solution

---

[36] The authors obtain the graph of r(x) from the graph of y = 1/x by shifting to the right by 3 units and then stretching by a factor of 2. The authors obtain the graph of s(x) from the graph of y = 1/x by shifting to the left by 2 units, reflecting across the x-axis, and then shifting upward 3 units. To rewrite s(x) the authors use long division, dividing the numerator 3x + 5 by the denominator x + 2.

31.

Solution

Graphing Rational Functions

These exercises ask you to write rational functions in standard form, factored form and compound fraction form, since each form allows you to access information about the functions' graph. Find the definitions of these forms in bold typeface in Section 6.6 of the textbook.

📖 **Read Section 6.6 Example 3**, in which the authors find the standard, factored and compound fraction forms of a rational function.

[37]*How do the authors obtain the compound fraction form?*
Answer in your own words:

---

[37] The authors find the highest power of x that appears in the numerator or denominator. Then they divide the numerator and denominator by that power. The computation for the numerator looks like this:

$$\frac{2x^2 + 7x - 4}{x^2} = \frac{2x^2}{x^2} + \frac{7x}{x^2} - \frac{4}{x^2} = 2 + \frac{7}{x} - \frac{4}{x^2}$$

📖 **Read the paragraph after Example 3**, in which the authors explain what information the standard form, factored form, and compound fraction form reveal about the graph.

[38] *To find the y-intercept, use the _____ (standard / factored / compound fraction) form. To find the x-intercept, use the _____ (standard / factored / compound fraction) form. To find the vertical asymptotes, use the _____ (standard / factored / compound fraction) form. To determine the end behavior (and find any horizontal asymptotes), use the _____ (standard / factored / compound fraction) form.*

✎ **Section 6.6 Exercise 13** Find the x- and y-intercepts of the rational function

$$t(x) = \frac{x^2 - x - 2}{x - 6}.$$

Solution

✎ **Section 6.6 exercise 23** Find all horizontal and vertical asymptotes of the graph of

$$s(x) = \frac{6x^2 + 1}{2x^2 + x - 1}.$$

Solution

📖 **Read Section 6.6 Example 4**, in which the authors construct the graph of a rational function from its formula.

[39] *The graphs in Figure 5 show the three steps to constructing the graph. What is on the graph in each step?*
Answer in your own words:

---

[38] To find the y-intercepts, use the standard form. To find the x-intercepts and the vertical asymptotes, use the factored form. To find the end behavior, use the compound fraction form.

[39] The first graph shows the x- and y-intercepts (red dots) and the horizontal asymptotes (dashed blue lines). The second graph shows also the graph of the function near the vertical asymptotes. In particular, since $y \to -\infty$ as $x \to -2^-$, the second graph shows a red line to the left of the asymptote at $x = -2$. The third graph shows the final answer, with the pieces from the second graph connected from left to right.

✐ **Section 6.6 Exercise 37** Find the intercepts and asymptotes, and then sketch a graph of the rational function $t(x) = \dfrac{4x - 8}{(x - 4)(x + 1)}$. Use a graphing calculator to confirm your answer.

Solution

CONTEXTS

Graphing Quotients of Linear Functions

These exercises ask you to model the real world using rational functions.

📖 **Read Section 6.6 Example 2**, in which the authors model the focal distance of a camera lens. Note that the introduction to this section presents these ideas.

[40] *In part (c), you are asked what happens to the variable _____ (x / y / F), as the variable _____ (x / y / F) increases. How do the authors answer?*
Answer in your own words:

_____

[40] Part (c) asks what happens to the variable y (the focusing distance) as the variable x (the distance from the object to the lens) increases. The authors find a formula for y as a function of x, and graph it. Then they use the graph to determine what happens to y as $x \to \infty$. In other words, they look on the graph for horizontal asymptotes as $x \to \infty$.

✏ **Section 6.6 Exercise 41 Focusing a Camera Lens** A camera has a lens of fixed focal length F. To focus on an object located a distance x from the lens, the film must be placed a distance y behind the lens, where F, x, and y are related by

$$\frac{1}{x} + \frac{1}{y} = \frac{1}{F}$$

(See Example 2.) Suppose the camera has a wide-angle 35-mm lens (✕).
(a) Express y as a function of x and graph the function.
(b) What happens to the focusing distance y as the object moves further away from the lens?
(c) What happens to the focusing distance y as the object moves closer to the lens?

Solution

For extra practice, use transformations of the graph of the reciprocal function $y = \frac{1}{x}$ to graph the rational function, as in Section 6.6 Example 1.

CONCEPTS

✓ **6.6 Exercises – Concepts – Fundamentals** Complete the Fundamentals section of the exercises at the end of Section 6.6. Compare your answers to those at the back of the textbook, and make corrections as necessary.

📖 **Read the Concept Check for Section 6.6**, in the Chapter Review at the end of Chapter 6.

6.6 Solutions to ✏ Exercises

✏ Section 6.6 Exercise 13

To find the y-intercept, we replace x by 0 in the standard form:

$$t(0) = \frac{0^2 - 0 - 2}{0 - 6} = \frac{1}{3}$$

So the y-intercept is $\frac{1}{3}$. To find the x-intercept, we replace $t(x)$ by 0 in the factored form.

$$\frac{(x-2)(x+1)}{x-6} = 0 \qquad\qquad \text{Replace } t(x) \text{ by 0 in factored form}$$

$$(x-2)(x+1) = 0 \qquad\qquad \text{The numerator is 0}$$

$$x - 2 = 0 \ \text{ or } \ x + 1 = 0 \qquad \text{Zero-Product Property}$$

$$x = 2 \ \text{ or } \ x = -1 \qquad\qquad \text{Solve}$$

So the x-intercepts are -1 and 2.

✏ Section 6.6 Exercise 23

The vertical asymptotes occur where the denominator is zero. The factored form is $s(x) = \frac{6x^2 + 1}{(2x-1)(x+1)}$. The denominator is zero when x is $\frac{1}{2}$ or -1. So the vertical asymptotes are the vertical lines $x = \frac{1}{2}$ and $x = -1$.

The horizontal asymptote is a horizontal line that the graph approaches as x gets large. Let's write the compound fraction form and see what happens as x gets large. In this case the fractions $\frac{1}{x^2}$, $\frac{1}{x}$, and $-\frac{1}{x^2}$ go to zero. So we have

$$s(x) = \frac{6 + \frac{1}{x^2}}{2 + \frac{1}{x} - \frac{1}{x^2}} \quad \rightarrow \quad \frac{6+0}{2+0+0} = 3$$

So the graph gets closer to 3 as x gets large. This means that the horizontal asymptote is the horizontal line $y = 3$.

✎ Section 6.6 Exercise 29

We can express s in terms of $f(x) = \dfrac{1}{x}$ as follows:

$$s(x) = \frac{3}{x+1} \qquad \text{Definition of } s$$

$$= 3\left(\frac{1}{x+1}\right) \qquad \text{Factor 3}$$

$$= 3f(x+1) \qquad \text{Because } f(x+1) = \frac{1}{x+1}$$

So the graph of s is obtained from the graph of f by shifting one unit to the left and stretching vertically by a factor of 3. Thus the graph of s has a vertical asymptote $x = -1$ and a horizontal asymptote of $y = 0$.

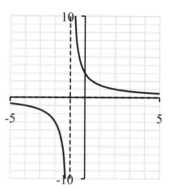

✎ Section 6.6 Exercise 31

We can express s in terms of $f(x) = \dfrac{1}{x}$ as follows:

$$s(x) = \frac{2x-3}{x-2} \qquad \text{Definition of } s$$

$$= 2 + \frac{1}{x-2} \qquad \text{Long Division } x-2\overline{)2x-3}$$

$$\begin{array}{r} 2 \\ x-2\overline{)2x-3} \\ \underline{2x-4} \\ 0x+1 \end{array}$$

$$= \frac{1}{x-2} + 2 \qquad \text{Rearrange terms}$$

$$= f(x-2) + 2 \qquad \text{Because } f(x-2) = \frac{1}{x-2}$$

So the graph of s is obtained from the graph of f by shifting two units to the right, and then shifting upward two units. Thus the graph of s has a vertical asymptote $x = 2$ and a horizontal asymptote of $y = 2$.

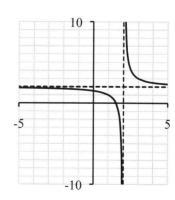

✎ Section 6.6 Exercise 37

To find the y-intercept, we replace x by 0 in the given form:

$$t(0) = \frac{4(0) - 8}{(0 - 4)(0 + 1)} = 2$$

So the y-intercept is 2. To find the x-intercept, we replace $t(x)$ by 0 in the factored form:

$$\frac{4x - 8}{(x - 4)(x + 1)} = 0 \qquad \text{Replace } t(x) \text{ by 0}$$

$$4x - 8 = 0 \qquad \text{The numerator is 0}$$

$$4x = 8 \qquad \text{Add 8}$$

$$x = 2 \qquad \text{Solve}$$

So the x-intercept is 2.

The vertical asymptotes occur where the denominator is zero. The denominator of $t(x) = \dfrac{4x - 8}{(x - 4)(x + 1)}$ is zero when x is 4 or -1. So the vertical asymptotes are the vertical lines $x = 4$ and $x = -1$.

A horizontal asymptote is a horizontal line that the graph approaches as x gets large. Let's write the compound fraction form and see what happens as x gets large.

$$t(x) = \frac{4x - 8}{(x - 4)(x + 1)} \qquad \text{Given}$$

$$= \frac{4x - 8}{x^2 - 3x - 4} \qquad \text{Standard form}$$

$$= \frac{\dfrac{4}{x} - \dfrac{8}{x^2}}{1 - \dfrac{3}{x} - \dfrac{4}{x^2}} \qquad \text{Compound fraction form}$$

In this case the fractions $4/x$, $8/x^2$, $-3/x$, and $-4/x^2$ go to zero. So we have

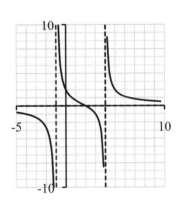

$$t(x) = \frac{\dfrac{4}{x} - \dfrac{8}{x^2}}{1 - \dfrac{3}{x} - \dfrac{4}{x^2}} \to \frac{0 + 0}{1 + 0 + 0} = 0$$

So the graph gets closer to 0 as x gets large. This means that the horizontal asymptote is the horizontal line $y = 0$.

✎ **Section 6.6 Exercise 41 Focusing a Camera Lens**

(a) We replace F by 35 in the equation and solve for y:

$$\frac{1}{x}+\frac{1}{y}=\frac{1}{35}$$ Replace F by 35

$$\frac{1}{y}=\frac{1}{35}-\frac{1}{x}$$ Subtract $1/x$

$$\frac{1}{y}=\frac{x-35}{35x}$$ Common denominator

$$y=\frac{35x}{x-35}$$ Take the reciprocal of each side

The last equation defines y as a rational function of x.

Let $f(x)=\frac{1}{x}$. Then we can express y as a function of f as follows:

$$y=\frac{35x}{x-35}$$ Definition of y

$$=35+\frac{1225}{x-35}$$ Long division $x-35\overline{)35x}$ ... $\begin{array}{r}35\\ 35x-35^2\\ \hline 35^2\end{array}$

$$=1225\left(\frac{1}{x-35}\right)+35$$ Factor 1225 and rearrange terms

$$=1225f(x-35)+35$$ Because $f(x-35)=\frac{1}{x-35}$

So the graph of y is obtained from the graph of f by shifting right by 35 units, stretching vertically by a factor of 1225, and then shifting upward by 35 units. Thus y has vertical asymptote $x=35$ and horizontal asymptote $y=35$.

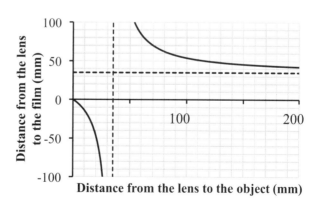

(b) From the graph we see that $y=35$ is a horizontal asymptote, so the focusing distance approaches 35 as x becomes large.

(c) From the graph we see that as the distance x gets close to 35, the focusing distance y tends to infinity. (Of course, x cannot be less than 35.)

7.1 Systems of Linear Equations in Two Variables

OBJECTIVES

☑ *Check the box when you can do the exercises addressing the objective, such as those included in this study guide, which are listed here.*   **Exercises**

The Substitution Method

☐ Solve a system of equations using the substitution method.   **17**

The Substitution Method

☐ Solve a system of equations using the elimination method.   **21, 29, 31, 51, 53, 55**

Graphical Interpretation: The Number of Solutions

☐ Identify inconsistent systems (i.e. systems with no solution).   **29**

☐ Identify a dependent system (i.e. one with infinitely many solutions). Express the infinite set of solutions in ordered pair form.   **31**

Applications: How Much Gold is in the Crown?

☐ Think about a problem by guessing a solution and checking if it is correct.   **51, 53, 55**

☐ Set up a system of equations that models the real world.   **51, 53, 55**

☐ Answer questions about the real world by solving systems of equations.   **51, 53, 55**

GET READY...

This section involves systems of linear equations, and allows you to relate the number of solutions a system has to the graphs of the equations. Review the graphs of linear equations in Section 2.3 of the textbook.

SKILLS

Systems of Equations and Their Solutions

This section is about systems of equations and their solutions. Find these terms in bold typeface in Section 6.6 of the textbook.

📖 **Read the subsection, 'Systems of Equations and Their Solutions,'** in which the authors define the key vocabulary of this section.

[1] *The solutions of Equation 1 appear on the graph as _____. The solutions of Equation 2 appear on the graph as _____. The solutions of the system are shown on the graph as _____.*

The Substitution Method

These exercises ask you to solve systems of equations, using the substitution method. Find its definition in a red box in Section 7.1 of the textbook.

---

[1] The solutions of Equation 1 appear on the graph as a red line. The solutions of Equation 2 appear on the graph as a blue line. The solutions of the system of equation appear on the graph as the point of intersection of the red and blue lines, (3, 1).

📖 **Read Section 7.1 Example 1**, in which the authors solve a system of equations using the substitution method.

[2] *The authors substitute y = _____ into Equation 2. The solutions of Equation 1 appear on the graph as _____. The solutions of Equation 2 appear on the graph as _____. The solutions of the system are shown on the graph as _____.*

✏ **Section 7.1 Exercise 17** Use the substitution method to solve the system of linear equations.

$$\begin{cases} x + y = 4 \\ -x + y = 0 \end{cases}$$

Solution

The Elimination Method

These exercises ask you to solve systems of equations using the elimination method. Find its definition in a red box in Section 7.1 of the textbook.

📖 **Read Section 7.1 Example 2**, in which the authors solve a system of equations using the elimination method.

[3] *The first step of the elimination method says to adjust the coefficients of the equations so that one variable will be eliminated when you add the equations. In this example, the given equations are already related in this way since the coefficient of _____ (x / y) is _____ in Equation 1 and _____ in Equation 2.*

---

[2] The authors substitute $y = 1 - 2x$ into Equation 2. Note that they write $(1 - 2x)$ when they make the substitution, including the parentheses. Always include parentheses when substituting negative or complicated expressions. The solutions of Equation 1 appear on the graph as a red line. The solutions of Equation 2 appear on the graph as a blue line. The solutions of the system of equation appear on the graph as the point of intersection of the red and blue lines $(-2, 5)$.

[3] The coefficient of y in Equation 1 is 2 and in Equation 2 is -2.

✐ **Section 7.1 Exercise 21** Use the elimination method to solve the system of linear equations.

$$\begin{cases} 2x - 3y = 9 \\ 4x + 3y = 9 \end{cases}$$

Solution

Graphical Interpretation: The Number of Solutions

These exercises ask you to determine how many solutions a system of linear equation has to the graphs of the equations. Find this summarized in a blue box in Section 7.1 of the textbook.

📖 **Read the introduction to the subsection, 'Graphical Interpretation: The Number of Solutions,'** in which the authors relate the solutions to a system of equations to the equations' graphs.

[4] *The solution to a system of two linear equations is the set of points at which two lines in the plane intersect. There are three possibilities for this intersection. What are they? Draw the graphs illustrating each case.*
Answer in your own words:

When the graphs of the two lines intersect at a single point, the system has _____ *(exactly one solution / no solution / infinitely many solutions). When the graphs do not intersect, the system has* _____ *(exactly one solution / no solution / infinitely many solutions), and is called* _____. *When the graphs of the two lines coincide, the system has* _____ *(exactly one solution / no solution / infinitely many solutions), and the system is called* _____.

---

[4] The two lines may intersect at a point, they may be parallel and not intersect at all, or the two lines may coincide. Graphs illustrating these can be found in the textbook in Section 7.1 Figure 4. When the graphs intersect at a point, the system has one solution. When the graphs do not intersect, the system has no solution, and is called inconsistent. When the graphs coincide, the system has infinitely many solutions and is called dependent.

📖 **Read Section 7.1 Example 3**, in which the authors investigate an inconsistent system.

[5]*In the original system, in the first equation, the coefficient on y is ____, and in the second equation, the coefficient on y is ____. The authors multiply the first equation by ____, and the second equation by ____, making the coefficient on y ____ in the first equation, and ____ in the second equation.*

📖 **Read Section 7.1 Example 4**, in which the authors investigate a dependent system.

[6]*In the original system, in the first equation, the coefficient on y is ____, and in the second equation, the coefficient on y is ____. The authors multiply the first equation by ____, and the second equation by ____, making the coefficient on y ____ in the first equation, and ____ in the second equation. If you write Equation 1 in slope-intercept form, you get _____; if you write Equation 2 in slope intercept form, you get _____. So if x = t, then y = _____, and all solutions to the system have the form (___, _____).*

✏ **Section 7.1 Exercise 29 and 31** Solve the system, or show that it has no solution. If the system has infinitely many solutions, express them in the ordered-pair form given in Example 4.

$$31. \begin{cases} 2x - 6y = 10 \\ -3x + 9y = -15 \end{cases}$$

Solution

---

[5] In the first equation, the coefficient on y is originally –2, and in the second equation, the coefficient on y is originally 3. The authors multiply the first equation by 3 and the second equation by 2, making the coefficient on y –6 in the first equation and 6 in the second equation.

[6] In the first equation, the coefficient on y is originally –6, and in the second equation, the coefficient on y is originally –8. The authors multiply the first equation by 4 and the second equation by 3, making the coefficient on y equal to –24 in both equations. If you write Equation 1 or 2 in slope intercept form, you get

$y = \frac{1}{2}x - 2$. So if $x = t$, then $y = \frac{1}{2}t - 2$, and all solutions to the system have the form $\left(t, \frac{1}{2}t - 2\right)$.

29. $\begin{cases} x + 4y = 8 \\ 3x + 12y = 2 \end{cases}$

Solution

Contexts

Applications: How Much Gold Is in the Crown?

These exercises ask you to set up and solve systems of linear equations that model the real world. Find the guidelines for modeling with systems of equations in a red box in Section 7.1 of the textbook.

📖 **Read the introduction to Section 7.1**, in which the authors describe the quandary Archimedes discovered in the bath.

[7]*When Archimedes discovered the solution, he shouted, "_____!"*

📖 **Read the introduction to the subsection, 'Applications: How Much Gold Is in the Crown?'** in which the authors explain how Archimedes found the volume of gold in the crown.

[8]*The density of a substance is its _____ divided by its _____.*

Hints and Tips 7.1: Thinking about the problem

To help you find the equations that you will solve to answer to a question about the real world, make up a possible answer, and then check if it is correct.

- Determine what quantities you are asked to find, and assign variables to them.
- Make up possible values for those variables. Include units on your guesses, to help you remember what they mean in context.
- Check if your guess is correct. To see how the given numbers in the problem contribute to the equations you seek, write out your calculations, and avoid doing computations in your head.
- Analyze your guess: was it too large or too small? What needs to change for it to be correct? Wrong guesses are often more illustrative than right ones.

---

[7] Eureka!

[8] The density of a substance is its weight divided by its volume.

📖 **Read Section 7.1 Example 5**, in which the authors find the weights of the gold and sliver in the crown.

[9] *What does this problem ask you to find?*
Answer in your own words:

<table>
<tr><td>

Following Hints and Tips 7.1, before beginning the solution, you can think about the problem by making up an answer, and checking if it is right. For example, suppose the weight of gold in the crown is x = 100 g, and the weight of the silver is y = 150 g.

The formula relating volume to density and weight is V = _____. The density of gold is D =_____ g/cm³, and the density of silver is D =_____ g/cm³. To check if the guess is right, complete the table.

The guess that the gold weighs 100 g and the silver weighs 150 g is _____ (correct / incorrect). How do you know?
Answer in your own words:

</td><td>

In words	*In numbers*
Weight of gold	
Weight of silver	
Total weight	
Volume of gold	$V = \dfrac{W}{D} = \dfrac{\boxed{} \; g}{\boxed{} \; g/cm^3}$
Volume of silver	$V = \dfrac{W}{D} = \dfrac{\boxed{} \; g}{\boxed{} \; g/cm^3}$
Total volume	

</td></tr>
</table>

[10] *In the system of equations, x represents _____ and y represents _____. So x/19.3 represents _____, and y/10.5 represents _____. When solving the problem, the authors get rid of the fractions by multiplying the Volume Equation by _____. Then the authors multiply the Weight Equation by _____. This makes the coefficient on x in the Weight Equation equal to _____, while the coefficient on x in the Volume Equation is _____.*

---

[9] This problem asks you to find the weight of the gold and the weight of the silver in the crown. The formula relating weight to volume and density is $V = W/D$. The density of gold is 19.3 g/cm³, and the density of silver is 10.5 g/cm³.

We fill out the table with our guesses. The weight of the gold is 100 g, and the weight of the silver is 150 g, so the total weight is 250 g.

The volume of the gold is (100 g)/(19.3 g/cm³) ≈ 5.18 cm³, and the volume of the silver is (150 g)/(10.5 g/cm³) ≈ 14.29 cm³. So the total volume comes to 19.47 cm³.

The guess is incorrect, since the total weight of the crown should be 235 g, and the total volume of the crown should be 14 cm³.

[10] The variable x represents the weight of the gold in the crown and the variable y represents the weight of the silver in the crown (both in grams). So, since $V = W/D$, the quantity $x/19.3$ represents the volume of gold in the crown, and $y/10.5$ represents the volume of silver in the crown (both in cm³). The authors get rid of fractions by multiplying the Volume Equation by (19.3)(10.5) = 202.65. The authors then multiply the Weight Equation by −10.5, making the coefficient on x in the Weight Equation equal to −10.5, while the coefficient on x in the Volume Equation is 10.5.

✎ **Section 7.1 Exercise 53 Nutrition** A researcher performs an experiment to test a hypothesis that involves the nutrients niacin and retinol. She feeds one group of laboratory rats a daily diet of precisely 32 units of niacin and 22,000 units of retinol. She uses two types of commercial pellet foods. Food A contains 0.12 unit of niacin and 100 units of retinol per gram. Food B contains 0.20 units of niacin and 50 units of retinol per gram. How many grams of each food does she feed this group of rats each day?

Solution

📖 **Read Section 7.1 Example 6**, in which the authors model the progress of a rowboat in a stream with a current.

[11] *What does this problem ask you to find?*
Answer in your own words:

Following Hints and Tips 7.1, before beginning the solution, you can think about the problem by guessing an answer and checking if it is correct. For example, suppose her rowing speed is x = 3 mi/hr, and the speed of the current is y = 2 mi/hr. To check if this is correct, calculate the distance she would go at with these speeds by completing the table.

In words	In numbers	In words	In numbers	In words	In numbers
Rowing speed		Speed upstream		Speed downstream	
Current speed					
		Time upstream		Time downstream	
		Distance upstream		Distance downstream	

The guess that the she rows at 3 mi/hr and the current flows at 2 mi/hr is _____ (correct / incorrect). How do you know?
Answer in your own words:

[12] *In the system of equations, x represents _____ and y represents _____. So $(x - y)\frac{3}{2}$ represents _____, and $(x + y)\frac{3}{4}$ represents _____. When solving the problem, the authors get rid of the fractions by multiplying the Equation 1 by _____, and Equation 2 by _____. After they apply the distributive law, the coefficient on y in Equation 1 equals _____, while the coefficient on y in the Equation 2 is _____.*

---

[11] This problem asks you to find the speed the woman rows and the speed of the current. We fill out the table with our guesses. The rowing speed is 3 mi/hr, and the current flows at 2 mi/hr.

Her speed upstream is $3 - 2 = 1$ mi/hr. Her time upstream is 1½ (or 3/2) hours, so with these speeds, her distance upstream is (3/2 hr)(1 mi/hr) = 3/2 mi.

Similarly, her speed down stream is $3 + 2 = 5$ mi/hr. Her time downstream is 45 minutes, which is ¾ hr, so with these speeds, her distance downstream is (3/4 hr)(5 mi/hr) = 15/4 = 3.75 mi.

The guess is incorrect, since the distance she travels upstream and the distance she travels downstream should both be 4 miles.

[12] The variable x represents the woman's rowing speed and the variable y represents the speed of the current (both in miles per hour). So $(x - y)\frac{3}{2}$ represents the distance she travels upstream and $(x + y)\frac{3}{4}$ represents the distance she travels downstream (both in miles). The authors get rid of fractions by multiplying Equation 1 by 2 and Equation 2 by 4. Then the coefficient on y in the Equation 1 equals -3, while the coefficient on y in Equation 2 is 3.

Section 7.1 Exercise 51 Airplane Speed A man flies a small airplane from Fargo to Bismarck, North Dakota—a distance of 180 mi. Because he is flying into a head wind, the trip takes him 2 hours. On the way back the, wind is still blowing at the same speed, so the return trip takes only 1 h 12 min. What is his speed in still air, and how fast is the wind blowing?

Solution

> Refer back to page 70 (see also pages 140-142)
> **Hints and Tips 1.8b: Thinking about the problem**

Following Hints and Tips 1.8b, think about these mixing problems by drawing a diagram, as you do in Section 2.6 Example 3.

📖 **Section 7.1 Example 7.1**, in which the authors determine the amount of alcohol solution needed to fortify wine.

[13] *What does this problem ask you to find?*
Answer in your own words:

Following Hints and Tips 1.8b and 7.1, before beginning the solution, you can think about the problem by making up an answer, checking if it is right and drawing a diagram. For example, suppose the vintner uses x = 700 L wine and y = 400 L of 70%-alcohol solution. Complete the table, adding the volume of each part to the diagram, to check if this guess is right.

		In words	In numbers
The part of the 70%-solution that is not alcohol	⎫	Volume of wine	_____
	⎬ Added	Volume of 70%-alcohol solution	_____
The part of the 70%-solution that is alcohol		Total volume of the mixture	_____
The part of the wine that is not alcohol	⎫	Volume of alcohol from the wine	_____
	⎬ Original	Volume of alcohol from the 70%-alcohol solution	_____
The part of the wine that is alcohol		Total alcohol in the mixture	_____

The guess that the vintner uses 700 L of wine and 400 L of solution is _____ (correct / incorrect). How do you know?
Answer in your own words:

[14] *In the system of equations, x represents _____ and y represents _____. Also, 0.1x represents _____, and 0.7y represents _____. When solving the problem, the authors get rid of the decimals by multiplying the Equation 1 by _____. The coefficient on x in Equation 1 and Equation 2 equals _____.*

---

[13] This problem asks you to find the volume of the wine and the volume of the 70%-alcohol solution. We fill out the table with our guesses. We guess that the vintner uses 700 L of wine, and 400 L of solution, so the total volume of the mixture is $700 + 400 = 1100$ L.

The wine is 10% alcohol, so it has $700(0.1) = 70$ L of alcohol. The solution is 70% alcohol, so it has $400(0.7) = 280$ L of alcohol. So the mixture has a total of $70 + 280 = 350$ L of alcohol.

The guess is incorrect, since the volume of the mixture should be 1000 L and the volume of alcohol should 16% alcohol, so the volume of alcohol in the mixture should be $1000(0.16)=160$ L.

[14] The variable x represents the volume of wine in the mixture and the variable y represents the volume of 70%-alcohol solution in the mixture (both in liters). So $0.1x$ represents the volume of alcohol in the wine and $0.7y$ represents the volume of alcohol in the solution (both in liters). The authors get rid of the decimals by multiplying Equation 1 by 10. Then the coefficients on c in both equations are 1.

🖎 **Section 7.1 Exercise 55 Mixture Problem** A biologist has two brine solutions, one containing 5% salt and another containing 20% salt. How many milliliters of each solution should she mix to obtain 1 liter of a solution that contains 14% salt?

Solution

CONCEPTS

✓ **7.1 Exercises – Concepts – Fundamentals** Complete the Fundamentals section of the exercises at the end of Section 7.1. Compare your answers to those at the back of the textbook, and make corrections as necessary.

📖 **Read the Concept Check for Section 7.1**, in the Chapter Review at the end of Chapter 7.

7.1 Solutions to ✏ Exercises

✏ Section 7.1 Exercise 17

We solve for y in the first equation.

$x + y = 4$	Equation 1
$y = 4 - x$	Subtract x from each side

Now we replace y by $4 - x$ in the second equation and solve for x.

$-x + y = 0$	Equation 2
$-x + (4 - x) = 0$	Replace y by $4 - x$
$4 - 2x = 0$	Simplify
$-2x = -4$	Subtract 4 from each side
$x = 2$	Divide each side by -2

Next we back-substitute $x = 2$ into the equation $y = 4 - x$.

$y = 4 - 2 = 2$	Back-substitute

Thus $x = 2$ and $y = 2$, so the solution is the ordered pair $(2, 2)$. If we graph the lines $x + y = 4$ and $-x + y = 0$, they would intersect at the point $(2, 2)$.

✏ Section 7.1 Exercise 21

Since the coefficients of the y-terms are the same, except that they have opposite signs, we can add the equations to eliminate y.

$2x - 3y = 9$	Equation 1
$4x + 3y = 9$	Equation 2
$6x \quad\quad = 18$	Add
$x = 3$	Divide each side by 6

Now we back-substitute $x = 3$ into one of the original equations and solve for y.

$2x - 3y = 9$	Equation 1
$2(3) - 3y = 9$	Replace x by 3
$6 - 3y = 9$	Simplify
$-3y = 3$	Subtract 6 from each side
$y = -1$	Divide each side by -3

The solution is $(3, -1)$. If we graph the lines $2x - 3y = 9$ and $4x + 3y = 9$, we would find they intersect at the point $(3, -1)$.

✎ Section 7.1 Exercise 29

We try to find a suitable combination of the two equations to eliminate the variable y. Multiplying the first equation by -3 gives

$$-3x - 12y = -24 \qquad\qquad -3 \times \text{Equation 1}$$
$$\underline{3x + 12y = \quad 2} \qquad\qquad \text{Equation 2}$$
$$0 = -22 \qquad\qquad\quad \text{Add}$$

Adding the two equations eliminates both x and y in this case, and we end up with $0 = -22$, which is obviously false. No matter what values we assign to x and y, we cannot make this statement true, so the system has no solution. If we graph the lines $x + 4y = 8$ and $3x + 12y = 2$, we would find they fail to intersect at all, since they are parallel.

✎ Section 7.1 Exercise 31

We try to find a suitable combination of the two equations to eliminate the variable y. Multiplying the first equation by 3 and the second equation by 2 gives

$$6x - 18y = \quad 30 \qquad\qquad 3 \times \text{Equation 1}$$
$$\underline{-6x + 18y = -30} \qquad\qquad 2 \times \text{Equation 2}$$
$$0 = 0 \qquad\qquad\quad \text{Add}$$

We see that the two equations in the original system are simply different ways of expressing the equation of one single straight line. The coordinates of any point on this line give a solution to the system, and there are infinitely many solutions. If we graph the lines $2x - 6y = 10$ and $-3x + 9y = -15$, we would find they coincide, and hence intersect at every point on the line.

We write the slope intercept form of the line.

$$2x - 6y = 10 \qquad\qquad \text{Equation 1}$$
$$-6y = 10 - 2x \qquad\qquad \text{Subtract } 2x \text{ from each side}$$
$$y = \frac{10 - 2x}{-6} \qquad\qquad \text{Divide each side by } -6$$
$$y = -\frac{5}{3} + \frac{1}{3}x \qquad\qquad \text{Simplify}$$

So for each real number t, the pair $x = t$ and $y = -\frac{5}{3}t + \frac{1}{3}$ is a solution to the system.

As ordered pairs, the solutions are of the form $\left(t, -\frac{5}{3}t + \frac{1}{3}\right)$, where t is any real number.

✐ **Section 7.1 Exercise 53 Nutrition**

Since the question asks about the amounts of each food the researcher feeds the rats, let x be the amount of Food A and y be the amount of Food B.

Thinking about the problem

Following **Hints and Tips 7.1**, we first think about the problem by making up an answer, and checking if it is correct. Suppose that she feeds the rats $x = 10$ g of Food A and $y = 20$ g of Food B. We calculate the retinol and niacin in the rats' diets in this case.

In words	*In numbers*	*In words*	*In numbers*	*In words*	*In numbers*
Food A	10 g	Niacin from Food A	(0.12 units/g)(10 g) = 1.2 units	Amount of retinol from Food A	(100 units/g)(10 g) = 1000 units
Food B	20 g	Niacin from Food B	(0.2 units/g)(20 g) = 4 units	Amount of retinol from Food B	(50 units/g)(20 g) = 1000 units
		Total niacin	5.2 units	Total retinol	2000 units

Our guess of $x = 10$ g of Food A and $y = 20$ g of Food B was nowhere near correct, since the rats should have a total of 32 units of niacin and 22,000 units of retinol.

From our work above, we see that we want two equations to hold.

$$\left(\begin{array}{c}\text{niacin from}\\ \text{Food A}\end{array}\right) + \left(\begin{array}{c}\text{niacin from}\\ \text{Food B}\end{array}\right) = 32 \text{ units}$$

$$\left(\begin{array}{c}\text{retinol from}\\ \text{Food A}\end{array}\right) + \left(\begin{array}{c}\text{retinol from}\\ \text{Food B}\end{array}\right) = 22,000 \text{ units}$$

Thus we want to find x and y so that

$$0.12x + 0.2y = 32 \qquad \text{Niacin Equation}$$
$$100x + 50y = 22,000 \qquad \text{Retinol Equation}$$

We solve the system by elimination. First to eliminate decimals, we multiply the Niacin Equation by 100.

$$12x + 20y = 3200 \qquad \begin{array}{l}100 \times \text{Niacin Equation}\\ = \text{new Niacin Equation}\end{array}$$
$$100x + 50y = 22,000 \qquad \text{Retinol Equation}$$

Then we multiply the new Niacin Equation by 5 and the Retinol Equation by –2, to eliminate the variable y.

$$\begin{array}{ll}60x + 100y = 16,000 & 5 \times \text{Niacin Equation}\\ \underline{-200x - 100y = -44,000} & -2 \times \text{Retinol Equation}\\ -140x = -28,000 & \text{Add}\\ x = 200 & \text{Divide each side by } -140\end{array}$$

So the amount of Food A is 200 g. Now we back-substitute $x = 200$ into one of the original equations. We choose the Retinol Equation, since it looks easier.

$100(200) + 50y = 22{,}000$	Replace x by 200 in the Retinol Equation
$20{,}000 + 50y = 22{,}000$	Simplify
$50y = 2000$	Subtract 20,000 from each side
$y = 40$	Divide each side by 50

So the diet includes 200 g of Food A and 40 g of Food B.

✓ **Check** Let's check that this is indeed a solution.

In words	In numbers	In words	In numbers	In words	In numbers
Food A	200 g	Niacin from Food A	(0.12 units/g)(200 g) = 24 units	Amount of retinol from Food A	(100 units/g)(200 g) = 20,000 units
Food B	40 g	Niacin from Food B	(0.2 units/g)(40 g) = 8 units	Amount of retinol from Food B	(50 units/g)(40 g) = 2000 units
		Total niacin	32 units	Total retinol	22,000 units

With 200 g of Food A and 40 g of Food B, the rats receive 32 units of niacin and 22,000 units of retinol, as required.

✒ Section 7.1 Exercise 51 Airplane Speed

Since the question asks for the speed of the airplane (in still air) and the speed of the wind, we let x be the airplane's speed, and y be the wind speed.

Thinking about the problem

Following **Hints and Tips 7.1**, we think about the problem by making up an answer and checking if it is correct. Suppose the airplane flies at 100 mi/hr, and the wind blows at 15 mi/hr. To check if this is right, we calculate the distance he could go in each direction with those speeds.

In words	In numbers	In words	In numbers	In words	In numbers
Airplane's speed	100 mi/h	Speed from Fargo to Bismarck	$100 - 15$ = 85 mi/h	Speed from Bismarck to Fargo	$100 + 15$ =115 mi/h
Wind speed	15 mi/h				
		Time from Fargo to Bismarck	2 h	Time from Bismarck to Fargo	1 h 12 min $=1\,\text{h}+\dfrac{12\ \text{min}}{60\ \text{min/h}}$ = 1.2 h
		Distance from Fargo to Bismarck	$d = st$ = (85 mi/h)(2 h) =170 mi	Distance from Bismarck to Fargo	$d = st$ = (115 mi/h)(1.2 h) = 138 mi

Our guess of 100 mi/hr for the plane's speed and 30 mi/hr for the wind speed is not right, since the distance from Fargo to Bismarck should be 180 mi.

From our work above, we see that we want two equations to hold:

$$\left(\begin{array}{c}\text{speed from} \\ \text{Fargo to Bismarck}\end{array}\right)\left(\begin{array}{c}\text{time from} \\ \text{Fargo to Bismarck}\end{array}\right)=\left(\begin{array}{c}\text{distance from} \\ \text{Fargo to Bismarck}\end{array}\right)$$

$$\left(\begin{array}{c}\text{speed from} \\ \text{Bismarck to Fargo}\end{array}\right)\left(\begin{array}{c}\text{time from} \\ \text{Bismarck to Fargo}\end{array}\right)=\left(\begin{array}{c}\text{distance from} \\ \text{Bismarck to Fargo}\end{array}\right)$$

Thus we want to find a speed of the plane x and a wind speed y so that

$(x - y)(2) = 180$	Fargo-to-Bismarck Equation
$(x + y)(1.2) = 180$	Bismarck-to-Fargo Equation

Before we solve, we apply the Distributive Property to both equations, and multiply the Bismarck-to-Fargo Equation by 10 to remove the decimals.

$2x - 2y = 180$	Fargo-to-Bismarck Equation
$12x + 12y = 1800$	$10 \times$ Bismarck-to-Fargo Equation = new Bismarck-to-Fargo Equation

We solve this system using elimination. We multiply the Fargo-to-Bismarck Equation by 6 to eliminate the y.

$12x - 12y = 1080$	$6 \times$ Fargo-to-Bismarck Equation
$\underline{12x + 12y = 1800}$	Bismarck-to-Fargo Equation
$24x = 2880$	Add
$x = 120$	Divide each side by 24

Thus the plane's speed is 120 mi/h. To find the wind speed, we back-substitute $x = 120$ into one of the original equations.

$2x - 2y = 180$	Fargo-to-Bismarck Equation
$2(120) - 2y = 180$	Replace x by 120
$240 - 2y = 180$	Simplify
$-2y = -60$	Subtract 240 from each side
$y = 30$	Divide each side by -2

Thus the plane speed is 120 mi/h, and the wind speed is 30 mi/h.

✓ **Check** We check that these speeds do meet the criteria in the following table.

In words	In numbers	In words	In numbers	In words	In numbers
Airplane's speed	120 mi/h	Speed from Fargo to Bismarck	120 − 30 = 90 mi/h	Speed from Bismarck to Fargo	120 + 30 =150 mi/h
Wind speed	30 mi/h	Time from Fargo to Bismarck	2 h	Time from Bismarck to Fargo	1.2 h
		Distance from Fargo to Bismarck	$d = st$ = (90 mi/h)(2 h) =180 mi	Distance from Bismarck to Fargo	$d = st$ = (150 mi/h)(1.2 h) = 180 mi

✐ Section 7.1 Exercise 55 Mixture Problem

The question asks for the volume of each brine solution to mix, so we let x be the volume of 5%-brine and y be the volume of 20%-brine.

Thinking about the problem

Following **Hints and Tips 1.8b and 7.1**, before beginning the solution, we think about the problem by making up an answer, checking if it is right, and drawing a diagram. For example, suppose the biologist uses x = 100 mL 5%-brine and y = 200 mL of 20%-alcohol solution. Complete the table, adding the volume of each part to the diagram, to check if this guess is right.

	In words	In numbers
The part of the 5%-brine that is water — 95 mL	Volume of 5%-brine	100 mL
The part of the 5%-brine that is salt — 5 mL (5% brine)	Volume of 20%-brine	200 mL
	Total volume of the mixture	300 mL
The part of the 20%-brine that is water — 160 mL	Volume of salt from the 5%-brine	(0.05)(100 mL) = 5 mL
The part of the 20%-brine that is salt — 40 mL (20% brine)	Volume of salt from the 20%-brine	(0.2)(200 mL) = 40 mL
	Total volume of salt in the mixture	45 mL

The guess of 100 mL of 5%-brine, and 200 mL of 20% brine is not right, since the total volume of the mixture is supposed to be 1 L, which is 1000 mL, and it is supposed to have (0.14)(1000 mL) = 140 mL of salt.

From our work above, we see that we want two equations to hold:

$$\left(\text{volume of 5\%-brine}\right) + \left(\text{volume of 20\%-brine}\right) = 1000 \text{ mL}$$

$$\left(\begin{array}{c}\text{volume of salt}\\\text{from the 5\%-brine}\end{array}\right) + \left(\begin{array}{c}\text{volume of salt}\\\text{from the 20\%-brine}\end{array}\right) = (0.14)(1000 \text{ mL}) = 140 \text{ mL}$$

Thus we want to find a volume of 5%-brine x and a volume of 20%-brine y such that

| $x + y = 1000$ | Total Mixture Equation |
| $0.05x + 0.2y = 140$ | Salt Equation |

Before we solve, we multiply the Salt Equation by 100 to clear the decimals.

| $x + y = 1000$ | Total Mixture Equation |
| $5x + 20y = 14{,}000$ | $100 \times$ Salt Equation |

Now we solve using elimination. We multiply the Total Mixture Equation by –20 to eliminate the y.

$-20x + -20y = -20{,}000$	Total Mixture Equation
$\underline{5x + 20y = 14{,}000}$	$100 \times$ Salt Equation
$-15x = -6000$	Add
$x = 400$	Divide each side by –15

Thus the amount of 5%-brine solution the biologist needs is 400 mL. To find y, back-substitute $x = 400$ into one of the original equations. We choose the Total Mixture Equation because it looks easier.

| $400 + y = 1000$ | Replace x by 400 in the Total Mixture Equation |
| $y = 600$ | Subtract 400 from each side |

Thus the biologist needs 400 mL of the 5%-brine solution and 600 mL of the 20%-brine solution to get the 1000 mL (or 1 L) of 14%-brine solution she needs.

✓ **Check** We verify that the resulting brine solution contains the correct percentage of salt in the table. Note that 14% of 1000 mL is 140 mL.

In words	*In numbers*
Volume of 5%-brine	400 mL
Volume of 20%-brine	600 mL
Total volume of the mixture	1000 mL
Volume of salt from the 5%-brine	(0.05)(400 mL) = 20 mL
Volume of salt from the 20%-brine	(0.2)(600 mL) = 120 mL
Total volume of salt in the mixture	140 mL

7.2 Systems of Linear Equations in Several Variables

OBJECTIVES

☑ *Check the box when you can do the exercises addressing the objective, such as those included in this study guide, which are listed here.*

	Exercises
Solving a Linear System	
☐ Identify systems that are written in triangular form, and solve them using back-substitution.	**5, 13, 17, 23, 27, 31**
☐ Use Gaussian elimination to rewrite a system in triangular form.	**13, 17, 23, 27, 31**
Inconsistent and Dependent Systems	
☐ Identify inconsistent systems of equations (i.e. those with no solution), from their triangular forms.	**23**
☐ Identify dependent systems of equations (i.e. one with infinitely many solutions), from their triangular forms. Express the infinite set of solutions as an ordered triple.	**27**
Modeling with Linear Systems	
☐ Think about a problem by guessing a solution and checking if it is correct.	**31**
☐ Set up a system of equations that models the real world.	**31**
☐ Answer questions about the real world by solving systems of equations.	**31**

SKILLS

Solving a Linear System

These exercises ask you to start with a system of linear equations and change it to an equivalent system in triangular form. Find a description of triangular form in Section 7.2 of the textbook.

^{15}*Look at Section 7.2 Exercises 5-30. The systems given in exercises ___, ___, ___, and ___, are in triangular form.*

To put the system in triangular form, the authors manipulate the equations in such a way that the solutions to the equations remain unchanged. Find a list of operations that lead to an equivalent system in a blue box in Section 7.2 of the textbook.

📖 **Read Section 7.2 Example 1**, in which the authors solve a system in triangular form.

^{16}*Since the system is in triangular form, the authors solve it using _____. First, the authors substitute _____ into Equation ___, and solve for the variable ___. Then the authors substitute _____ and _____ into Equation ____, and solve for the variable ___. They express their final answer as $(x, y, z) = ($___, ___, ___$)$.*

---

[15] The systems given in exercises 5, 6, 7, and 8 are in triangular form.

[16] The authors solve the system using back-substitution. First the authors substitute $z = 3$ into Equation 2, and solve for y. Then the authors substitute $y = -1$ and $z = 3$ into Equation 1 and solve for x. They express their final answer as $(x, y, z) = (2, -1, 3)$.

✐ **Section 7.2 Exercise 5** Use back-substitution to solve the triangular system.

$$\begin{cases} x - 2y + 4z = 3 \\ \quad\quad y + 2z = 7 \\ \quad\quad\quad\quad z = 2 \end{cases}$$

Solution

📖 **Read Section 7.2 Example 2**, in which the authors solve a system of equations in three variables.

[17]*In the first two steps, the authors eliminate the variable ___ from the second and third equations. Then they eliminate the variable ___ from the third equation. To finish solving the system, the authors use*

_____.

✐ **Section 7.2 Exercise 13** Find the solution of the linear system.

$$\begin{cases} x - y - z = 4 \\ 2y + z = -1 \\ 2x + 3y - 2z = 8 \end{cases}$$

Solution

---

[17] First the authors eliminate the variable x from the second and third. Then the authors eliminate y from the third equation. To finish solving the system, the authors use back-substitution.

Inconsistent and Dependent Systems

These exercises ask you to determine when a system is inconsistent or dependent. Find the definitions of inconsistent and dependent in bold typeface in Section 7.2 of the textbook. Also see the subsection of 7.1 entitled 'Graphical Interpretation: The Number of Solutions.'

📖 **Read Section 7.2 Example 3**, in which the authors determine that a system is inconsistent (has no solution).

[18]*How do the authors determine that the system has no solution?*
Answer in your own words:

📖 **Read Section 7.2 Example 4**, in which the authors determine that a system is dependent (has infinitely many solutions). Read also the paragraph after Example 4.

[19]*How do the authors determine that the system has infinitely many solutions?*
Answer in your own words:

To write the solution as an ordered triple, the authors solve the second equation for y. They get
$y =$ _____ . *Then they substitute this into the first equation, and solve for x. They get x =* _____ .

So if z = t, then x = _____ , *and y =* _____ . *So the ordered triple (x, y, z) = (*_____ , _____ , _____ *)*
describes all possible solutions to the system.

Give an example of numbers x, y, and z that solve the system of equations, other than those listed in the table after Example 4. Show your work below.

---

[18] The authors first put the system into triangular form. Then they note that the third equation is $0 = 2$, which is false. No matter what values you assign to x, y, and z, the third equation will never be true, so the system has no solution.

[19] The authors first put the system into triangular form. Then they note that the third equation is $0 = 0$, which is true, but tells us nothing. This leaves only two equations. You could solve the two equations for x and y in terms of z, but z could take on any value, so there are infinitely many solutions.

To write the solutions as an ordered triple, the authors solve Equation 2 for y. They get $y = 2z + 2$. Then they substitute $y = 2z + 2$ into Equation 1, and solve for x. They get $x = -3z$. So if $z = t$, then $x = -3t$, and $y = 2t + 2$. So the ordered triple $(x, y, z) = (-3t, 2t + 2, t)$ describes all possible solutions to the system. For example, by taking $t = 100$, we see that $x = -3(100) = -300$, $y = 2(100) + 2 = 202$ and $z = 100$. So the triple, $(x, y, z) = (-300, 202, 100)$ is a solution of this equation. You can choose any number for t to determines an example of a solution to this system.

✎ **Section 7.2 Exercise 17, 23, 27** Solve the system, or show that it has no solution. If the system has infinitely many solutions, express them in the ordered triple form given in Example 4.

17. $\begin{cases} x \quad\ \ - 4z = 1 \\ 2x - y - 6z = 4 \\ 2x + 3y - 2z = 8 \end{cases}$

Solution

23. $\begin{cases} x + 2y - z = \ \ 1 \\ 2x + 3y - 4z = -3 \\ 3x + 6y - 3z = \ \ 4 \end{cases}$

Solution

✏ **Section 7.2 Exercise 31 Finance** An investor has $100,000 to invest in three types of bonds: short-term, intermediate-term and long-term. Short-term bonds pay 4% annually, intermediate-term bonds pay 5%, and long-term bonds pay 6%. The investor wishes to realize a total annual income of 5.1% with equal amounts invested in short- and intermediate-term bonds. How much should she invest in each type of bond?

Solution

CONCEPTS

✓ **7.2 Exercises – Concepts – Fundamentals** Complete the Fundamentals section of the exercises at the end of Section 7.2. Compare your answers to those at the back of the textbook, and make corrections as necessary.

📖 **Read the Concept Check for Section 7.2**, in the Chapter Review at the end of Chapter 7.

7.2 Solutions to ✐ Exercises

✐ Section 7.2 Exercise 5

From the last equation, we know that $z = 2$. We back-substitute this into the second equation and solve for y.

$$y + 2(2) = 7 \qquad \text{Back-substitute } z = 2 \text{ into Equation 2}$$

$$y = 3 \qquad \text{Solve for } y$$

Now we back-substitute $y = 3$ and $z = 2$ into the first equation and solve for x.

$$x - 2(3) + 4(2) = 3 \qquad \text{Back-substitute } y = 3 \text{ and } z = 2 \text{ into Equation 1}$$

$$x = 1 \qquad \text{Solve for } x$$

The solution of the system is $x = 1$, $y = 3$, and $z = 2$. We can also write the solution as the ordered triple $(1, 3, 2)$.

✐ Section 7.2 Exercise 13

To change this to a triangular system, we need to eliminate the x- and y- terms from the third equation. We begin by eliminating the x-term. We multiply Equation 1 by -2 and add the result to Equation 3, to create a new equation 3.

$$-2x + 2y + 2z = -8 \qquad \text{Equation } 1 \times (-2)$$

$$\underline{2x + 3y - 2z = 8} \qquad \text{Equation 3}$$

$$5y = 0 \qquad \text{Equation } 1 \times (-2) + \text{Equation 3} = \text{new Equation 3}$$

Now the new Equation 3 is equivalent to $y = 0$, and you could greatly simplify your system by back-substituting at this point. However, to practice creating a triangular system, we continue as planned. Our system is now

$$\begin{cases} x - y - z = 4 \\ 2y + z = -1 \\ 5y = 0 \end{cases}$$

We use the second equation to eliminate the y from the third equation (don't worry, the z will reappear). To get $-5y$ in the second equation, we multiply by $-5/2$.

$$-5y - \frac{5}{2}z = \frac{5}{2} \qquad \text{Equation } 2 \times \left(-\frac{5}{2}\right)$$

$$\underline{5y \phantom{- \frac{5}{2}z} = 0} \qquad \text{Equation 3}$$

$$-\frac{5}{2}z = \frac{5}{2} \qquad \text{Equation } 2 \times \left(-\frac{5}{2}\right) + \text{Equation 3} = \text{new Equation 3}$$

With this Equation 3, the system is now in triangular form, but it will be easier to work with if we divide the last equation by $-\dfrac{5}{2}$ on each side. Our new, triangular system is

$$\begin{cases} x - y - z = 4 \\ \quad\; 2y + z = -1 \\ \qquad\qquad z = -1 \end{cases}$$

We use back-substitution to solve the triangular system. From the third equation we get $z = -1$. We back-substitute $z = -1$ into the second equation and solve for y.

$2y + (-1) = -1$	Replace z by -1 in Equation 2
$y = 0$	Solve for y

Now we back-substitute $y = 0$ and $z = -1$ into the first equation and solve for x.

$x - 0 - (-1) = 4$	Replace y by 0 and z by -1 in Equation 1
$x = 3$	Solve for x

The solution of the system is $x = 3$, $y = 0$ and $z = -1$. We can also write the solution as the ordered triple $(3, 0, -1)$.

✎ Section 7.2 Exercise 17

To put this system in triangular form, we begin by eliminating the x from the second and third equations.

$-2x \qquad + 8z = -2$	Equation 1 $\times$ (–2)
$\underline{2x - y - 6z = \;\; 4}$	Equation 2
$-y + 2z = \;\; 2$	Equation 1 $\times$ (–2) + Equation 2 = new Equation 2

$-2x \qquad + 8z = -2$	Equation 1 $\times$ (–2)
$\underline{2x + 3y - 2z = \;\; 8}$	Equation 3
$3y + 6z = \;\; 6$	Equation 1 $\times$ (–2) + Equation 3 = new Equation 3

This gives us a new, equivalent system:

$$\begin{cases} x \qquad\;\; - 4z = 1 \\ \quad -y + 2z = 2 \\ \quad\;\; 3y + 6z = 6 \end{cases}$$

We now use the second equation to eliminate the y from the third equation.

$-3y + 6z = 6$	Equation 2 × 3
$\underline{3y + 6z = 6}$	Equation 3
$12z = 12$	Equation 2 × 3 + Equation 3 = new Equation 3

With this as the third equation, the system is now in triangular form, but it will be easier to work with if we divide the last equation by 12 on each side. Our new, triangular system is

$$\begin{cases} x \qquad\; -4z = 1 \\ \quad -y + 2z = 2 \\ \qquad\qquad z = 1 \end{cases}$$

We use back-substitution to solve the triangular system. From the third equation, we get $z = 1$. We back-substitute $z = 1$ into the second equation, and solve for y.

$-y + 2(1) = 2$	Replace z by 1 in Equation 2
$y = 0$	Solve for y

Now we back-substitute $z = 1$ into the first equation, and solve for x.

$x - 4(1) = 1$	Replace z by 1 in Equation 1
$x = 5$	Solve for x

The solution of the system is $x = 5$, $y = 0$ and $z = 1$. We can also write the solution as the ordered triple $(5, 0, 1)$.

✐ Section 7.2 Exercise 23

To put this system in triangular form, we begin by eliminating the x-term from the second and third equation.

$-2x - 4y + 2z = -2$	Equation 1 × (−2)
$\underline{2x + 3y - 4z = -3}$	Equation 2
$-y - 2z = -5$	Equation 1 × (−2) + Equation 2 = new Equation 2

$-3x - 6y + 3z = -3$	Equation 1 × (−3)
$\underline{3x + 6y - 3z =\;\; 4}$	Equation 3
$0 = 1$	Equation 1 × (−3) + Equation 3 = new Equation 3

The new third equation is $0 = 1$, which is false. No matter what values we assign to x, y, and z, the third equation will never be true. This means that the system has no solution.

✐ Section 7.2 Exercise 27

To put the system in triangular form, we begin by eliminating the x from the second and third equations.

$-x - \ y + z = \ 0$	Equation 1 $\times (-1)$
$\underline{x + 2y - 3z = -3}$	Equation 2
$y - 2z = -3$	Equation 1 $\times (-1)$ + Equation 2 = new Equation 2

$-2x - 2y + 2z = \ 0$	Equation 1 $\times (-2)$
$\underline{2x + 3y - 4z = -3}$	Equation 3
$y \ - 2z = -3$	Equation 1 $\times (-2)$ + Equation 3 = new Equation 3

This gives us the new, equivalent system

$$\begin{cases} x + y - \ z = \ 0 \\ \quad\ \ y - 2z = -3 \\ \quad\ \ y - 2z = -3 \end{cases}$$

We now use the second equation to eliminate the y in the third equation.

$-y + 2z = \ 3$	Equation 2 $\times (-1)$
$\underline{y - 2z = -3}$	Equation 3
$0 = 0$	Equation 2 $\times (-1)$ + Equation 3 = new Equation 3

The new third equation is true, but it gives us no new information, so we can drop it from the system. Only two equations are left. We can use them to solve for x and y in terms of z.

To find the complete solution of the system, we begin by solving for y in terms of z, using the second equation.

$y - 2z = -3$	Equation 2
$y = -3 + 2z$	Add $2z$ to each side.

Then we solve for x in terms of z using the first equation.

$x - 3 + 2z - z = 0$	Replace y by $-3 + 2z$ in Equation 1
$x - 3 + z = 0$	Simplify
$x = 3 - z$	Subtract $-3 + z$ from each side

To describe the complete solution, we let t represent any real number. The solution is

$$x = 3 - t$$
$$y = -3 + 2t$$
$$z = t$$

We can also write the solution as the ordered triple $(3 - t, -3 + 2t, t)$.

🖋 Section 7.2 Exercise 31 Finance

Since the question asks for the amounts she should invest in each of the three bonds, let x be the amount she invests in the short term bond, y be the amount she invests in the intermediate term bond, and z be the amount she invests in the long term bond.

Thinking about the problem

Following **Hints and Tips 7.1**, we think about the problem by guessing an answer, and checking if it is correct. For example, suppose the investor puts $x = \$25,000$ into the short-term bond, $y = \$35,000$ into the intermediate-term bond and $z = \$40,000$ into the long-term bond. We complete the table to see if our guess is correct.

In words	*In numbers*	*In words*	*In numbers*
Amount invested in the short-term bond	$25,000	Interest per year from the short-term bond	(0.04)(25,000) = \$1000
Amount invested in the intermediate term bond	$35,000	Interest per year from the intermediate term bond	(0.05)(35,000) = \$1750
Amount invested in the long term bond	$40,000	Interest per year from the long-term bond	(0.06)(40,000) = \$2400
Total investment	$100,000	Total interest per year	$5150

The investor wants the total invested to be $100,000 (our guess satisfies this requirement). From the investment, she wants to have 5.1% total annual interest, which on $100,000 is $(0.051)(100,000) = \$5,100$ (our guess has $5150, instead). The investor also specifies that she wants equal amounts invested in the short- and intermediate-term bonds, which our guess fails to have.

From our work above, we see that we want the following three equations to hold.

$$\left(\begin{array}{c}\text{investment in the}\\ \text{short-term bond}\end{array}\right)+\left(\begin{array}{c}\text{investment in the}\\ \text{intermediate-term bond}\end{array}\right)+\left(\begin{array}{c}\text{investment in the}\\ \text{long-term bond}\end{array}\right)=\$100,000$$

$$\left(\begin{array}{c}\text{interest from the}\\ \text{short-term bond}\end{array}\right)+\left(\begin{array}{c}\text{interest from the}\\ \text{intermediate-term bond}\end{array}\right)+\left(\begin{array}{c}\text{interest from the}\\ \text{long-term bond}\end{array}\right)=\$5,100$$

$$\left(\begin{array}{c}\text{investment in the}\\ \text{short-term bond}\end{array}\right)=\left(\begin{array}{c}\text{investment in the}\\ \text{intermediate-term bond}\end{array}\right)$$

Thus we want to find the amounts x, y, and z that solve the following system.

$$x + y + z = 100,000 \qquad \text{Equation 1}$$
$$0.04x + 0.05y + 0.06z = 5100 \qquad \text{Equation 2}$$
$$x = y \qquad \text{Equation 3}$$

Multiplying the second equation by 100 to get rid of the decimals, and rewriting the third equation, gives the following system, which we solve using Gaussian elimination.

$x + \ y + \ z = 100,000$	Equation 1
$4x + 5y + 6z = 510,000$	Equation 2
$x - \ y \qquad = 0$	Equation 3

We begin by eliminating the x from the second and third equations.

$x + \ y + \ z = 100,000$	Equation 1
$y + 2z = \ 110,000$	$-4 \times$ Equation 1 + Equation 2 = new Equation 2
$-2y - \ z = -100,000$	$-1 \times$ Equation 1 + Equation 3 = new Equation 3

Now we eliminate the y from the third equation.

$x + \ y + \ z = 100,000$	Equation 1
$y + 2z = 110,000$	Equation 2
$3z = \ 120,000$	$2 \times$ Equation 2 + Equation 3 = new Equation 3

Now that the system is in triangular form, we use back-substitution to find the solution. The third equation tells us $3z = 120,000$, so $z = 40,000$. We back-substitute $z = 40,000$ into the second equation and solve for y.

$y + 2(40,000) = 110,000$	Replace z by 40,000 in Equation 2
$y = 30,000$	Solve for y

We back-substitute $y = 30,000$ and $z = 40,000$ into the first equation and solve for x.

$x + 30,000 + 40,000 = 100,000$	Replace y by 30,000 and z by 40,000 in Equation 1
$x = 30,000$	Solve for x

This means the investor should invest $30,000 in the short- and intermediate-term bonds and $40,000 in the long-term bond.

✓ **Check** We check that this choice indeed satisfies the criteria in the following table.

In words	*In numbers*	*In words*	*In numbers*
Amount invested in the short-term bond	$30,000	Interest per year from the short-term bond	(0.04)(30,000) = $1200
Amount invested in the intermediate term bond	$30,000	Interest per year from the intermediate term bond	(0.05)(30,000) = $1500
Amount invested in the long term bond	$40,000	Interest per year from the long-term bond	(0.06)(40,000) = $2400
Total investment	$100,000	Total interest per year	$5100

7.3 Using Matrices to Solve Systems of Equations

OBJECTIVES

☑ *Check the box when you can do the exercises addressing the objective, such as those included in this study guide, which are listed here.*

Exercises

The Augmented Matrix of a Linear System

☐ Write a system of linear equations as an augmented matrix, and convert an augmented matrix back into a system of linear equations.

1, 11, 15, 21, 25, 27, 41

Elementary Row Operations

☐ Use elementary row operations to write a system of equations in triangular form, and use back-substitution to find the solution.

11

Row-Echelon Form

☐ Use a graphing calculator to find the row-echelon form of the augmented matrix of a system of linear equations, and then use back-substitution to solve the system.

15

Reduced Row-Echelon Form

☐ Use a graphing calculator to find the reduced row-echelon form of the augmented matrix of a system of linear equations, and then translate the matrix back into equations to solve the system.

21, 25, 27, 41

☐ Model the real world using a system of linear equations, and solve the system using matrices.

41

Inconsistent and Dependent Systems

☐ Identify inconsistent systems of equations from their augmented matrices in reduced row-echelon form.

25

☐ Identify dependent systems of equations from their augmented matrices in reduced row-echelon form.

27

☐ Express the complete solution of a depend system as an ordered triple.

27

SKILLS

Matrices

These exercises ask you to express systems of equations as matrices. Find the definition of a matrix in a blue box in Section 7.3 of the textbook.

[22]*The dimensions of the matrix in Section 7.3 Exercise 5 are ___ × ___.*

---

[22] The dimensions are the number of rows by the number of columns, so in Exercise 5, the dimensions are 3×2.

The Augmented Matrix of a Linear System

These exercises ask you to find the augmented matrix that represents a system of equations. Find the definition of an augmented matrix in bold typeface in Section 7.3 of the textbook.

📖 **Read Section 7.3 Example 1**, in which the authors represent a system by an augmented matrix.

[23] *Since there is no y in Equation 2 in this system, the coefficient of y in that equation is _____. In Equation 3, the coefficient on ___ is 0.*

✎ **Section 7.3 Exercise 1** Write the augmented matrix of the following system of equations.

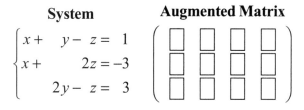

$$\begin{cases} x + & y - z = & 1 \\ x + & 2z = & -3 \\ & 2y - z = & 3 \end{cases}$$

Elementary Row Operations

These exercises ask you to perform elementary row operations to transform a matrix that represents a system of equations into another matrix that represents an equivalent system of equations. Find the Elementary Row Operations in a blue box in Section 7.3 of the textbook.

📖 **Read Section 7.3 Example 2**, in which the authors use elementary row operations to solve a linear system of equations.

[24] *Why did the authors choose to multiply the first row by –3 and add it to the third row, in the first step? Answer in your own words:*

---

[23] Since there is no y in Equation 2, the coefficient of y in that equation is 0. In Equation 3, the coefficient on x is 0.

[24] The authors want a 0 in the first column of the third row. They multiply the first row by –3 and add the result to the third row to get –3+3=0 in the first column of the new third row.

✏ **Section 7.3 Exercise 11** The system of linear equations has a unique solution. Find the solution using elementary row operations as in Example 2.

$$\begin{cases} x - 2y + z = 1 \\ y + 2z = 5 \\ x + y + 3z = 8 \end{cases}$$

Solution

Row-Echelon Form

These exercises ask you to find the row-echelon form of a matrix using a calculator. Find the definition of row-echelon form in a blue box in Section 7.3 of the textbook.

📖 **Read the introduction to the subsection, 'Row-Echelon Form,'** in which the authors give examples of a matrix that is and a matrix that is not in row-echelon form.

[25] *In the example that is not in row-echelon form, what one elementary row operation could you do to put the matrix in row-echelon form?* _____

---

[25] To put the matrix in row-echelon form, interchange Row 1 and Row 2.

□ **Read Section 7.3 Example 3**, in which the authors solve a system of equations using a calculator.

[26]*What are the steps for finding the solution of a system of equations using a calculator?*
Answer in your own words:

✎ **Section 7.3 Exercise 15** Use a graphing calculator to put the augmented matrix of the system into row-echelon form. Then solve the system (as in Example 3).

$$\begin{cases} x + 3y - 2z = -2 \\ x + 4y - 3z = -3 \\ x + 3y - z = 0 \end{cases}$$

Solution

Reduced Row-Echelon Form

These exercises ask you to solve systems of equations by first putting them in reduced row-echelon form. Find the definition of reduced row-echelon form in a blue box in Section 7.3 of the textbook.

[27]*Is this matrix in reduced row-echelon form?* _____ *(Yes / No) Explain how you know.*
Answer in your own words:

$$\begin{bmatrix} 1 & 0 & 3 & -2 \\ 0 & 1 & 4 & 3 \\ 0 & 0 & 0 & 0 \end{bmatrix}$$

---

[26] First write the system as an augmented matrix. Then enter the matrix into your calculator and use the ref (row-echelon form) command to get the row-echelon form of the matrix. Then write the corresponding system of equations, which is in triangular form. Use back-substitution to solve the system, from its triangular form.

[27] The matrix is in reduced row-echelon form, since there are leading 1's in columns 1 and 2, and the other entries in those columns are 0's. In column 3, there is no leading 1, so the entries can be any number.

📖 **Read Section 7.3 Example 4**, in which the authors solve a system, using the reduced row-echelon form of the augmented matrix for the system.

[28]*Why is reduced row-echelon form better for solving the system of equations than the row-echelon form?*
Answer in your own words:

✎ **Section 7.3 Exercise 21** Use a graphing calculator to put the augmented matrix of the system into reduced row-echelon form, then solve the system (as in Example 4).

$$\begin{cases} x + 2y - z = 9 \\ 2x \quad - z = -2 \\ 3x + 5y + 2z = 22 \end{cases}$$

Solution

Inconsistent and Dependent Systems

These exercises ask you to find the leading variables from the reduced row-echelon form of the augmented matrix representing a system of equations to determine if the system has a unique solution, infinitely many solutions or no solution. Find the definition of a leading variable in bold typeface and a description of the row-echelon forms for each of these cases in a blue box in Section 7.3 of the textbook.

📖 **Read Section 7.3 Example 5**, in which the authors determine that a system has no solution by looking at its reduced row-echelon form.

[29]*How can you tell that the system has no solution from the reduce row-echelon form of the augmented matrix in this example?*
Answer in your own words:

---

[28] When you write the system of equations corresponding to reduced row-echelon form, you do not need to back substitute to find the solution(s), as you do with (non-reduced) row-echelon form.

[29] The last row of the matrix is [0 0 0 1]. If we translate the last row back into equation form, we get $0x + 0y + 0z = 1$, or $0 = 1$, which is false. No matter what values we pick for x, y, and z, the last equation will never be a true statement. This means that the system has no solution.

📖 **Read Section 7.3 Example 6**, in which the authors determine that a system has infinitely many solutions by looking at its reduced row-echelon form.

[30]*Give two examples of numbers x, y, z such that the triple (x, y, z) is a solution to the system. Example 1: x = ___, y = ___, and z = ___. Example 2: x = ___, y = ___, and z = ___.*

How did you find these examples?
Answer in your own words:

✎ **Section 7.3 Exercises 25 and 27** Use a graphing calculator to put the augmented matrix of the system into reduced row-echelon form. Determine whether the system is inconsistent or dependent. If it is dependent, find the complete solution.

25. $\begin{cases} x + y + z = 2 \\ \quad\ \ y - 3z = 1 \\ 2x + y + 5z = 0 \end{cases}$

Solution

---

[30] To find an example, choose a value for t, and substitute it into the formula for the complete solution. For example, if $t = 1$, then we get $x = 7(2) - 5 = 9$, $y = 3(1) + 1 = 4$ and $z = 1$, so $(x, y, z) = (9, 4, 1)$. For another example, let $t = 0$. Then we get $(x, y, z) = (-5, 1, 0)$. Make up your own solution.

27. $\begin{cases} 2x - 3y - 9z = -5 \\ x + 3z = 2 \\ -3x + y - 4z = -3 \end{cases}$

Solution

CONTEXTS

Reduced Row-Echelon Form

These exercises ask you to solve system of linear equations that model the real world.

Refer back to page 411

Hints and Tips 7.1: Thinking about the problem

📖 **Read Section 7.3 Example 7**, in which the authors model the amount of potassium, protein and vitamin D in a combination of commercial diet foods.

[31] *What does this problem ask you to find?*
Answer in your own words:

Following Hints and Tips 7.1, before beginning to set up the equations, you can think about the problem by making up an answer and checking if it is right. For example, suppose the subject eats $x = 4$ ounces of MiniCal, $y = 3$ ounces of LiquiFast, and $z = 2$ ounces of SlimQuick. The MiniCal has ____ mg of potassium per ounce, the LiquiFast has ____ mg of potassium per ounce, and the SlimQuick has ____ mg of potassium per ounce. Complete the table to check if this guess has the right amount of potassium.

In words	In numbers	In words	In numbers
Amount of MiniCal	_____	Potassium from MiniCal	_____
Amount of LiquiFast	_____	Potassium from Liquifast	_____
Amount of SlimQuick	_____	Potassium from SlimQuick	_____
		Total Potassium	_____

How do you know that the guess $x = 4$, $y = 3$ and $z = 2$ are wrong?
Answer in your own words:

[32] *In the system of equations for Example 7, x represents _____, y represents _____, and z represents _____. So in the first equation, 50x represents _____, 75y represents _____, and 10z represents _____. The constant 500 on the right hand side of the first equation represents _____. What do the authors do to retrieve the solution from the reduced row-echelon form of the matrix?*
Answer in your own words:

---

[31] This problem asks you to find how much of each diet food should the subject eat every day; that is how much MiniCal, how much LiquiFast and how much SlimQuick. MiniCal has 50 mg of potassium per ounce, Liquifast has 75 mg of potassium per ounce, and SlimQuick has 10 mg potassium per ounce. Our guess is 4 oz. of MiniCal, 3 oz. of LiquiFast and 2 oz of SlimQuick. With these quantities of diet foods, the subject gets is $4(50) = 200$ mg potassium from MiniCal, $75(3) = 225$ mg potassium from LiquiFast, and $2(10) = 20$ mg potassium from SlimQuick. So on that diet, the patient gets a total of $200 + 225 + 20 = 445$ mg potassium. So the guess that $x = 4$, $y = 3$, and $z = 2$ is wrong, since the patient is supposed to have a diet that includes 500 mg of potassium.

[32] The variable x represents the amount MiniCal in the patient's diet, the variable y represents the amount of Liquifast in the patient's diet, and the variable z represents the amount of SlimQuick in the patient's diet. So, $50x$ represents the potassium the patient gets from MiniCal, $75y$ represents the potassium the patient gets from LiquiFast, and $10z$ represents the potassium the patient gets from SlimQuick. The constant 500 (which is in milligrams) represents the total potassium in the patient's diet. To get the solution, the authors change the reduced row-echelon form of the matrix back into a system of equations. They get $x = 5$, $y = 2$, and $z = 10$, which is the solution to the system.

✎ **Section 7.3 Exercise 41 Nutrition** A doctor recommends that a patient take 50 mg each of niacin, riboflavin, and thiamin daily to alleviate a vitamin deficiency. In his medicine chest at home the patient finds three brands of vitamin pills. The amounts of the relevant vitamins per pill are given in the table. How many pills of each type should he take every day to get 50 mg of each vitamin?

Solution

Vitamin (mg)	VitaMix	Vitron	VitaPlus
Niacin	5	10	15
Riboflavin	15	20	0
Thiamin	10	10	10

CONCEPTS

✓ **7.3 Exercises – Concepts – Fundamentals** Complete the Fundamentals section of the exercises at the end of Section 7.3. Compare your answers to those at the back of the textbook, and make corrections as necessary.

📖 **Read the Concept Check for Section 7.3**, in the Chapter Review at the end of Chapter 7.

7.3 Solutions to ✐ Exercises

✐ **Section 7.3 Exercise 1**

<table>
<tr><th colspan="2">System</th><th>Augmented Matrix</th></tr>
</table>

$$\begin{cases} x + y - z = 1 \\ x + 2z = -3 \\ 2y - z = 3 \end{cases} \qquad \begin{bmatrix} 1 & 1 & -1 & 1 \\ 1 & 0 & 2 & -3 \\ 0 & 2 & -1 & 3 \end{bmatrix}$$

✐ **Section 7.3 Exercise 11**

Our goal is to eliminate the x-term from the second equation and the x- and y-term from the third equation. In the augment matrix, this means using elementary row operations to get 0's in the second and third row of column 1 and in the third row of column 2.

We begin with the augmented matrix of the system.
$$\begin{bmatrix} 1 & -2 & 1 & 1 \\ 0 & 1 & 2 & 5 \\ 1 & 1 & 3 & 8 \end{bmatrix}$$

Add $(-1) \times$ Row 1 to Row 3
$$\begin{bmatrix} 1 & -2 & 1 & 1 \\ 0 & 1 & 2 & 5 \\ 0 & 3 & 2 & 7 \end{bmatrix}$$

Add $(-3) \times$ Row 2 to Row 3
$$\begin{bmatrix} 1 & -2 & 1 & 1 \\ 0 & 1 & 2 & 5 \\ 0 & 0 & -4 & -8 \end{bmatrix}$$

Multiply Row 3 by $-\dfrac{1}{4}$
$$\begin{bmatrix} 1 & -2 & 1 & 1 \\ 0 & 1 & 2 & 5 \\ 0 & 0 & 1 & 2 \end{bmatrix}$$

This is the augmented matrix for the triangular system below, which is equivalent to the original system of equations.
$$\begin{cases} x - 2y + z = 1 \\ y + 2z = 5 \\ z = 2 \end{cases}$$

We use back-substitution to find the solution. From the third equation, we have $z = 2$. We substitute $z = 2$ into the second equation and solve for y.

$$y + 2(2) = 5 \qquad \text{Replace } z \text{ by 2 in Equation 2}$$
$$y = 1 \qquad \text{Solve for } y$$

Now we back-substitute $y = 1$ and $z = 2$ into the first equation and solve for x.

$$x - 2(1) + 2 = 1 \qquad \text{Replace } y \text{ by 1 and } z \text{ by 2 in Equation 1}$$
$$x = 1 \qquad \text{Solve for } x$$

Thus the solution is $x = 1$, $y = 1$, and $z = 2$. Written as an ordered triple, the solution is $(1, 1, 2)$.

✎ **Section 7.3 Exercise 15**

First we write the augmented matrix of the system.

$$\begin{bmatrix} 1 & 3 & -2 & -2 \\ 1 & 4 & -3 & -3 \\ 1 & 3 & -1 & 0 \end{bmatrix}$$

Using the ref command on a graphing calculator, we put the matrix into row-echelon form:

$$\begin{bmatrix} 1 & 3 & -2 & -2 \\ 0 & 1 & -1 & -1 \\ 0 & 0 & 1 & 2 \end{bmatrix}$$

This is the augmented matrix for the triangular system below, which is equivalent to the original system of equations.

$$\begin{cases} x + 3y - 2z = -2 \\ y - z = -1 \\ z = 2 \end{cases}$$

We use back-substitution to find the solution. From the third equation, we have $z = 2$. We substitute $z = 2$ into the second equation and solve for y.

$$y - 2 = -1 \qquad \qquad \text{Replace } z \text{ by 2 in Equation 2}$$
$$y = 1 \qquad \qquad \text{Solve for } y$$

Now we back-substitute $y = 1$ and $z = 2$ into the first equation and solve for x.

$$x + 3(1) - 2(2) = -2 \qquad \qquad \text{Replace } y \text{ by 1 and } z \text{ by 2 in Equation 1}$$
$$x = -1 \qquad \qquad \text{Solve for } x$$

Thus the solution is $x = -1$, $y = 1$, and $z = 2$. Written as an ordered triple, the solution is $(-1, 1, 2)$.

✎ **Section 7.3 Exercise 21**

First we write the augmented matrix of the system.

$$\begin{bmatrix} 1 & 2 & -1 & 9 \\ 2 & 0 & -1 & -2 \\ 3 & 5 & 2 & 22 \end{bmatrix}$$

Using the rref command on a graphing calculator, we put the matrix into reduced row-echelon form:

$$\begin{bmatrix} 1 & 0 & 0 & -1 \\ 0 & 1 & 0 & 5 \\ 0 & 0 & 1 & 0 \end{bmatrix}$$

This is the augmented matrix for the triangular system below, which is equivalent to the original system of equations.

$$\begin{cases} x = -1 \\ y = 5 \\ z = 0 \end{cases}$$

Hence we immediately arrive at the solution $(-1, 5, 0)$.

✐ Section 7.3 Exercises 25

First we write the augmented matrix of the system.

$$\begin{bmatrix} 1 & 1 & 1 & 2 \\ 0 & 1 & -3 & 1 \\ 2 & 1 & 5 & 0 \end{bmatrix}$$

Using the rref command on a graphing calculator, we put the matrix into reduced row-echelon form:

$$\begin{bmatrix} 1 & 0 & 4 & 0 \\ 0 & 1 & -3 & 0 \\ 0 & 0 & 0 & 1 \end{bmatrix}$$

If we translate the last row back into equation form, we get $0x + 0y + 0z = 1$, or $0 = 1$, which is false. No matter what values we pick for x, y, and z, the last equation will never be a true statement. This means that the system has no solution.

✐ Section 7.3 Exercises 27

First we write the augmented matrix of the system.

$$\begin{bmatrix} 2 & -3 & -9 & -5 \\ 1 & 0 & 3 & 2 \\ -3 & 1 & -4 & -3 \end{bmatrix}$$

Using the rref command on a graphing calculator, we put the matrix into reduced row-echelon form:

$$\begin{bmatrix} 1 & 0 & 3 & 2 \\ 0 & 1 & 5 & 3 \\ 0 & 0 & 0 & 0 \end{bmatrix}$$

The third row corresponds to the equation $0 = 0$. This equation is always true, no matter what values are used for x, y, and z. Since the equation adds no new information about the variables, we can drop it from the system. So the last matrix corresponds to the system

$$\begin{cases} x + 3z = 2 \\ y + 5z = 3 \end{cases}$$

We solve for the leading variables are x and y, in terms of the nonleading variable z.

$$x = 2 - 3z \qquad \text{Solve for } x \text{ in Equation 1}$$

$$y = 3 - 5z \qquad \text{Solve for } y \text{ in Equation 2}$$

To obtain the complete solution, we let t represent any real number, and we express x, y, and z in terms of t.

$$x = 2 - 3t$$

$$y = 3 - 5t$$

$$z = t$$

We can also write the solution as the ordered triple $(2 - 3t, 3 - 5t, t)$, where t is any real number.

✎ Section 7.3 Exercise 41 Nutrition

Since the question asks for the number pills of each type that the patient should take, let x be the number of VitaMix pills, y be the number of Vitron pills, and z be the number of VitaPlus pills.

Thinking about the problem

Following **Hints and Tips 7.1**, before setting up the equations, we think about the problem by making up an answer and checking if it is right. For example, suppose that the patient takes $x = 2$ VitaMix pills, $y = 1$ Vitron pill and $z = 4$ VitaPlus pills. We fill out the table to see if the patient gets the right amount of Niacin with that combination.

In words	*In numbers*	*In words*	*In numbers*
Number of VitaMix pills	2 pills	Niacin from VitaMix	(5 mg/pill)(2 pills) = 10 mg
Number of Vitron pills	1 pill	Niacin from Vitron	(10 mg/pill)(1 pill) = 10 mg
Number of VitaPlus pills	4 pills	Niacin from VitaPlus	(15 mg/pill)(4 pills) = 60 mg
		Total Niacin	80 mg

The patient should bet 50 mg of Niacin, so this guess is not right.

From our work above, we see that we have these equations

$$\begin{pmatrix} \text{Niacin} \\ \text{from VitaMix} \end{pmatrix} + \begin{pmatrix} \text{Niacin} \\ \text{from Vitron} \end{pmatrix} + \begin{pmatrix} \text{Niacin} \\ \text{from VitaPlus} \end{pmatrix} = \begin{pmatrix} \text{Total amount} \\ \text{of niacin} \end{pmatrix}$$

$$\begin{pmatrix} \text{Riboflavin} \\ \text{fromVitaMix} \end{pmatrix} + \begin{pmatrix} \text{Riboflavin} \\ \text{from Vitron} \end{pmatrix} + \begin{pmatrix} \text{Riboflavin} \\ \text{from VitaPlus} \end{pmatrix} = \begin{pmatrix} \text{Total amount} \\ \text{of riboflavin} \end{pmatrix}$$

$$\begin{pmatrix} \text{Thiamin} \\ \text{fromVitaMix} \end{pmatrix} + \begin{pmatrix} \text{Thiamin} \\ \text{from Vitron} \end{pmatrix} + \begin{pmatrix} \text{Thiamin} \\ \text{from VitaPlus} \end{pmatrix} = \begin{pmatrix} \text{Total amount} \\ \text{of thiamin} \end{pmatrix}$$

Thus, we want to find x, y and z that solve the following system.

$5x + 10y + 15z = 50$ Niacin Equation

$15x + 20y \qquad = 50$ Riboflavin Equation

$10x + 10y + 10z = 50$ Thiamin Equation

We begin by writing the augmented matrix for this system.

$$\begin{bmatrix} 5 & 10 & 15 & 50 \\ 15 & 20 & 0 & 50 \\ 10 & 10 & 10 & 50 \end{bmatrix}$$

Using the rref command on a graphing calculator, we put the matrix into reduced row-echelon form:

$$\begin{bmatrix} 1 & 0 & 0 & 2 \\ 0 & 1 & 0 & 1 \\ 0 & 0 & 1 & 2 \end{bmatrix}$$

This is the augmented matrix for the triangular system below, which is equivalent to the original system of equations.

$$\begin{cases} x = 2 \\ y = 1 \\ z = 2 \end{cases}$$

Thus the patient should take 2 VitaMix pills, 1 Vitron pill and 2 VitaPlus pills to get 50 mg each of Niacin, Riboflaven and Thiamin.

✓ **Check** We verify that the patient gets the right amount of niacin with these doses in the following table.

In words	In numbers	In words	In numbers
Number of VitaMix pills	2 pills	Niacin from VitaMix	(5 mg/pill)(2 pills) = 10 mg
Number of Vitron pills	1 pill	Niacin from Vitron	(10 mg/pill)(1 pill) = 10 mg
Number of VitaPlus pills	2 pills	Niacin from VitaPlus	(15 mg/pill)(2 pills) = 30 mg
		Total Niacin	50 mg

7.4 Matrices and Data in Categories

OBJECTIVES

☑ *Check the box when you can do the exercises addressing the objective, such as those included in this study guide, which are listed here.*

Exercises

Organizing Categorical Data in a Matrix

☐ Read information from a matrix of categorical data. — **7, 11, 13, 15, 17**

☐ Calculate the total number of data points in each category, from a matrix of categorical data. — **7, 17**

☐ Calculate proportions for a matrix of categorical data. Organize proportions in a matrix. — **11, 15**

Adding Matrices

☐ Add matrices of data. — **13**

☐ Interpret the sum of two matrices of categorical data. — **13**

Scalar Multiplication of Matrices

☐ Calculate a scalar multiple of a matrix. — **15**

☐ Compare proportions in a matrix of categorical data to proportions of a scalar multiple of the matrix. — **15**

Multiplying a Matrix Times a Column Vector

☐ Multiply a matrix by a column vector. — **17**

☐ Interpret the entries in the product of a matrix of categorical data and an appropriate column vector. — **17**

CONTEXTS

📖 **Read the introduction to Section 7.4**, in which the authors discuss categorical data.

[33]*Give an example of categorical data.*
Answer in your own words:

Organizing Categorical Data in a Matrix

📖 **Read Section 7.4 Example 1**, in which the authors organize information about the hair and eye color of a class of college algebra students.

[34]*In the matrix, to count the number of students with the particular eye-color, you add the values in the corresponding _____ (column / row), and to count the number of students a particular hair-color, you add the values in the corresponding _____ (column / row).*

---

[33] An example of categorical data is a list of responses to a multiple choice survey question. For example, the survey could ask the eye color and hair color of each of its respondents. The list of responses is an example of categorical data.

[34] To count the students with a particular hair-color, add the numbers in the corresponding column. To count the students with a particular eye-color, add the numbers in the corresponding row.

✎ **Section 7.4 Exercise 7** A Cincinnati car dealership conducted a customer satisfaction survey, and the results are tabulated in the data matrix A.

(a) How many of the customers who had a bill of less than $50 gave a fair service rating?
(b) How many customers had a bill above $100?
(c) How many customers gave an excellent service rating?
(d) How customers were surveyed?

Solution

Cincinnati dealership
Amount of bill

Less than $50 ↓	$50-$100 ↓	More than $100 ↓		
10	8	5	← Excellent	
5	7	6	← Good	**Service rating**
1	3	5	← Fair	

📖 **Read Section 7.4 Example 2**, in which the authors calculate the proportions, for example of students in a college algebra class with dark hair and brown eyes.

[35] *To calculate the (3, 3) entry in the matrix P, the authors divide _____ by _____.*

✎ **Section 7.4 Exercise 11** A tasting booth at Joe's Specialty Foods offers pesto pizza on Monday, spinach ravioli on Tuesday, and macaroni and cheese on Wednesday. The sales distribution for these products is tabulated in matrix A.

(a) Express each entry in the matrix as a proportion of the column total. Call the resulting matrix P.
(b) What does the (1,3) entry in P represent? What about the (3,3) entry?

Solution

Specialty food

Pesto pizza ↓	Spinach ravioli ↓	Macaroni and cheese ↓		
50	20	15	← Monday	
40	75	20	← Tuesday	**Day**
35	60	100	← Wednesday	

---

[35] To calculate the (3, 3) entry in the matrix, the authors divide 1 (the number of red haired green eyed students) by 4 (the total number of red haired students).

Adding Matrices

These exercises ask you to interpret the entries in the sum of two matrices that hold real-world categorical data.

📖 **Read Section 7.4 Example 3**, in which the authors organize information about the hair and eye color of a class of history students.

[36] *There are _____ students in the college algebra class with blond hair and blue eyes, and there are _____ students in the history class with blond hair and blue eyes, so the total number of students with blond hair and blue eyes is _____. The number of students with blond hair and blue eyes in each class is the (_____, _____) entry in the matrices A and B. So the total number of students with blond hair and blue eyes is the (_____, _____) entry in the matrix A + B.*

✏ **Section 7.4 Exercise 13 (extended) Service Ratings** The following questions refer to the matrix from Section 7.4 Exercise 8, given below, and the matrix in Section 7.4 Exercise 7, included in this study guide, above.

(a) Find a data matrix for the results of the survey at the Cincinnati dealership (in Exercise 7 above) combined with the results of the survey at the Akron dealership (below).
(b) What is the total number of customers (from both surveys) who had a bill between $50 and $100 and gave an excellent service rating?
(c) What does the (1, 1) entry in each of the three matrices represent?

Solution

Akron dealership

Amount of bill

Less than $50	$50-$100	More than $100	
↓	↓	↓	
9	15	12	← Excellent
8	10	6	← Good
5	2	7	← Fair

Service rating

---

[36] There are 6 blond-haired blue-eyed students in the college algebra class, and 10 blond-haired blue-eyed students in the history class, so in total, there are 16 blond-haired blue-eyed students. This information is the (2, 2) entry in each matrix.

Scalar Multiplication of Matrices

In these exercises, you perform scalar multiplication on a matrix holding real-world data. Find the definition of scalar multiplication in bold typeface in Section 7.4 of the textbook.

📖 **Read Section 7.4 Example 4**, in which the authors multiply the entries of the matrix holding information about the hair and eye-color of the college algebra students by a scalar.

[37] *If the (2,1) entry in the matrix A is 4, then the (2,1) entry in the matrix 3A is ____.*

✎ **Section 7.4 Exercise 15 (extended) Sales** The following questions refer to the matrix A, given in Section 7.4 Exercise 11, included in this study guide, above.

(a) Find the scalar product 4A.
(b) In this new data matrix, what proportion of the pesto pizza sales occurs on Wednesday?
(c) In the matrix A, what proportion of the pesto pizza sales occurs on Wednesday?

Solution

Multiplying a Matrix Times a Column Matrix

These exercises ask you to multiply a matrix by a column matrix. Find the definition of this product in the introduction to this subsection of Section 7.4 in the textbook.

📖 **Read Section 7.4 Example 5**, in which the authors organize information about the revenue a farmer earns from selling produce at a market.

[38] *On Thursday, the farmer earns ____ by selling oranges, ____ by selling broccoli, and ____ by selling beans. So his total revenue on Thursday is ____. This total is found in the (___, ___) entry of the column matrix AC.*

---

[37] If the (2, 1) entry in the matrix A is 4, then the (2, 1) entry in the matrix $3A$ is 12.

[38] On Thursday, the farmer earns $50(0.90) = \$45$ by selling oranges, $20(1.20) = \$24$ by selling broccoli, and $10(1.50) = \$15$ by selling beans. So his Thursday total is $\$45 + \$24 + \$15 = \$84$. This total is the (1, 1) entry in the column matrix AC.

✎ **Section 7.4 Exercise 17 Car-Manufacturing Profits** A specialty-car manufacturer has plants in Auburn, Biloxi, and Chattanooga. Three models are produced, with daily production given in matrix A. The profit (in dollars) per car is tabulated by model in matrix C.

(a) Find the product matrix AC.

(b) Assuming that all cars produced are sold, what is the daily profit from the Biloxi plant?

(c) What is the total daily profit (from all three plants)?

Cars produced each day

Model K   Model R   Model W
↓   ↓   ↓

$$\begin{bmatrix} 12 & 10 & 0 \\ 4 & 4 & 20 \\ 8 & 9 & 12 \end{bmatrix}$$

← Auburn
← Biloxi **City**
← Chattanooga

Profit (dollars per car)

$$C = \begin{bmatrix} 1000 \\ 2000 \\ 1500 \end{bmatrix}$$

← Model K
← Model R
← Model W

Solution

CONCEPTS

✓ **7.4 Exercises – Concepts – Fundamentals** Complete the Fundamentals section of the exercises at the end of Section 7.4. Compare your answers to those at the back of the textbook, and make corrections as necessary.

📖 **Read the Concept Check for Section 7.4**, in the Chapter Review at the end of Chapter 7.

7.4 Solutions to ✐ Exercises

✐ Section 7.4 Exercise 7

(a) This is the entry in the "less than $50" column and the "fair" row (the (1, 3) entry in the matrix). We see that there is one customer who spent less than $50 and gave the dealership a fair rating.

(b) We add the numbers in the "More than $100" column: $5 + 6 + 5 = 16$. Sixteen customers had a bill above $100.

(c) We add the numbers in the "Excellent" row: $10 + 8 + 5 = 23$. Twenty-three customers gave the dealership an excellent rating.

(d) We add all the entries in the matrix:

$$10 + 8 + 5 + 5 + 7 + 6 + 1 + 3 + 5 = 50$$

Fifty customers were surveyed.

✐ Section 7.4 Exercise 11

(a) The total of the first column of the given matrix is $50 + 40 + 35 = 125$. So the entries in the first column of the matrix P are

$$\frac{50}{125} = 0.4 \qquad \frac{40}{125} = 0.32 \qquad \frac{35}{125} = 0.28$$

The other entries in the matrix P are calculated similarly.

Specialty food

	Pesto pizza ↓	Spinach ravioli ↓	Macaroni and cheese ↓		
$P =$	0.4	0.13	0.11	← Monday	
	0.32	0.48	0.15	← Tuesday	**Day**
	0.28	0.39	0.74	← Wednesday	

(b) The (1, 3) entry in P tells us that of the sales of macaroni and cheese in those three days, 11% occurred on Monday. The (3, 3) entry in P tells us that of the macaroni and cheese sales in those three days, 74% occurred on Wednesday. Since the tasting booth served macaroni and cheese on Wednesday, it makes sense that the largest number of macaroni and cheese sales were on that day.

Section 7.4 Exercise 13 (extended) Service Ratings

(a) To combine the results, we add the Cincinnati matrix to the Akron matrix. The total in each category is simply the sum of the corresponding entries in the matrices A and B.

$$\begin{bmatrix} \text{Cincinnati} \\ \text{matrix} \end{bmatrix} + \begin{bmatrix} \text{Akron} \\ \text{matrix} \end{bmatrix} = \begin{bmatrix} 10 & 8 & 5 \\ 5 & 7 & 6 \\ 1 & 3 & 5 \end{bmatrix} + \begin{bmatrix} 9 & 15 & 12 \\ 8 & 10 & 6 \\ 5 & 2 & 7 \end{bmatrix} = \begin{bmatrix} 19 & 23 & 17 \\ 13 & 17 & 12 \\ 6 & 5 & 12 \end{bmatrix}$$

(b) From the (1, 2) entry in the sum matrix, we see that 23 customers from these dealerships had a bill between \$50 and \$100 and gave an excellent service rating.

(c) The (1, 1) entry in the Cincinnati matrix is the number of customers of the Cincinnati dealership with a bill less than \$50, who gave an excellent service rating. Similarly, the (1, 1) entry in the Akron matrix is the number of customers of the Akron dealership with a bill less than \$50, who gave an excellent service rating. The (1, 1) entry in the sum matrix is the total number of customers at both locations with a bill less than \$50, who gave an excellent service rating.

Section 7.4 Exercise 15 (extended) Sales

(a) To perform the scalar multiplication, we multiply each entry in the data matrix by 4.

$$4A = 4 \begin{bmatrix} 50 & 20 & 15 \\ 40 & 75 & 20 \\ 35 & 60 & 100 \end{bmatrix} = \begin{bmatrix} 200 & 80 & 60 \\ 160 & 300 & 80 \\ 140 & 240 & 400 \end{bmatrix}$$

(b) The total in the "pesto pizza" column in the new matrix is $200 + 160 + 140 = 500$. So the proportion of the pesto pizza sales that occur on Wednesday in this matrix is $\frac{140}{500} = 0.28$.

(c) The total in the "pesto pizza" column in the original matrix is $50 + 40 + 35 = 125$. So the proportion of the pesto pizza sales that occur on Wednesday in this matrix is $\frac{35}{125} = 0.28$, as well.

Section 7.4 Exercise 17 Car-Manufacturing Profits

(a) The product matrix AC given by the following.

$$AC = \begin{bmatrix} 12 & 10 & 0 \\ 4 & 4 & 20 \\ 8 & 9 & 12 \end{bmatrix} \begin{bmatrix} 1000 \\ 2000 \\ 1500 \end{bmatrix} = \begin{bmatrix} 12(1000)+10(2000)+0(1500) \\ 4(1000)+4(2000)+20(1500) \\ 8(1000)+9(2000)+12(1500) \end{bmatrix} = \begin{bmatrix} 32,000 \\ 42,000 \\ 44,000 \end{bmatrix}$$

(b) The (2,1) entry of the matrix is calculated as follows:

$$\begin{pmatrix} 4 \text{ model K cars} \\ \text{sold in Biloxi} \end{pmatrix}\left(1000\,\frac{\text{dollars}}{\text{model K car}}\right) + \begin{pmatrix} 4 \text{ model R cars} \\ \text{sold in Biloxi} \end{pmatrix}\left(2000\,\frac{\text{dollars}}{\text{model L car}}\right)$$

$$+ \begin{pmatrix} 20 \text{ model W cars} \\ \text{sold in Biloxi} \end{pmatrix}\left(1500\,\frac{\text{dollars}}{\text{model W car}}\right) = \$42{,}000$$

This means that the (2, 1) entry of matrix AC is the sum of Biloxi's daily profit from the sales of each of the three models. So the total profit in Biloxi is $42,000.

(c) We add the three entries in the column matrix AC to get the total profit for the day: $32{,}000 + 42{,}000 + 44{,}000 = 118{,}000$. So the daily profit from all three plants is $118,000.

7.5 Matrix Operations: Getting Information from Data

OBJECTIVES

☑ *Check the box when you can do the exercises addressing the objective, such as those included in this study guide, which are listed here.*

Exercises

Addition, Subtraction, and Scalar Multiplication

☐ Add, subtract or perform scalar multiplication of two matrices.   **15**

☐ Determine if it is possible to add or subtract two matrices.   **15**

Multiplying Matrices

☐ Multiply matrices by hand.   **13**

☐ Multiply matrices on a graphing calculator.   **13**

☐ Determine if it is possible to multiply two matrices.   **11, 13**

Getting Information from Categorical Data

☐ Organize information about the real world with matrices and use matrix multiplication to get information from the model.   **27**

SKILLS

Addition, Subtraction, and Scalar Multiplication

These exercises ask you perform the operations matrix operations you defined in context in Section 7.4.

📖 **Read Section 7.5 Example 1**, in which the authors perform addition, subtraction and scalar multiplication of matrices.

[39]*Matrix A has dimensions ____ × ____. Matrix C has dimensions ____ × ____. Why is there no answer for part (c)?*
Answer in your own words:

---

[39] Matrix A is 3×2, and matrix C is 2×3. You cannot add matrices with different dimensions.

✎ **Section 7.5 Exercise 15** Matrices B, C and F are defined as follows. Perform the indicated matrix operations(s), or explain why it cannot be performed.

$$B = \begin{bmatrix} 3 & \frac{1}{2} & 5 \\ 1 & -1 & 3 \end{bmatrix}, \quad C = \begin{bmatrix} 2 & -\frac{5}{2} & 0 \\ 0 & 2 & -3 \end{bmatrix}, \quad F = \begin{bmatrix} 1 & 0 & 0 \\ 0 & 1 & 0 \\ 0 & 0 & 1 \end{bmatrix}$$

(a) $B + C$ (b) $B + F$

Solution

Matrix Multiplication

These questions ask you to multiply matrices. Find the definition of the product of two matrices in a blue box in Section 7.5 of the textbook.

📖 **Read Section 7.5 Example 2**, in which the authors multiply matrices.

[40] *You can multiply an $m \times n$ matrix by a $p \times q$ matrix provided that ____ (m / n) equals ____ (p / q). In that case, the resulting matrix has dimensions ____ × ____. To find c_{23}, the (2, 3) entry in AB, the authors find the inner product of the 2nd _____ (row / column) of A and the 3rd _____ (row / column) of B.*

📖 **Read Section 7.5 Example 3**, in which the authors use a graphing calculator to perform matrix multiplication.

✎ **Section 7.5 Exercises 11 and 13** Perform the matrix operation by hand, or explain why the operation is not defined. If the operation is defined, verify your answer by doing the calculation on your graphing calculator.

$$11. \begin{bmatrix} 2 & 6 \\ 1 & 3 \\ 2 & 4 \end{bmatrix} \begin{bmatrix} 1 & -2 \\ 3 & 6 \\ -2 & 0 \end{bmatrix}$$

Solution

---

[40] You can multiply an $m \times n$ matrix by a $p \times q$ matrix, provided that $n = p$. In that case the resulting matrix has dimensions $m \times q$. To find the (2, 3) entry in AB, find the inner product of the second row of A and the third column of B.

13.
$$\begin{bmatrix} 1 & 2 \\ -1 & 4 \end{bmatrix}\begin{bmatrix} 1 & -2 & 3 \\ 2 & 2 & -1 \end{bmatrix}$$

Solution

CONTEXTS

Getting Information from Categorical Data

📖 **Read Section 7.4 Example 5 again,** paying particular attention to the meaning of the entries in the matrix AC.

[41] *What does the (2, 1) entry of the matrix AC represent in context?*
Answer in your own words:

📖 **Read Section 7.5 Example 4,** in which the authors model a farmer's revenue from selling oranges, broccoli and beans at a market.

[42] *What does the (2, 1) entry in the matrix AC represent in context?*
Answer in your own words:

---

[41] The (2, 1) entry is the sum of Friday's revenue from the sales of oranges, broccoli and beans.

[42] The (2, 1) entry is the sum of Friday's revenue from the sales of oranges, broccoli and beans in week 1.

✎ **Section 7.5 Exercise 27 Fast-Food Sales** A small fast-food chain with restaurants in Santa Monica, Long Beach, and Anaheim sells only hamburgers, hotdogs, and milk shakes. On a certain day, sales were distributed according to matrix A. The price of each item (in dollars) is given by matrix C.

(a) Calculate the product matrix CA.

(b) What is the total revenue from the Long Beach restaurant?

(c) What is the total revenue (from all three restaurants)?

Number of items sold

$$A = \begin{bmatrix} 4000 & 1000 & 3500 \\ 400 & 300 & 200 \\ 700 & 500 & 9000 \end{bmatrix} \begin{matrix} \leftarrow \text{Hamburgers} \\ \leftarrow \text{Hot dogs} \\ \leftarrow \text{Milkshakes} \end{matrix}$$

Santa Monica, Long Beach, Anaheim (columns); Item (rows)

Item

$$C = \begin{bmatrix} 0.90 & 0.80 & 1.10 \end{bmatrix}$$

Hamburger, Hotdog, Milkshake

Solution

CONCEPTS

✓ **7.5 Exercises – Concepts – Fundamentals** Complete the Fundamentals section of the exercises at the end of Section 7.5. Compare your answers to those at the back of the textbook, and make corrections as necessary.

📖 **Read the Concept Check for Section 7.5**, in the Chapter Review at the end of Chapter 7.

7.5 Solutions to ✐ Exercises

✐ Section 7.5 Exercise 15

(a) Since B and C are matrices with the same dimensions (2×3), we an add them.

$$B + C = \begin{bmatrix} 3 & \frac{1}{2} & 5 \\ 1 & -1 & 3 \end{bmatrix} + \begin{bmatrix} 2 & -\frac{5}{2} & 0 \\ 0 & 2 & -3 \end{bmatrix} = \begin{bmatrix} 5 & -2 & 5 \\ 1 & 1 & 0 \end{bmatrix}$$

(b) $B + F$ is not defined because we cannot add matrices with different dimensions, and B is 2×3, while F is 3×3.

✐ Section 7.5 Exercises 11

The product is not defined since the dimensions of each matrix is 3×2. When we write $(2 \times 3)(2 \times 3)$, we see that the inner two numbers are not the same, so the rows and columns won't match up when we try to calculate the product.

✐ Section 7.5 Exercises 13

The first matrix has dimensions 2×2, and the second has dimensions 2×3. When we write $(2 \times 2)(2 \times 3)$, we see the inner two numbers are the same, so the product is defined. The resulting matrix will have dimension 2×3 (from the outer two numbers).

$$\begin{bmatrix} 1 & 2 \\ -1 & 4 \end{bmatrix} \begin{bmatrix} 1 & -2 & 3 \\ 2 & 2 & -1 \end{bmatrix} = \begin{bmatrix} 1\cdot1+2\cdot2 & 1\cdot-2+2\cdot2 & 1\cdot3+2\cdot-1 \\ -1\cdot1+4\cdot2 & -1\cdot-2+4\cdot2 & -1\cdot3+4\cdot-1 \end{bmatrix} = \begin{bmatrix} 5 & 2 & 1 \\ 7 & 10 & -7 \end{bmatrix}$$

✐ Section 7.5 Exercise 27 Fast-Food Sales

(a) The first matrix has dimensions 1×3, and the second has dimensions 3×3. When we write $(1 \times 3)(3 \times 3)$, we see the inner two numbers are the same, so the product is defined. The resulting matrix will have dimension 1×3 (from the outer two numbers).

$$CA = \begin{bmatrix} 0.90 & 0.80 & 1.10 \end{bmatrix} \begin{bmatrix} 4000 & 1000 & 3500 \\ 400 & 300 & 200 \\ 700 & 500 & 9000 \end{bmatrix} = \begin{bmatrix} 4690 & 1690 & 13{,}970 \end{bmatrix}$$

(b) The $(1, 2)$ entry in the matrix CA is obtained as follows:

$$\left(0.90\frac{\text{dollar}}{\text{hamburger}}\right)\left(\begin{array}{c}1000 \text{ hamburgers} \\ \text{sold in Long Beach}\end{array}\right) + \left(0.80\frac{\text{dollar}}{\text{hotdog}}\right)\left(\begin{array}{c}300 \text{ hotdogs} \\ \text{sold in Long Beach}\end{array}\right)$$
$$+ \left(1.10\frac{\text{dollar}}{\text{milkshake}}\right)\left(\begin{array}{c}500 \text{ milkshakes} \\ \text{sold in Long Beach}\end{array}\right) = \$1690$$

Thus the total revenue from the Long Beach restaurant on that day is $1690.

(c) The total revenue from all three locations is the sum of the entries in the matrix CA: $4690 + 1690 + 13{,}970 = 20{,}350$. Thus the total revenue on that day is $20,350.

7.6 Matrix Equations: Solving a Linear System

OBJECTIVES

☑ *Check the box when you can do the exercises addressing the objective, such as those included in this study guide, which are listed here.*

Exercises

The Inverse of a Matrix

☐ Verify that given matrices are inverses. **3**

☐ Use a graphing calculator to find the inverse of a matrix, if it exists. **7, 13, 15, 27**

Matrix Equations

☐ Solve a system of equations by first writing it as a matrix equation. **19, 23, 27**

Modeling with Matrix Equations

☐ Use a linear system of equations to organize information about the real world. **27**

SKILLS

The Inverse of a Matrix

These exercises ask you to find a solution for a system of equations by first finding the inverse of the coefficient matrix. This works because the inverse of the matrix multiplied by the matrix is the identity matrix. Find the definitions of the identity matrix and the inverse of a matrix in blue boxes in Section 7.6 of the textbook.

📖 **Read the introduction to the subsection, 'The Inverse of a Matrix,'** in which the authors introduce the identity matrix and the inverse of a matrix.

[43] *The identity matrix I_3 has ___ rows and ___ columns. In general the identity matrix I_n has ___ rows and ___ columns. If you multiply the identity matrix I_3 by a 3×3 matrix A, then you get $I_3A =$___. If A and B are 3×3 matrices and A is the inverse of B, then $AB =$ ____. The matrix A^{-1} ___ (is / is not) equal to $1/A$.*

📖 **Read Section 7.6 Example 1,** in which the authors determine whether or not two matrices are inverses.

[44] *To check if A and B are inverses, calculate _____ and _____. If A and B are inverses, then your answer for both should be ____.*

---

[43] The identity matrix I_3 has 3 rows and 3 columns; I_n has n rows and n columns. If you multiply, $I_3A = A$. If A is the inverse of B, then $AB = I_3$. If A is a matrix, then it does not make any sense to divide by it; division makes sense for numbers, not matrices. So for a matrix A, A^{-1} does not equal $1/A$, since $1/A$ is meaningless.

[44] To check if A and B are inverses, calculate AB and BA. If they are inverses, you should get the identity matrix I_2 as your answer for both calculations.

✎ **Section 7.6 Exercise 3** Calculate the products AB and BA to verify that matrix B is the inverse of matrix A.

$$A = \begin{bmatrix} 4 & 1 \\ 7 & 2 \end{bmatrix} \qquad B = \begin{bmatrix} 2 & -1 \\ -7 & 4 \end{bmatrix}$$

Solution

📖 **Read Section 7.6 Example 2**, in which the authors find the inverse of a matrix using a graphing calculator.

[45] *Multiply the given matrix A by the inverse that the authors found on the calculator. You get AA^{-1} = _____.*

✎ **Section 7.6 Exercise 7 and 13** Use a calculator that can perform matrix operations to find the inverse of the matrix, if it exists.

7. $\begin{bmatrix} -3 & -5 \\ 2 & 3 \end{bmatrix}$

Solution

13. $\begin{bmatrix} 2 & 4 & 1 \\ -1 & 1 & -1 \\ 1 & 4 & 0 \end{bmatrix}$

Solution

---

[45] Since A^{-1} is the inverse of A, the product $AA^{-1} = I_3$.

Matrix Equations

These exercises ask you to represent a system of equations as a matrix equation, and then use the inverse of the coefficient matrix to solve the matrix equation. Find the definition of a coefficient matrix of a system of equations in bold typeface in Section 7.6 of the textbook.

📖 **Read the introduction to the subsection, 'Matrix Equations,'** in which the authors solve a system of equations using matrices.

[46] *If a system of equations has three equations with three variables, then in the matrix equation, AX = B, the coefficient matrix A has dimension ___ × ___, the variable matrix X has dimension ___ × ___, and the answer vector B has dimension ___ × ___.*

📖 **Read Section 7.6 Examples 3 and 4**, in which the authors solve a system of equations using matrices.

[47] *What steps to the authors take to solve the systems of equations?*
Answer in your own words:

✎ **Section 7.6 Exercise 19 and 23** Solve the system of linear equations by expressing the system as a matrix equation and using the inverse of the coefficient matrix. Use the inverses you found in Exercises 7 and 13, found in this study guide above.

19. $\begin{cases} -3x - 5y = 4 \\ 2x + 3y = 0 \end{cases}$

Solution

---

[46] The matrix A has dimension 3×3, and the matrices X and B have dimension 3×1.

[47] First, the authors write the system of equations as a matrix equation $AX = B$. Then they use a calculator to find the inverse A^{-1} of the coefficient matrix A, using a graphing calculator. They then calculate the product $A^{-1}B$. The solution to the system are the entries in the column matrix $A^{-1}B$.

23. $\begin{cases} 2x + 4y + z = 7 \\ -x + y - z = 0 \\ x + 4y = -2 \end{cases}$

Solution

📖 **Read Section 7.6 Example 5**, in which the authors consider a matrix that does not have an inverse.

[48]*A matrix with no inverse is called a* _____ *matrix.*

✎ **Section 7.6 Exercise 15** Use a calculator that can perform matrix operations to find the inverse of the matrix, if it exists.

$$\begin{bmatrix} 1 & 2 & 3 \\ 4 & 5 & -1 \\ 1 & -1 & -10 \end{bmatrix}$$

Solution

CONTEXTS

Modeling with Matrix Equations

These exercises ask you to model the real world with a system of linear equations and solve it using a matrix equation.

📖 **Read Section 7.6 Example 6**, in which the authors model the diets of a hamster and a gerbil.

[49]*What does the (2, 3) entry in the coefficient matrix A represent? What does the (1, 3) entry in the vector B represent?*
Answer in your own words:

---

[48] A matrix without an inverse is called a singular matrix.

[49] The (2, 3) entry of the matrix A is 10. This entry represents the amount of fat (in mg) in one gram of Rodent Chow. The (1, 3) entry in the vector B is 440. This entry represents the amount of carbohydrates (in mg) that a hamster needs each day.

✎ **Section 7.6 Exercise 27 Sales Commissions** An encyclopedia saleswoman works for a company that offers three different grades of bindings for its encyclopedias: standard, deluxe, and leather. For each set that she sells, she earns a commission based on the set's binding grade. One week she sells one standard, one deluxe, and two leather sets and makes $675 in commission. The next week she sells two standard, one deluxe, and one leather set for a $600 commission. The third week she sells one standard, two deluxe, and one leather set, earning $625 in commission.

(a) Let x, y, and z represent the commission she earns on standard, deluxe, and leather sets, respectively. Translate the given information into a system of equations.

(b) Express the system of equations you found in part (a) as a matrix equation of the form $AX = B$.

(c) Find the inverse of the coefficient matrix A and use it to solve the matrix equation in part (b). How much commission does the saleswoman earn on a set of encyclopedias in each grade of binding?

Solution

CONCEPTS

✓ **7.6 Exercises – Concepts – Fundamentals** Complete the Fundamentals section of the exercises at the end of Section 7.6. Compare your answers to those at the back of the textbook, and make corrections as necessary.

📖 **Read the Concept Check for Section 7.6**, in the Chapter Review at the end of Chapter 7.

7.6 Solutions to ✎ Exercises

✎ Section 7.6 Exercise 3

We perform matrix multiplication to show that $AB = I$ and $BA = I$.

$$AB = \begin{bmatrix} 4 & 1 \\ 7 & 2 \end{bmatrix} \begin{bmatrix} 2 & -1 \\ -7 & 4 \end{bmatrix} = \begin{bmatrix} 4\cdot2+1\cdot-7 & 4\cdot-1+1\cdot4 \\ 7\cdot2+2\cdot-7 & 7\cdot-1+2\cdot4 \end{bmatrix} = \begin{bmatrix} 1 & 0 \\ 0 & 1 \end{bmatrix}$$

$$BA = \begin{bmatrix} 2 & -1 \\ -7 & 4 \end{bmatrix} \begin{bmatrix} 4 & 1 \\ 7 & 2 \end{bmatrix} = \begin{bmatrix} 2\cdot4-1\cdot7 & 2\cdot1-1\cdot2 \\ -7\cdot4+4\cdot7 & -7\cdot1+4\cdot2 \end{bmatrix} = \begin{bmatrix} 1 & 0 \\ 0 & 1 \end{bmatrix}$$

✎ Section 7.6 Exercise 7

A graphing calculator gives the following inverse of this matrix.

$$\begin{bmatrix} -3 & -5 \\ 2 & 3 \end{bmatrix}^{-1} = \begin{bmatrix} 3 & 5 \\ -2 & -3 \end{bmatrix}$$

✎ Section 7.6 Exercise 13

A graphing calculator gives the following inverse of this matrix.

$$\begin{bmatrix} 2 & 4 & 1 \\ -1 & 1 & -1 \\ 1 & 4 & 0 \end{bmatrix}^{-1} = \begin{bmatrix} -4 & -4 & 5 \\ 1 & 1 & -1 \\ 5 & 4 & -6 \end{bmatrix}$$

✎ Section 7.6 Exercise 19

This system is equivalent to a matrix system $AX = B$, where A, X, B are given by:

$$A = \begin{bmatrix} -3 & -5 \\ 2 & 3 \end{bmatrix} \qquad X = \begin{bmatrix} x \\ y \end{bmatrix} \qquad B = \begin{bmatrix} 4 \\ 0 \end{bmatrix}$$

Using a calculator, we find (see Exercise 7 above) that

$$A^{-1} = \begin{bmatrix} 3 & 5 \\ -2 & -3 \end{bmatrix}$$

So because $X = A^{-1}B$, we have

$$\begin{bmatrix} x \\ y \end{bmatrix} = \begin{bmatrix} 3 & 5 \\ -2 & -3 \end{bmatrix} \begin{bmatrix} 4 \\ 0 \end{bmatrix} = \begin{bmatrix} 12 \\ -8 \end{bmatrix}$$

So the solution to the system is $x = 12$, $y = -8$, or $(12, -8)$.

🖉 Section 7.6 Exercise 23

This system is equivalent to a matrix system $AX = B$, where A, X, B are given by:

$$A = \begin{bmatrix} 2 & 4 & 1 \\ -1 & 1 & -1 \\ 1 & 4 & 0 \end{bmatrix} \qquad X = \begin{bmatrix} x \\ y \\ z \end{bmatrix} \qquad B = \begin{bmatrix} 7 \\ 0 \\ -2 \end{bmatrix}$$

Using a calculator, we find (see Exercise 7 above) that

$$A^{-1} = \begin{bmatrix} -4 & -4 & 5 \\ 1 & 1 & -1 \\ 5 & 4 & -6 \end{bmatrix}$$

So because $X = A^{-1}B$, we have

$$X = \begin{bmatrix} -4 & -4 & 5 \\ 1 & 1 & -1 \\ 5 & 4 & -6 \end{bmatrix} \begin{bmatrix} 7 \\ 0 \\ -2 \end{bmatrix} = \begin{bmatrix} -38 \\ 9 \\ 47 \end{bmatrix}$$

So the solution to the system is $x = -38$, $y = 9$, and $z = 47$, or $(-38, 9, 47)$.

🖉 Section 7.6 Exercise 15

When we try to find the inverse of this matrix on a calculator, we get an error, because this matrix does not have an inverse. On a TI-83 calculator, the error message says ERR:SINGULAR MAT indicating that the matrix is singular.

🖉 Section 7.6 Exercise 27 Sales Commissions

We record the information given in this exercise in a table for easy reference:

	Number with standard binding sold	Number with deluxe binding sold	Number with leather binding sold
Week 1	1	1	2
Week 2	2	1	1
Week 3	1	2	1

Her commission in each week is given by

$$\left(\text{Total Commission}\right) = \left(\begin{array}{c}\text{Commission from sales} \\ \text{with standard binding}\end{array}\right) + \left(\begin{array}{c}\text{Commission from sales} \\ \text{with deluxe binding}\end{array}\right) + \left(\begin{array}{c}\text{Commission from sales} \\ \text{with leather binding}\end{array}\right)$$

Since x, y, and z represent the commission she earns on standard, deluxe, and leather sets, respectively, we have

$$\left(\text{Total Commission}\right) = \left(\begin{array}{c}\text{Number with} \\ \text{standard binding sold}\end{array}\right)(x) + \left(\begin{array}{c}\text{Number with} \\ \text{deluxe binding sold}\end{array}\right)(y) + \left(\begin{array}{c}\text{Number with} \\ \text{leather binding sold}\end{array}\right)(z)$$

Applying this to each of the weeks listed in the table above, we have the following system of equations.

$1x + 1y + 2z = 675$ Week 1 Equation
$2x + 1y + 1z = 600$ Week 2 Equation
$1x + 2y + 1z = 625$ Week 3 Equation

(b) This system is equivalent to a matrix system $AX = B$, where A, X, B are given by:

$$A = \begin{bmatrix} 1 & 1 & 2 \\ 2 & 1 & 1 \\ 1 & 2 & 1 \end{bmatrix} \qquad X = \begin{bmatrix} x \\ y \\ z \end{bmatrix} \qquad B = \begin{bmatrix} 675 \\ 600 \\ 625 \end{bmatrix}$$

(c) Using a calculator, we find

$$A^{-1} = \begin{bmatrix} -1/4 & 3/4 & -1/4 \\ -1/4 & -1/4 & 3/4 \\ 3/4 & -1/4 & -1/4 \end{bmatrix}$$

So because $X = A^{-1}B$, we have

$$X = \begin{bmatrix} -1/4 & 3/4 & -1/4 \\ -1/4 & -1/4 & 3/4 \\ 3/4 & -1/4 & -1/4 \end{bmatrix} \begin{bmatrix} 675 \\ 600 \\ 625 \end{bmatrix} = \begin{bmatrix} 125 \\ 150 \\ 200 \end{bmatrix}$$

So the solution to the system is $x = 125$, $y = 150$, and $z = 200$. Thus she earns $125 commission on sales of encyclopedias with standard binding, $150 commission on sales with deluxe binding and $200 on sales with leather binding.